Studies in Global Science Fiction

Studies in Global Science Fiction (edited by Anindita Banerjee, Rachel Haywood Ferreira, and Mark Bould) is a brand-new and first-of-its-kind series that opens up a space for Science Fiction scholars across the globe, inviting fresh and cutting-edge studies of both non-Anglo-American and Anglo-American SF literature. Books in this series will put SF in conversation with postcolonial studies, critical race studies, comparative literature, transnational literary and cultural studies, among others, contributing to ongoing debates about the expanding global compass of the genre and the emergence of a more diverse, multinational, and multi-ethnic sense of SF's past, present, and future. Topics may include comparative studies of selected (trans)national traditions, SF of the African or Hispanic Diasporas, Indigenous SF, issues of translation and distribution of non-Anglophone SF, SF of the global south, SF and geographic/cultural borderlands, and how neglected traditions have developed in dialogue and disputation with the traditional SF canon.

More information about this series at
http://www.palgrave.com/gp/series/15335

Zachary Kendal • Aisling Smith
Giulia Champion • Andrew Milner
Editors

Ethical Futures and Global Science Fiction

palgrave
macmillan

Editors
Zachary Kendal
Monash University
Clayton, VIC, Australia

Aisling Smith
Monash University
Clayton, VIC, Australia

Giulia Champion
University of Warwick
Coventry, UK

Andrew Milner
Monash University
Clayton, VIC, Australia

ISSN 2569-8826 ISSN 2569-8834 (electronic)
Studies in Global Science Fiction
ISBN 978-3-030-27892-2 ISBN 978-3-030-27893-9 (eBook)
https://doi.org/10.1007/978-3-030-27893-9

Cover illustration: Mint Images / Getty

This Palgrave Macmillan imprint is published by the registered company Springer Nature Switzerland AG.
The registered company address is: Gewerbestrasse 11, 6330 Cham, Switzerland

Preface

In a time of global political and economic uncertainty, with countries around the world facing environmental, social and financial crises, the question of what kind of future we want to create is more relevant than ever. Central to this speculation lies the question of ethics—the kinds of values this future will reflect and how it will shape our engagement with others, whether human, animal or the environment. This book explores the ethical dimensions of the reconfigurations of our future found in global science fiction (SF) literatures, where authors have worked through the kinds of utopian futures we want to embrace and the dystopian alternatives we must strive to avoid. The studies collected here also show the influence of various strands of ethical criticism, from rhetorical and narratological studies, to poststructuralist and deconstructionist approaches, to a host of social and cultural theories, including ecocritical, postcolonial, Marxist, feminist and intersectional studies.

Early fan histories of SF tended to trace the genre back only so far as the inter-war American "pulp" fiction magazines, especially Hugo Gernsback's *Amazing* and John W. Campbell's *Astounding*. More recently, more scholarly and more European accounts have tended to range across the entire Western literary heritage, preferred starting points including both Lucian's *Alēthēs Historia* (second century CE) and Thomas More's *Utopia* (1516). If fan histories were too parochially American, scholarly histories tended to occlude what was actually most distinctive about modern SF, its preoccupation with science as a practical activity productive of new technologies. This preoccupation was part of the Romantic reaction against the effects of the Industrial Revolution, which registered in the first place in Britain. Hence, Brian Aldiss's famous decision to trace the "origins of the species" to Mary Shelley's *Frankenstein* (1818). Whatever the starting point, however, the focus tended to fall disproportionately on British, French and American variants of the genre. This collection also starts in the Anglosphere with Zachary Kendal's close reading of Isaac Asimov's *Foundation* trilogy and Joshua Bulleid's detailed account of vegetarianism in nineteenth-century British and American utopianism. And subsequent chapters by Rachel Fetherston and Jacqueline Dutton do indeed pay due credit to the rich tradition of French SF. But we have sought to achieve a more fully global range, so that the collection also addresses texts from Algeria, Australia, Canada, China, Egypt, Germany, Haiti, India, Jamaica, Macedonia, Mexico, Russia and South Africa. In keeping with common practice in the discipline of Comparative Literature, we have wherever possible included original-language quotations for all non-Anglophone primary texts, followed by English translations in parentheses.

The book opens with three diverse essays on "Ethics and the Other," with each examining the approaches to ethics and alterity in different SF traditions, while also providing a historical introduction to the collection. Zachary Kendal's chapter draws on Emmanuel Levinas's writing on ethics to contrast the approaches to totality and infinity in Asimov's *Foundation* trilogy (1951–1953) and Yevgeny Zamyatin's *Мы* (*We*) (1921). Kendal suggests that Asimov's commitment to the knowability of all things, including the other person, ultimately left the trilogy open to accusations of both vulgar Marxism and outright fascism, whereas Zamyatin's focus on the irrational and the unknowability of the other person reflects a more ethical approach to the other. Sreejata Paul then explores Begum Rokeya Sakhawat Hossain's early-twentieth-century utopias *Sultana's Dream* (1905) and *পদ্মরাগ* (*Padmarag*) (1924), finding that these texts advocate for radical feminist ethics and a feminine ethics of care, respectively. In doing so, Paul finds that the texts explore the issues facing women in colonial Bengal and how Hossain worked through ethical issues around feminism and femininity. Finally, Joshua Bulleid's chapter turns the reader's attention to animal ethics and the treatment of the animal other, as he traces the theme of vegetarianism through English and American utopian traditions. Bulleid finds a firmly established trend of vegetarian utopians grounded in concerns over the ethical treatment of animals.

In Part II the collection moves to consider humankind's treatment of the environmental other. The rise of climate fiction, or what Daniel Bloom terms "cli-fi," and of science-fictional representations of the consequences of climate change, along with the growth of ecocriticism and scholarly discourse on ecology, now define this as a central ethical theme in its own right. Part II begins with Andrew Milner's chapter surveying global trends in cli-fi, establishing climate change as an issue of ethical significance and examining the different responses to it in SF from Australia, Britain, Canada, China, France, Germany, the United States and South Africa. Rachel Fetherston's chapter follows, a comparative study of Michel Houellebecq's *Les Particules élémentaires* (1998) and Margaret Atwood's *MaddAddam* trilogy (2003–2013), which develops the notion of "ecological posthumanism." Thomas Moran then turns to Chinese SF in his examination of environmental activism and willed human extinction in Liu Cixin's Hugo Award-winning 三体 (*The Three-Body Problem*) (2006). Finally, Giulia Champion examines the ecological and social deteriorations in Haitian and Mexican visions of ecopocalypse developed, respectively, in Jacques Roumain's *Gouverneurs de la rosée* (1944) and Homero Aridjis's *La Leyenda de los soles* (1993) and *¿En quién piensas cuando haces el amor?* (1995).

The chapters in Part III explore the evolution of the relation between SF as a literary genre and colonialism, beginning with Bill Ashcroft's investigation of the legacy of imperialism in SF and closing on Lara Choksey's exploration of African futurism and decolonial praxis. This shows the crucial tension existing between the postcolonial SF literatures found throughout this section and their inextricable nexus with historical (re)visions, since futuristic narratives from the Global South must often address the social and environmental destruction and uneven development created by imperial practices. The chapters in this section tackle critical ethical issues in reconstructing postcolonial social and political spaces as well as in reclaiming SF as a literary genre. Ashcroft examines the ethics of power and empire in his exploration of Octavia Butler's *Xenogenesis* (or *Lilith's Brood*) trilogy (1987–1989), looking at the novel's inversion of colonial dominance and tackling discussions surrounding postcolonial SF. Challenging colonial pasts is also crucial in Nudrat Kamal's chapter, which examines Amitav Ghosh's *The Calcutta Chromosome* (1995). Kamal focuses on Ghosh's conceptualisation of the posthuman, postcolonial cyborg and its ramifications for postcolonial ethics. Finally, Choksey closes this section by exploring Assia Djebar's *Ombre sultane* (1987), Nalo Hopkinson's *Midnight Robber* (2000) and the intersection between political participation and gender, bringing forth women's role in decolonial praxis and Pan-Africanism, focusing on care and feminist ethics.

Finally, Part IV of the collection considers the political dimensions of SF, with four chapters further extending the themes broached in earlier essays. Jacqueline Dutton's chapter examines the failure of French secular state ethics to offer utopian visions of the future of the "City of Light." Dutton compares the futures imagined by Louis-Sébastien Mercier in *L'An 2440: Rêve s'il en fut jamais* (1771) and Michel Houellebecq in *Soumission* (2015), discussing how these texts conceptualise the relationships between ethics, religion and the state. Kalina Maleska then considers the emergence of dystopian literature in contemporary Macedonia through recent SF novels. Maleska evaluates how Sanja Mihajlović Kostadinovska's *517* (2015) and Tomislav Osmanli's *Бродот. Конзархија* (The Ship. Consarchy) (2016) respond directly to Macedonian politics and the ethics of nationalist policies, treating Vlada Urošević's *Дворскиот поет во апарат за летање* (The Court Poet in a Flying Machine) (1996) as a precursor to this nascent genre. Next, Anna Madoeuf and Delphine Pagès-El Karoui compare two visions of future Cairo, exploring the state's role in economic, spatial and social inequality in Jamil Nasir's *Tower of Dreams* (1999) and Ahmed Khaled Towfik's *Utopia* (2008). This matter of economic inequality is central to the book's closing chapter by Nick Lawrence, who explores the dearth of utopian visions of ethical post-capitalist futures, despite the pressure of ongoing global economic crisis and political tension. Lawrence reflects on the Western utopian tradition in earlier literature and ultimately affirms the necessity of utopian thought for negotiating the problems of the contemporary world.

Clayton, VIC, Australia — Zachary Kendal
Clayton, VIC, Australia — Aisling Smith
Coventry, UK — Giulia Champion
Clayton, VIC, Australia — Andrew Milner

Acknowledgements

The editors of this book would like to thank the Monash Warwick Alliance, a partnership between Monash University and the University of Warwick, whose funding and support made this international collaboration possible. We also thank the colleagues, friends and family who provided advice, support and encouragement while we worked on this project, particularly Evie Kendal and Verity Burgmann.

ACKNOWLEDGMENTS

[illegible]

Praise for *Ethical Futures and Global Science Fiction*

"A brilliantly curated, critically-rigorous, and discerning book. It stands apart among even the best of the emerging canon of global sf studies in the range of national literatures, historical periods, and subject areas covered, and in the reach and depth of its transdisciplinary approach. This is a collection to which students and scholars in the field will return again and again."

—Terry Harpold, Associate Professor, *University of Florida, USA*

"This volume makes an important contribution to the theoretical debates on the question of ethics for a sustainable future. Contributors investigate global science fiction literature, explore alterity and climate change and analyse the political dimensions of utopian and dystopian narratives in different geographical and cultural contexts. It is a valuable collection for both established and new scholars from different disciplines, providing original and up to date interdisciplinary perspectives on the intrinsically global qualities of science fiction."

—Raffaella Baccolini, *Professor of English Literature at the University of Bologna, Italy, and co-author of* Utopia Method Vision *(2007) and* Dark Horizons: Science Fiction and the Dystopian Imagination *(2003)*

"This book makes a valuable contribution to our understanding of how science fiction responds to contemporary ethical and political dilemmas. Four well-chosen sections frame insights into SF narratives on encounters with the other, environmental ethics, post-colonial challenges and socio-political inequality. These concerns speak directly to the felt problems of the present moment. At the same time, the collection is rooted in a rich understanding of SF's long history of engagement with social, political and ethical issues. The global and especially postcolonial focus of this collection is a particularly welcome to SF criticism which has only recently turned to fully address the US and Eurocentric history of the genre and its histories. We need to listen much more closely to indigenous and postcolonial stories and theories in the face of Western modernity's impasses. *Ethical Futures and Global Science Fiction* puts these issues and positions into a complex and subtle conversation with reflections on the spectre of climate crisis and the persistence of global capitalism and its inequities. The collection reiterates the importance of speculative fiction to necessary thinking about probable, possible and desirable futures, and brings important new texts and voices to our attention."

—Lisa Garforth, *Senior Lecturer in Sociology at Newcastle University, UK, and author of* Green Utopias: Environmental Hope Before and After Nature *(2017)*

Contents

Part I Ethics and the Other 1

1 Science Fiction's Ethical Modes: Totality and Infinity in Isaac Asimov's *Foundation* Trilogy and Yevgeny Zamyatin's *Мы* (*We*) 3
Zachary Kendal

2 Inversion and Prolepsis: Begum Rokeya Sakhawat Hossain's Feminist Utopian Strategies 29
Sreejata Paul

3 Better Societies for the Ethical Treatment of Animals: Vegetarianism and the Utopian Tradition 49
Joshua Bulleid

Part II Environmental Ethics 75

4 Eutopia, Dystopia and Climate Change 77
Andrew Milner

5 Evolving a New, Ecological Posthumanism: An Ecocritical Comparison of Michel Houellebecq's *Les Particules élémentaires* and Margaret Atwood's *MaddAddam* Trilogy 99
Rachel Fetherston

6 The Perverse Utopianism of Willed Human Extinction: Writing Extinction in Liu Cixin's *The Three-Body Problem* *(三体)* 119
Thomas Moran

7 Ecopocalyptic Visions in Haitian and Mexican Landscapes of Exploitation 141
Giulia Champion

Part III Postcolonial Ethics 163

8 Postcolonial Science Fiction and the Ethics of Empire 165
Bill Ashcroft

9 The Postcolonial Cyborg in Amitav Ghosh's *The Calcutta Chromosome* 187
Nudrat Kamal

10 Wagering the Future: Split Collectives and Decolonial Praxis in Assia Djebar's *Ombre sultane* and Nalo Hopkinson's *Midnight Robber* 211
Lara Choksey

Part IV Ethics and Global Politics 233

11 Rewriting France's Future: From Louis-Sébastien Mercier's Pre-Revolutionary Projections to Michel Houellebecq's Islamic Agendas via Secular State Ethics 235
Jacqueline Dutton

12 The Appearance of Dystopian Fiction in Macedonia and its Ethical Concerns 261
Kalina Maleska

13 Cairo in 2015 and in 2023: The Dreadful Fates of the Egyptian Capital in Jamil Nasir's *Tower of Dreams* and Ahmed Khaled Towfik's *Utopia* 283
Anna Madoeuf and Delphine Pagès-El Karoui

14 Post-Capitalist Futures: A Report on Imagination 303
Nick Lawrence

Index 329

Notes on Contributors

Bill Ashcroft is a renowned critic and theorist, founding exponent of postcolonial theory and co-author of *The Empire Writes Back*, the first text to offer a systematic examination of the field of postcolonial studies. He is author and co-author of 21 books and over 200 articles and chapters, variously translated into six languages, and he is on the editorial boards of ten international journals. His latest work is *Utopianism in Postcolonial Literatures.* He is Emeritus Professor at the University of New South Wales and is a fellow of the Australian Academy of the Humanities.

Joshua Bulleid is a PhD researcher at Monash University, whose thesis project investigates the representation of vegetarianism in science fiction and utopian literature.

Giulia Champion is completing her doctoral thesis at the University of Warwick. Her research investigates postcolonial literature in original languages and aims to theorise literary cannibalism as a set of practices through the world ecology framework and historical materialism.

Lara Choksey is a postdoctoral research associate at the Wellcome Centre for Cultures and Environments of Health at the University of Exeter. Her main research interests are biology and literature, critical race and decolonial studies, and modern and contemporary fiction. Her book *Narrative in the Age of the Genome: Genetic Worlds* is forthcoming with Bloomsbury.

Jacqueline Dutton is Associate Professor of French Studies at the University of Melbourne. She has published widely on utopian studies, including a chapter on "non-Western" utopian traditions for *The Cambridge*

Companion to Utopian Literature (2010) and a book on utopianism in the work of JMG Le Clézio, *Le Chercheur d'or et d'ailleurs* (2003). She is a leading researcher on French culture and identity, teaching courses on travel writing, food and wine, cinema and literature. She is currently writing a cultural history of the Bordeaux, Burgundy and Champagne wine regions and co-editing (with Peter J. Howland) *Wine, Terroir and Utopia* (forthcoming).

Rachel Fetherston is co-founder and publications manager of the nature connection charity Remember The Wild. She is also a PhD candidate in Literary Studies and Social Science at Deakin University, where she is investigating the representation of the nonhuman in Australian ecofiction and the impact of such fiction on readers' connection to nature.

Nudrat Kamal is Lecturer of Social Sciences and Liberal Arts at the Institute of Business Administration, Karachi. She has an MA in Comparative Literature from Stony Brook University, New York, which she attended as a Fulbright scholar. Her primary research traces diasporic aesthetics in Urdu and Anglophone literature on the South Asian Partition of 1947, focusing particularly on the gendered experiences of violence and dislocation. She is also interested in the intersections of postcolonial theory and science fiction and fantasy, and has written extensively on literature, film and popular culture for publications such as *Dawn*, *The Express Tribune*, *Newsline* and *Truthdig*.

Zachary Kendal is a librarian in Rare Books at Monash University Library. He was recently an editor-in-chief of *Colloquy: Text, Theory, Critique* and is completing a PhD in Literary and Cultural Studies at Monash University, researching ethics and literary representation in science fiction.

Nick Lawrence teaches American literature, world literature and critical theory at the University of Warwick. A member of the Warwick Research Collective, he is co-author of *Combined and Uneven Development: Towards a New Theory of World-Literature* (2015). He is currently working on *Post-Capitalist Aesthetics*, a monograph charting contemporary dystopian landscapes in tandem with cultural prefigurations of a world order beyond the present era of neoliberal capitalism.

Anna Madoeuf is Professor of Geography at the Université de Tours, in charge of EMAM (Middle East & North Africa Research Team) in the CITERES laboratory (Cities, Territories, Spaces & Societies). Her key publications include *Explorer le temps au Liban et au Proche-Orient*, co-edited with Sylvia Chiffoleau, Elie Dannaoui and Souad Slim (2017); *Lire les villes: Panoramas du monde urbain conremporain*, co-edited with Raffaele Cattedra (2012); and *Les Pèlerinages au Maghreb et au Moyen-Orient*, co-edited with Sylvia Chiffoleau (2010).

Kalina Maleska is a Macedonian writer, a literary critic and Professor of Literature at Saints Cyril and Methodius University in Skopje. Her works include three collections of short stories, two novels and a play. Her stories have been translated into several languages and published in various magazines and anthologies, including the anthology *Best European Fiction 2018*. Her literary criticism has also been published in international journals. She translates literary works from English into Macedonian and vice versa, including translations of *Tristram Shandy* by Laurence Stern, *Huckleberry Finn* by Mark Twain and *Selected Stories* by Ambrose Bierce.

Andrew Milner is Emeritus Professor of English and Comparative Literature at Monash University and Honorary Professor at the University of Warwick. He is the author of numerous books, including, most recently, *Locating Science Fiction* (2012), *Again, Dangerous Visions: Essays in Cultural Materialism* (2018) and, with J. R. Burgmann, *Science Fiction and Climate Change* (2020).

Thomas Moran is a PhD candidate at Monash University writing on the death of cinema and the aesthetics of extinction. His master's thesis was titled *The Surreal Realist Cinema of Jia Zhangke*.

Delphine Pagès-El Karoui is Senior Lecturer in Geography at INALCO (National Institute of Oriental Languages and Civilisations), USPC (Sorbonne Paris Cité University). Following her doctorate on secondary cities in the Nile Delta, her current research addresses Egyptian migrations (transnational networks and diasporas in Europe and the Gulf, including imaginaries in literature and cinema), the spatial dimensions of Arab revolutions and its interactions with migrations, and urban diversity and cosmopolitanism in Gulf cities. With Stéphane Sawas (INALCO), she has coordinated the research programme Imaginaires Migratoires (2015–2017), which explores cinematic and literary imaginaries of migration.

Sreejata Paul holds a BA and an MA in English Literature from Jadavpur University and an MPhil from Christ University. At present, she is enrolled in a dual-badged PhD programme at the Indian Institute of Technology (Bombay) and Monash University. Her doctoral work focuses on Muslim women writers from colonial Bengal. She is also Associate Editor for *Colloquy: Text, Theory, Critique* journal and a teaching assistant at Indian Institute of Technology Bombay-Monash Research Academy.

Aisling Smith is a teaching associate in Literary Studies at Monash University and Deakin University. Her PhD examined affect theory and the works of David Foster Wallace. She is also a creative writer, a former editor-in-chief of *Colloquy: Text, Theory, Critique* and an editor of the *Verge: Chimera* (2017) anthology.

PART I

Ethics and the Other

CHAPTER 1

Science Fiction's Ethical Modes: Totality and Infinity in Isaac Asimov's *Foundation* Trilogy and Yevgeny Zamyatin's *Мы* (*We*)

Zachary Kendal

Introduction

Can genres have an intrinsic predisposition to a particular politics? Critics who ask this question of science fiction (SF) often arrive at different answers. Carl Freedman, for example, posits that SF is a "privileged and paradigmatic genre" for Marxism and critical theory, having "the deepest … affinity with the rigors of dialectical thinking," which would explain the genre's attraction to Marxist critics such as Darko Suvin, Raymond Williams, and Fredric Jameson.[1] By contrast, Aaron Santesso claims the genre leans in a more regressive direction, replete with "fascist energies and ideas."[2] In this chapter, I ask whether SF has a predisposition to a particular ethical outlook, exploring the dominant ethical modes of the SF genre. As with the question of politics, which resurfaces throughout this essay, how this question of ethics is answered depends on how SF is defined and, centrally, the significance given to the American SF pulp magazines, which are central to Santesso's

Z. Kendal (✉)
Monash University, Clayton, VIC, Australia

Z. Kendal et al. (eds.), *Ethical Futures and Global Science Fiction*, Studies in Global Science Fiction,
https://doi.org/10.1007/978-3-030-27893-9_1

study but excised from Freedman's. However, establishing a systematic definition of SF today—one that would include works generally considered SF, exclude (if desired) those of other genres, and navigate a growing body of "post-genre" or "slipstream" works that blend different conventions—would be impossible. It is thus more accurate to speak of SF's multitude of literary traditions, which provide it with what Samuel R. Delany called SF's "historical, theoretical, stylistic, and valuative plurality."[3] Rather than seek a single answer to this question of SF's ethics, I will examine two classic SF works and the traditions they represent: Isaac Asimov's original *Foundation* trilogy (1951–1953; serialised 1942–1950), one of the most iconic and well-loved series of the American pulps and SF's "golden age," and Yevgeny Zamyatin's *Мы* (*We*, 1921), a highly influential dystopian novel from an Eastern European SF tradition. I argue that the genre SF that developed in the American pulps was dominated by themes and modes of representation best described as *totalising*, but that works such as *Мы* demonstrate the potential for SF to engage with a more ethical discourse.[4]

I thus begin by exploring Emmanuel Levinas's critique of totalisation, which he finds violent and reductive in its attempts to reduce the world and its inhabitants to finite and knowable concepts. I then turn to Asimov's *Foundation* trilogy, focusing primarily on the series' representation of "psychohistory," for an example of this totalising project finding a utopian expression in pulp SF. Returning to Levinas, I then consider his writings on the idea of infinity and the disruption of totality in the face-to-face encounter, concepts integral to his understanding of ethics and responsibility. This leads into a Levinasian reading of Zamyatin's *Мы*, which demonstrates a philosophical orientation contrary of that of Asimov's *Foundation* trilogy, focusing on the power of the encounter with the unknowable other to break apart totalising ideologies. Finally, I will argue that the literary form of each text underpins their different ethical orientations, with Asimov's straightforwardly "pulpish" mode supporting his series' totalising approach, and Zamyatin's disruptive modernist form allowing him to present singular and unknowable characters.

Totality and Ethics

Writing in the decades following World War II, Levinas saw the need to drastically rethink the foundations of a philosophical tradition that had been complicit in the rise of fascism. This resulted in his blistering critique of the concept of totality, which he associates with Western philosophy as a whole. He describes this philosophy as "une tentative de synthèse universelle, une réduction de toute l'expérience, de tout ce qui est sense, à une totalité où la conscience embrasse le monde, ne laisse rien d'autre hors d'elle" ("an attempt at universal synthesis, a reduction of all experiences, of all that is reasonable, to a totality wherein consciousness embraces the world, leaving nothing other outside of itself").[5] Through integration into a comprehensible totality, what is *other* is stripped of its infinite alterity (its difference) in order to become *finite* and *known*—to be conquered, cognitively. Such an approach finds no resistance in objects, which can be integrated into the totality "sans jamais mettre en question la liberté du moi" ("without ever putting into question the freedom of the I").[6] It is when this approach is extended to the other person, singular and unknowable, that its unethical orientation is revealed. The totalisation of the Other necessitates the erasure of their alterity, dissolving them into a concept and subsuming them into a totality, making it an act of violence against the other's very otherness. Levinas thus identifies totalisation as a kind of imperialism, associating it with "toute la civilisation occidentale de propriété, d'exploitation, de tyrannie politique et de guerre" ("the whole Western civilization of property, exploitation, political tyranny, and war").[7] It is also foundational to fascism, which depends on reductive totalisation to create the neat and all-encompassing narratives and identities that would separate "us" from "them."

John Clute has argued that genre SF is governed by "two essential assumptions": first, that the world and its people "can be seen as wholly as required, and described accurately, in words"; and second, "that the 'world' ... does in the end have a story which can be told."[8] This two-fold belief that the world is knowable and representable reflects genre SF's often dogmatic scientific positivism, which validates only empirical and conclusive scientific knowledge and would extend "hard" science approaches into areas of social science and philosophy. The result is a generic disposition to totalising themes, narratives, and literary forms. Recasting this in Levinasian terms, genre SF is committed to the idea that the universe, and the others who inhabit it, are ultimately totalisable, and that this totality can be adequately communicated in writing.

Isaac Asimov's *Foundation* Trilogy

The totalising tendency that dominated genre SF finds powerful expression in Asimov's *Foundation* trilogy. Originally published as a series of short stories and novellas in *Astounding Science-Fiction* between 1942 and 1950, then re-worked into the novels *Foundation* (1951), *Foundation and Empire*

(1952), and *Second Foundation* (1953), the *Foundation* trilogy remains one of the most influential series of the pulp SF era. Although it was awarded a one-off special Hugo Award for Best All-Time Series in 1966, critics have often questioned the reasons behind its enormous success.[9] This question frames studies by James Gunn and Charles Elkins, but even Asimov asked the question when re-reading the original trilogy in preparation for *Foundation's Edge* (1982):

> I read [the stories] with mounting uneasiness. I kept waiting for something to happen, and nothing ever did. All three volumes, all the nearly quarter of a million words, consisted of thoughts and of conversations. No action. No physical suspense.
>
> What was all the fuss about, then? Why did everyone want more of that stuff?[10]

Gunn posits that the series' success lies in its commitment to "rationalism," as evidenced in its posing, then solving, different problems.[11] Elkins attributes it to the deterministic "concept of history" that gives the stories an unstoppable momentum towards an apparently utopian future.[12] Although there is an element of truth in each of these conclusions, I wish to propose another, not unrelated, answer: the trilogy's success lies in its readers' attraction to the series' utopian vision of absolute totalisation—a vision of total cognitive domination over the Other and the systematic elimination of difference.

Set in the distant future, the *Foundation* trilogy chronicles the decline and fall of a galaxy-wide empire and the struggles for power and survival that follow. It focuses primarily on a Foundation of scientists, established on the galaxy's periphery by Hari Seldon, intended to survive the Galactic Empire's collapse and minimise the duration of the ensuing "dark ages." According to Seldon's plan, this First Foundation would eventually work with a Second Foundation—the nature and location of which is shrouded in mystery—to establish an even more powerful and far-reaching Second Empire. Seldon can predict and manipulate events on a galactic scale through his mastery of *psychohistory*, a discipline that transforms psychology and the social sciences into empirical sciences governed by mathematical laws. "The future isn't nebulous," we are told, "it's been calculated out by Seldon and charted."[13] The *Foundation* stories are at pains to emphasise the concreteness and objectivity of psychohistory, "*that branch of mathematics which deals with the reactions of human conglomerates to fixed social and economic stimuli*."[14] The powers of psychohistory are demonstrated in the series' earlier stories: each time the Foundation faces a crisis, such as invasion by neighbouring kingdoms, we find that the threat was anticipated by Seldon, who orchestrated events such that the Foundation would be victorious. An apparent caveat to the powers of psychohistory is that it "cannot predict the future of a single man with any accuracy," but even this is undermined throughout the series, as Seldon and, later, the psychohistorians of the Second Foundation, prove capable of using psychohistory to manipulate specific individuals.[15]

In "The Mule," however, Seldon's plan seems to unravel when a mutant with psionic powers begins a campaign of galactic domination, eventually seizing control of the Foundation. A holographic recording of Seldon, played during the invasion, reveals that he had not predicted this turn of events. The nature of this tyrant—the Mule—is significant: as his powers are the result of a random genetic mutation, he is apparently "beyond calculation" in ways that ordinary people are not.[16] In the trilogy's final volume, however, the Mule is undone by the Second Foundation, which we find consists of psychohistorians operating in secret to re-work Seldon's equations when new data comes to light. Even the most unique and unexpected individual, then, appears powerless against Seldon's plan.

The deterministic nature of psychohistory has seen it likened to a vulgar form of Marxism. Early Marxism sought to expose the immutable laws governing socioeconomic development, often maintaining that the totality of social relations could be understood in terms of class relations. Georg Lukács articulated the importance of totalisation for Marxism thus: "Die Kategorie der Totalität, die allseitige, bestimmende Herrschaft des Ganzen über die Teile ist das Wesen der Methode, die Marx von Hegel übernommen und originell zur Grundlage einer ganz neuen Wissenschaft umgestaltet hat." ("The category of totality, the all-pervasive supremacy of the whole over the parts is the essence of the method which Marx took over from Hegel and brilliantly transformed into the foundations of a wholly new science.")[17] Whereas totalisation has unavoidably negative implications for Levinas and many poststructuralist thinkers, for certain strains of Marxism it is a utopian ambition, as it is throughout Asimov's *Foundation* trilogy. Donald Wollheim thus posits that "psychohistory is the science that Marxism never became," while Andrew Milner and Robert Savage note that "Marx's scientific socialism, yet another model for psychohistory, seems merely amateurish by comparison."[18] Although Marxists of the Frankfurt School, such as Theodor Adorno, tended to be more critical of the notion of totality and the reductive aspects of "scientific socialism," the Marxism that most widely circulated 1930s and 1940s America followed a more totalising and deterministic strain—what Elkins calls "the vulgar, mechanical, debased version of Marxism."[19] Elkins argues that Asimov "takes this brand of Marxism to its logical end," resulting in both a "pervading fatalism" and a cynical resentment of the proletarian masses, who must always remain ignorant of Seldon's plan.[20] This is demonstrated in the early story "The Encyclopedists," where it is revealed that although Seldon established the First Foundation with the ostensible goal of creating a galactic encyclopedia that would record all human knowledge, this was only a ruse to facilitate his manipulation of multiple generations of scientists.[21] This impulse—setting down in words all human knowledge—can in turn be recognised as another totalising endeavour, and one that eventually gets realised, as we know from reading excerpts from the *Encyclopedia Galactica* throughout the novels. If some form of Marxism is at play here, it would appear to be one that detests the working class and is resolutely authoritarian. And whereas the pre-determined end-point of history from certain Marxist perspectives is global communism, Seldon's anticipated utopia is a new galaxy-wide empire.[22]

Another strand of criticism sees a very different politics at work in the *Foundation* trilogy, finding within many of the tropes and dispositions of fascism and the far-right. Donald M. Hassler, for example, suggests that Asimov's early work may have been unwittingly influenced by "fascist propaganda with its colorful art deco and its grand notions of the heroic and the large futurism of Empire and dominance."[23] Likewise, when reviewing *Foundation's Edge*, David Langford finds Asimov struggling with "certain implications of his original trilogy being slightly distasteful" and sounding "rather like fascism."[24] Far from being intentional on Asimov's part, I suggest that this underlying fascistic inflection is the result of a work philosophically committed to totalisation and written within the politically and socially regressive traditions of American pulp SF. Santesso argues that fascist politics underlie much genre SF, as "certain foundational tropes and traditions of the genre carry the DNA of fascism … to the extent that even liberal, progressive authors working within the genre's more refined strains often (inadvertently) employ fascistic tropes and strategies."[25] Among the tropes identified are the fear of alien invasion, which thinly veils nationalism and fear of the other; a militaristic inflection, coming through preoccupation with advanced weaponry and war; the presence of a heroic "strongman" leader, reflecting Nietzsche's "Übermensch"; and a utopian technological optimism that suggests scientific progress will solve all our problems, rendering social change unnecessary.

A critical reading of Asimov's *Foundation* trilogy reveals the extent to which these tropes permeate the stories. The fear of the other, for example, is embodied in the Mule, who is dangerous *because* of his difference—his mutation. This theme of genetic purity is connected to the fascist trope of the master race, which also pervades the trilogy. Seldon is said to have selected the "youngest" and "strongest" individuals to populate his Foundations, and this rhetoric is intensified when the Second Foundation's psychohistorians are introduced as "a higher subdivision of Man" that is "inherently able" to master complex mathematics.[26] Although they are also referred to as "supermen," there is a greater and more powerful scientific "superman" at work throughout the series: the infallible Hari Seldon.[27] As the founder of the Foundations and perfecter of psychohistory, Seldon is worshipped across the galaxy, his name even coming to stand in for "God"—"for Seldon's sake," characters awkwardly proclaim.[28] Then there is the ultimate, utopian goal of Seldon and his Foundations: the establishment of a Second Empire that would achieve total galactic domination, with Second Foundation psychohistorians destined to rule as "a ready-made ruling class," and the leaders of the First Foundation providing a political infrastructure.[29] With the centrality of imperialism, authoritarianism, master races, and supermen, it is unsurprising that critics have long pointed to the trilogy's fascist overtones.

No doubt Asimov would be horrified by the suggestion that his work shares ideas and philosophical orientations with fascism, and I am certainly not suggesting that Asimov was himself fascist. He was, in fact, an ardent liberal, born into a Russian Jewish family that emigrated to the US when he was a boy, and his memoirs demonstrate his disdain for fascism. As a teenager in Brooklyn, he joined the New York Futurians, an anti-fascist fan organisation that wanted to recruit SF for a broader political movement. Counting among their members several high-profile SF authors, including Wollheim, Frederik Pohl, and Cyril Kornbluth, the Futurians wanted to rid pulp SF of its fascist dispositions, including its militarism, nationalism, heroic strongmen, and uncritical technological utopianism.[30] Asimov's early work was, however, heavily influenced by the editorial oversight of John W. Campbell Jr. and the genre SF conventions that crystallised in Campbell's *Astounding*. Asimov conceived of *Foundation* as a science-fictional retelling of Edward Gibbon's *The History of the Decline and Fall of the Roman Empire* (1776–1789), but Campbell played a major role in shaping the finished product. As Asimov noted in a 1979 interview:

> psychohistory originated in a discussion between myself and Campbell, as so many of the things in my early science-fiction stories did. ... There is some reference to symbolic logic in the first story and that was more or less forced on me by John Campbell ... [who felt] that symbolic logic, further developed, would so clear up the mysteries of the human mind as to leave human actions predictable.[31]

The idea of psychohistory was consistent with what Wollheim identified as Campbell's "mechanistic approach to psychology, sociology, and history," which contributed to his preoccupation with psionics and attraction to L. Ron Hubbard's dianetics.[32] In Levinasian terms, this can be recognised as an overriding desire to overcome the other's unknowability. Campbell also influenced the *Foundation* stories in other ways—for example, his insistence that any representation of aliens establish human beings as superior was distasteful to Asimov, who avoided the issue by writing an all-human galaxy.[33] Asimov described Campbell as "an idiosyncratic conservative ... somewhere to the right of Attila the Hun in politics," but remained one of his most outspoken apologists.[34] Campbell's far-right politics and racist tendencies were well known,[35] with Michael Moorcock calling him "an out-and-out fascist," under whose reign *Astounding* became "a crypto-fascist deeply philistine magazine."[36] Asimov acknowledged that his relationship with Campbell and eagerness to be published in *Astounding* meant he "caught the Campbell flavor," but failed to realise that this flavour, which infused much pulp SF, was often fascist in its orientation, and violent and totalising in its approach to the Other.[37]

Becoming preoccupied with the trilogy's politics, however, can distract from the underlying philosophy that gives rise to its pseudo-Marxist determinism and fascist strains: commitment to the project of totalisation. Asimov did not set out to write communist or fascist propaganda, but he was writing strictly and unquestioningly within the West's totalising philosophical framework. This goes deeper than politics, right to the core of ethics—to the encounter with, and representation of, the Other. Seldon's psychohistory is totalisation *par excellence*. No human individual can stand in the way of this mathematical system in which "all is taken into account."[38] Thus Damien Broderick identifies Asimov's *Foundation* trilogy as "the most explicit instance" of genre SF's tendency towards "the bleakest mechanical determinism and ... the obliteration of the volitional subject."[39] This obliteration is presented in utopian terms, as the reader is invited to celebrate every success of Seldon's plan and behold the incredible power of psychohistory. But the ethical implications of a society based on totalisation are apparent: the individual's alterity is erased and they become only subservient proletarian workers; oppressive social hierarchies assert themselves; authoritarian control is handed to "strongman" leaders and master races; and imperialism and colonisation are glorified. Although such totalising themes dominated pulp SF, and indeed continued to influence the genre SF that followed, SF nonetheless has enormous potential to disrupt totalising approaches and explore ethical encounters with others—a potential that I will explore in relation to Zamyatin's *Мы*.

Infinity and the Face-to-Face Encounter

Let us briefly return to Levinas to consider how totalisation can be disrupted by infinity and the face-to-face encounter. In *Totalité et Infini* (*Totality and Infinity*) (1961), Levinas posits that the idea of infinity is engendered in me by the face of the other person, which "détruit à tout moment, et déborde l'image plastique qu'il me laisse" ("at each moment destroys and overflows the plastic image it leaves me").[40] The face, here, is not reducible to what one can see and touch, but is every aspect the other's countenance and speech, their very presence. In its expression, the face can break apart any image formed of it: "Le visage est présent dans son refus d'être contenu. Dans ce sens il ne saurait être compris, c'est-à-dire englobé." ("The face is present in its refusal to be contained. In this sense it cannot be comprehended, that is, encompassed.")[41] The face-to-face encounter attests to the other's radical separation from the self and their "dimension de l'intériorité" ("dimension of interiority"), which "refuse au concept et résiste à la totalisation" ("declines the concept and withstands totalization").[42] This refusal to be contained, wherein the Other always exceeds the idea I hold of them, gives rise to the idea of infinity—that which cannot be subsumed into the finite totality.

For Levinas, ethics originates in this calling into question of my totalising powers. The face-to-face encounter demands that we consider the Other in their irreducibility, acknowledging their "singularité absolue" ("absolute singularity") as "irreprésentable" ("unrepresentable").[43] As Diane Perpich notes, the face is beyond all representation: "*there is simply no way to do justice to the singularity of a face in a description.*"[44] This presents a problem for literature, for how can we write something that is necessarily beyond representation? For SF, which often deals with truly alien and out-of-this-world others, the problem is even greater. Too often, genre SF allowed its totalising orientation to extend to these unrepresentable Others, which were described in clear and unambiguous prose. Ursula K. Le Guin took pulp SF to task for its failure to develop complex characters, finding the genre replete with "captains and troopers, and aliens and maidens and scientists, and emperors and robots and monsters—all signs, all symbols, statements, effigies, allegories, everything between the Stereotype and the Archetype"—but never singular, unknowable characters.[45] Thus, we find Asimov's *Foundation* trilogy filled with character types, but devoid of fully formed characters.

Other SF traditions, however, found ways to utilise the intellectual and theoretical resources of SF to undertake more ethical explorations of otherness. When examining the "unknowability thesis" in SF, Fredric Jameson points to the Polish author Stanisław Lem, whose *Solaris* (1961) offers a compelling study in "the impossibility of understanding the Other"—a position Jameson finds "implacably negative and skeptical," albeit with a "concomitant ethical imperative."[46] *Solaris* focuses on the unknowability of the alien other—the living ocean of Solaris, which scientists try and fail to comprehend—and reveals something of the radical alterity of the human other in the process. Zamyatin's *Мы*, however, foregrounds the face-to-face encounter and its power to disrupt totalising philosophies and instil a sense of ethics and responsibility. So I turn now to Zamyatin, whose unknowable characters in *Мы* led Le Guin to proclaimed him "the author of the first science fiction *novel*."[47]

Yevgeny Zamyatin's *МЫ*

Whereas Asimov's *Foundation* trilogy envisages a utopian potential for the mathematically rigorous totalisation of all things, Zamyatin's *Мы* offers a dire assessment of a society governed by reductive mathematical formulae and the erasure of otherness. Written in Russia between 1919 and 1921, *Мы* did not see publication in the Soviet Union until 1988, well after the author's death in Paris in 1937. The novel was first published in English translation in 1924, with a full Russian-language version not appearing in print until 1952.[48] It now stands alongside Aldous Huxley's *Brave New World* (1932) and George Orwell's *Nineteen Eighty-Four* (1949) as one of the most influential dystopian novels of the twentieth century. Although Huxley's and Orwell's novels each contain themes present in *Мы*—the question of happiness versus freedom, for example, and wariness of authoritarian politics—Zamyatin's text offers a more concerted attack on scientific positivism and the totalising impulse. Indeed, Daniel Walker identifies *Мы* as one of the earliest SF novels to clearly oppose scientific positivism, as it presents "a dramatization of what happens when an entire population becomes ruled by scientism, when a state practises scientism as its religion, creed, and political philosophy."[49] The rationalism that, according to Gunn, provided *Foundation* with its sense of wonder, here takes on purely negative connotations, with wonder instead being found in the *irrational*.

Мы is set in the One State, a totalitarian metropolis that formed after a devastating Two-Hundred Year War. The One State is separated from the outside world, including all plants and animals, by the Green Wall, a massive glass wall encompassing the city. Everything about life in the One State is strictly regulated and reduced to comprehensible formulae. Workers are referred to as "ciphers" ("нумера," or "numbers"), having alphanumeric designators instead of proper names. The Table of Hours governs each cipher's movements and activities, with every minute of the day accounted for, from eating, working, and resting to their mandatory physical regimen at the Taylor Exercise Hall—only two Personal Hours each day remain. Even love, we are told, has been "побеждена, то есть организована, математизирована" ("conquered, i.e., organized and mathematicized") as ciphers can acquire tickets for sexual encounters with any other cipher they choose.[50] Everything in the One State is designed to erase individuality. Even personal accommodations are not only uniform but made entirely of glass, ensuring full transparency into others' lives and aiming at the eradication of interiority.

Мы follows D-503, a senior mathematician and engineer for the One State, who sees his rigid, mechanised society as a utopia of efficiency and rationality. D-503 is the designer of the Integral, a spaceship intended to bring "математически безошибочное счастье" ("mathematically infallible happiness") (One State ideology) to the universe.[51] The novel begins with D-503 in loyal service to the One State, subscribing wholeheartedly to its ideals and looking forward to the day that Personal Hours are eradicated from the Table of Hours. He finds value in his work as a mathematician specifically for its totalising objectives, declaring that

> работа высшего, что есть в человеке,—рассудка—сводится именно к непрерывному ограничению бесконечности, к раздроблению бесконечности на удобные, легко переваримые порции—дифференциалы. В этом именно божественная красота моей стихии—математики.

> (the worthiest human efforts are those intellectual pursuits that specifically seek the uninterrupted delimiting of infinity, the reduction of infinity into convenient, easily digestible portions—into differentials. The divine beauty of my medium—mathematics—is exactly that.)[52]

This attempted integration of the infinite into a finite totality provides the philosophical foundation of D-503's dystopian society. As Natasha Randall notes, "mathematics travels through *We* almost as an allegorical super-text."[53] Unlike Asimov's trilogy, which praises psychohistory for its mathematical rigour, Zamyatin's novel offers a critique of the impulse towards finality and certainty that mathematics is taken to represent. The disruption to D-503's totalising approach comes in his encounters with others, which ultimately lead him to betray the One State and join a resistance movement, the MEPHI.

The most consequential of these encounters is with I-330, a woman D-503 meets on a regulated walk with his occasional lover, O-90. Upon meeting I-330, D-503 is disturbed by his inability to quantify (and totalise) her the way he would anything or anyone else: "какой-то странный раздражающий икс, и я никак не могу его поймать, дать ему цифровое выражение" ("there was a kind of strange and irritating X to her, and I couldn't pin it down, couldn't give it numerical expression").[54] True to the mathematical language that infuses *Мы*, I-330 is, to D-503, an "X," an *unknown*, and his subsequent encounters with her leave him distressed at the *irrationality*, the "$\sqrt{-1}$," that has entered his life.[55] In one of these early encounters, D-503 is particularly troubled by I-330's inaccessible interiority. He describes her eyes lowering "как шторы" ("like blinds") and later laments: "Там, за шторами, в ней происходило что-то такое—не знаю что, что выводило меня из терпения" ("behind those blinds, inside her, something was going on—I don't know what—and it exasperated me").[56] These encounters mark the beginnings of the breakdown of D-503's faith in the knowability of things and the validity of extending his scientific mode of thinking to the Other.

It is also through conversations with I-330 that D-503 finds his dependence on uniformity and finality drawn into question. Upon finding that MEPHI is planning a revolution by hijacking the Integral, D-503 is horrified. Having been taught the revolution that created the One State was final—the last revolution—D-503 finds the idea of another incomprehensible. I-330, however, appeals to his mathematical reasoning: "назови мне последнее число" ("What is the final number?"), she asks, to which he must concede, "число чисел—бесконечно" ("the number of numbers is infinite"). Likewise, I-330 explains, "Последней [революцию]—нет, революции—бесконечны. Последняя—это для детей: детей бесконечность пугает, а необходимо—чтобы дети спокойно спали по ночам" ("there isn't a final [revolution]. Revolutions are infinite. Final things are for children because infinity scares children and it is important that children sleep peacefully at night").[57] When D-503 objects, arguing that the One State will bring uniformity throughout the universe, I-330 emphasises the importance of *difference*: "равномерно, повсюду! … Тебе, математику,—разве не ясно, что только разности—разности—температур, только тепловые контрасты—только в них жизнь" ("Uniform, all over! … To you, a mathematician, isn't it clear that it's the differences—the differences—between temperatures, it's in thermal contrast that life lies").[58] As well as illustrating the importance of difference in bringing progressive social change, this also serves as a metaphor for the infinite difference between individuals, thwarted by the One State's search for uniformity.

Soon after meeting I-330, D-503 finds himself facing similar disruptive encounters with those closest to him, including O-90 and his friend R-13. "Милая О… Милый R…" ("Sweet O… Dear R…"), D-503 laments, "в нем есть тоже что-то, не совсем мне ясное" ("there is something about them that is … not totally clear to me").[59] During a conversation in which R-13 begins to voice, obliquely, his dissatisfaction with the One State, D-503 imagines his head as an impenetrable, opaque suitcase: "Я смотрел на его крепко запертый чемоданчик и думал: что он сейчас там перебирает—у себя в чемоданчике?" ("I watched his tightly locked little suitcase and thought: what is he now mulling over in that little suitcase of his?")[60] D-503, however, avoids engaging with R-13 further, opting to preserve his image of him as a loyal member of the Institute of State Poets and Writers. D-503 then takes solace that he will later be visited by O-90: "Завтра придет ко мне милая О, все будет просто, правильно и ограничено, как круг" ("Tomorrow, sweet O will come to me and everything will be as simple, correct, and delimited as a circle").[61] But this kind of thinking cannot withstand the face-to-face encounter. When he finally sees O-90 again, D-503 is taken aback by what he perceives as her irrationality—her love of him, jealousy of I-330, and desire to bear an unsanctioned child. His assumptions about O-90 and R-13—the mental images he held of them—turn out to be wholly inadequate, revealing the limits of his totalising powers.

"La relation éthique" ("The ethical relationship"), states Levinas, "met en question le moi" ("puts the I in question").[62] Although Levinas's writings on ethics and the face appear limited to encounters with humans, a clear example of this calling into question in *Мы* comes in D-503's face-to-face encounter with an animal. In the One State, all animals have been exiled to the "диким зеленым океаном" ("wild, green ocean") beyond the Green Wall—to the "неразумного, безобразного мира деревьев, птиц, животных" ("irrational, chaotic world of the trees, birds, animals").[63] Later, as he walks alongside the Green Wall, trying to clear his mind of X's and $\sqrt{-1}$'s, D-503 finds himself face-to-face with one of these wild animals:

> Сквозь стекло на меня—туманно, тускло—тупая морда какого-то зверя, желтые глаза, упорно повторяющие одну и ту же непонятную мне мысль. Мы долго смотрели друг другу в глаза—в эти шахты из поверхностного мира в другой, заповерхностный. И во мне копошится: "А вдруг он, желтоглазый,—в своей нелепой, грязной куче листьев, в своей невычисленной жизни—счастливее нас?"
>
> (Through the glass—foggy and dim—I saw the stupid muzzle of some kind of beast, his yellow eyes, obstinately repeating one and the same incomprehensible thought at me. We looked at each other for a long time, eye to eye, through the mineshaft from the surface world to that other world, beyond the surface. But a thought swarmed in me: what if he, this yellow-eyed being—in his ridiculous, dirty bundle of trees, in his uncalculated life—is happier than us?)[64]

This encounter has D-503 immediately concerned with the animal's inaccessible interiority, as he was with I-330, O-90, and R-13, even making him question the value of his own life. For Levinas, this drawing-into-question of the self is core to the face-to-face encounter, founding our responsibility to and for the Other.

These encounters all lead D-503 to a more ethical orientation towards the Other. According to Levinas, the face of the Other, in calling me into question, "me rappelle à mes obligations et me juge" ("summons me to my obligations and judges me").[65] It is this summons to responsibility that establishes the other person as my neighbour and makes the face-to-face encounter distinctly *ethical*. Over the course of the novel, D-503 gradually aligns himself with MEPHI, a resistance movement seeking the liberation of ciphers from their restrictive, mechanised lives and a return to nature. An example of D-503's sense of responsibility for the Other comes towards the end of the novel in his final encounter with O-90. Having previously avoided situations that could force him to empathise with others, here D-503 slows down and attempts to show compassion for the desperate O-90, despite his prevailing confusion, offering to help her find a way to escape the One State so she and her child can live together.

Such an ethical orientation means breaking with totalising modes of thinking and embracing the irreducible alterity of others and, therefore, the idea of infinity. At the end of the novel, with riots having broken out across the city, D-503 encounters an exasperated fellow mathematician who is desperate to find comfort in the finite in the tumult of revolution:

> успокойтесь … Все это вернется, неминуемо вернется. Важно только, чтобы все узнали о моем открытии. Я говорю об этом вам первому: я вычислил, что бесконечности нет! … Да, да, говорю вам: бесконечности нет. Если мир бесконечен, то средняя плотность материи в нем должна быть равна нулю. А так как она не нуль—это мы знаем,—то, следовательно, Вселенная—конечна … Вы понимаете: все конечно, все просто, все—вычислимо; и тогда мы победим философски,—понимаете?
>
> (Be calm. … Everything will return, inevitably return. The only important thing is that everyone finds out about my discovery. I am telling you this first: I have calculated that there is no infinity! … If the world is infinite—then the average density of matter in it must be exactly zero. And since it is not zero—this we know—then, consequently, the universe is finite … You understand: everything is finite, everything is simple, everything is calculable. Then we will conquer, in the philosophical sense—do you understand?)[66]

Faced with the unthinkable—a mass uprising against the One State—this man seeks comfort in the idea that everything is finite and knowable. This prevents him from encountering the others around him in an ethical way, facing their infinite alterity and responsibility. Such totalising modes of thinking, however, no longer satisfy D-503, who asks: "а там, где кончается ваша конечная Вселенная? Что там—дальше?" ("Where does your finite universe end? What is there—and what comes next?")[67] If finality is for children, D-503 has grown up.

Writing from within the increasingly mechanised social structures of post-revolutionary Russia, Zamyatin, unlike Asimov, was all too aware of the oppressive and authoritarian tendencies of a society governed by totalising approaches to the Other. As Randall notes, *Мы* responds quite directly to its context of War Communism, rapid industrialisation, and post-revolutionary utopianism. For example, Russia's Proletarian Culture movement, established in 1917 "to engender a new proletarian cultural universal," is an obvious source of the One State's approach to art and the individual.[68] But *Мы* is not simply an attack on Zamyatin's contemporary Russia and reading it as an anti-socialist allegory is overly reductive. Zamyatin was a socialist, after all, and an active Bolshevik during his student days, although he was deeply critical of what he saw as the brutal authoritarianism of the Bolshevik leadership after the 1917 October Revolution.[69] The totalising systems critiqued in *Мы* also extend far beyond Zamyatin's Russian context. A clear progenitor of the One State's mechanisation, for example, is Taylorism, the system of scientific management developed in the US by Frederick Winslow Taylor in the late-nineteenth century, directly referenced in *Мы*'s Taylor Exercise Hall. The One State's glass structures reflect London's Crystal Palace, an iron and glass megastructure created to host the Great Exhibition of the Works of Industry of All Nations in 1851. As Peter G. Stillman notes, by Zamyatin's time the Crystal Palace had come to symbolise "Western rationalism's clarity, transparency, and ability to use modern technology to conquer nature and satisfy human needs."[70] The One State thus reflects the trajectories of all modern cities, whether communist or capitalist, towards the destruction of individuality, mechanisation of society, and authoritarianisation of politics, making *Мы* a timeless cautionary dystopia about the ethical implications of a society governed by totalisation and the cognitive conquering of the other.

Science Fiction and the Unenglobable Literary Space

In both Asimov's *Foundation* trilogy and Zamyatin's *Мы*, the text's approach to totalisation and the Other is mirrored in its written form, highlighting the ethical significance of different modes of literary representation. Reflecting on his *Foundation* trilogy, Asimov boasted: "I generally manage to tie up all the loose ends into one neat little bow-knot at the end of my stories, no matter how complicated the plot might be."[71] This tendency towards closure is evident throughout the *Foundation* stories, with each developing a problem and then solving it (usually by invoking psychohistory), carefully explaining this solution to the reader in a comprehensive summation akin to a classic detective story's parlour room scene. The trilogy ends, for example, with an extended (and somewhat contrived) conversation between the leader of the Second Foundation and his student, in which the true goings-on of earlier events are clearly laid out for the reader. Narrative comprehensibility is also reinforced by the trilogy's seemingly omniscient and objective third-person narrator, one-dimensional characters, adherence to linear narrative structures, and clear and unambiguous writing. Much of this was required by Campbell, who controlled form as well as content.[72] Gunn describes Asimov's writing as "rarely more than adequate," while Elkins calls it "a watered-down idiom of … the banal, pseudo-factual style of the mass-circulation magazines."[73] There is no literary experimentation here, no glimpse of what Roland Barthes might call "writerly" form—this was never Asimov's goal: "I don't want to write poetically," he explained, "I only want to write clearly."[74] In this, Asimov adheres to the conventions of genre SF, which typically opts for a realist literary form to communicate scientific ideas and deliver the closed narratives its readers expect.

Concern for narrative's complicity in totalisation underlies Levinas's antipathy towards popular literature. In "La réalité et son ombre" ("Reality and Its Shadow") (1948), Levinas accuses art of being *inhumain* and *monstrueux* ("inhuman and monstrous") for its attempted reduction of the Other to static, lifeless images.[75] Literature does not escape this condemnation, for Levinas is adamant that narrative "n'ébranle pas la fixité de l'image" ("does not shatter the fixity of images"), describing characters in a novel as "êtres enfermés, prisonniers" ("beings that are shut up, prisoners"), forever unable to escape their clearly delimited fates.[76] This critique has particular resonance for the deterministic *Foundation* stories, in which everything happens "by psycho-historical necessity."[77] This fixity is the totalising dimension of narrative that does violence to the Others it represents—as Levinas writes in *Autrement qu'être* (*Otherwise Than Being*) (1974): "L'inénarrable!—autrui perdant dans la narration son visage de prochain" ("The unnarratable other loses his face as a neighbor in narration").[78] It is precisely this immobilisation of the face in representation that Levinas finds monstrous. Yet, the kind of narrative Levinas critiques is distinctly a straightforward, realist narrative form—it is irreversible, settled, closed; its characters knowable; its language transparent. No other form would have worked in Asimov's *Foundation* trilogy, at least not without shaking the philosophical foundations that underpin psychohistory.

In Zamyatin's *Мы*, however, an open and disruptive literary form complements the text's focus on infinity, irrationality, and the face-to-face encounter. *Мы* is presented as the diary of D-503, written to be placed aboard the Integral and carry his description of the "математически совершенной жизни Единого Государства" ("mathematically perfect life of the One State") to alien life.[79] Early records (chapters) are rigid and mechanised, but they become fragmentary and disordered as the narrator's totalising worldview breaks down. The keywords (or *конспекта*, abstract) that open and organise each record start out clear and descriptive, but soon D-503 is unable to categorise his experiences—the keywords of record 27 read: "Никакого конспекта—нельзя" ("No Keywords of Any Kind Are Possible").[80] D-503's intention is to record events with absolute clarity and certainty, and when he finds himself troubled by unknowns, he believes he is "обязанным" ("duty bound") as a writer, to explain them:

> вообще неизвестное органически враждебно человеку, и homo sapiens—только тогда человек в полном смысле этого слова, когда в его грамматике совершенно нет вопросительных знаков, но лишь одни восклицательные, запятые и точки.
>
> (all unknowns are … man's natural enemy. *Homo sapiens* is only man, in the fullest sense of the word, when his grammar contains no question marks, only exclamation marks, commas, and periods.)[81]

This kind of unambiguous prose and narrative closure proves impossible, however, in a world of unknowable Others. At the end of the novel, D-503 is returned to the service of the One State having forcibly received the "Великой Операции" ("Great Operation"), a lobotomy-like procedure that removes an individual's "фантазия" ("imagination"), or what D-503 calls his "душа" ("soul"). D-503 is thus confident that the MEPHI uprising will fail, but the One State is in retreat and the revolution seems to have popular support. Although D-503 provides the final record with the keywords "Факты" ("Facts") and "Я уверен" ("I Am Certain"), this certainty is ultimately withheld from the reader.[82] As Zamyatin wrote in 1923, "этого ограниченного, неподвижного мира нет, он—условность, абстракция, нереальность. И потому реализм—нереален" (the "finite, fixed world … is a convention, an abstraction, an unreality. And therefore Realism—be it 'socialist' or 'bourgeois'—is unreal").[83] "Ах, если бы и в самом деле это был только роман" ("If only this was really some sort of novel"), D-503 laments, but such a realist form cannot be maintained when faced with infinity and the untotalisable Other.[84] Zamyatin made this dense writerly form feel essential: "Старых, медленных, дормезных описаний нет: лаконизм" ("The old, slow, creaking descriptions are a thing of the past"), he wrote soon after completing *Мы*, "но огромная заряженность, высоковольтность каждого слова … и синтаксис—эллиптичен, летуч" ("every word must be supercharged, high-voltage. ... And hence, syntax becomes elliptic, volatile").[85] Indeed, *Мы* achieves an superfluity of meaning in its prose, providing the reader with a text open to myriad interpretive lenses.

The ambiguity and openness of *Мы* resonates strongly with the kind of writing Levinas values, and even attributes ethical qualities. "Reality and Its Shadow" ends with Levinas acknowledging the value of modern(ist) literature, which he notes achieves an awareness of the limitations of literary representation, and in an earlier essay on Proust he praises the openness and indeterminacy of *In Search of Lost Time*.[86] Levinas finds the clearest demonstration of literature's potential to disrupt totalisation in the fragmentary and often opaque writing of Maurice Blanchot, which he claims opens up the "non-englobable espace littéraire" ("unenglobable literary space") through its self-interruption and resistance to closure.[87] To use the language of Levinas's later work, this is literature that strives to *unsay* its Said, returning discourse to the infinitely ambiguous—and infinitely meaningful—realm of the Saying. Similarly, when Zamyatin writes of disruptive face-to-face encounters, these are realised through a form that itself challenges totalising modes of thinking, with the literary form becoming commensurate to the ethical content it seeks to convey.

Conclusion

The drive to cognitively conquer the unknown, and the unknowable Other, is at the core of both Asimov's utopian vision in the *Foundation* trilogy and Zamyatin's dystopian One State in *Мы*. Asimov embraced the formal and thematic conventions of pulp SF, particularly those of Campbell's *Astounding*, and the *Foundation* trilogy illustrates how, when an author is working within a genre's narrow confines, a particular politics—and a particular *ethics*—can assert itself. The totalising imperative of Western philosophy is central to Asimov's trilogy and out of it arises an inflection towards oppressive, totalitarian politics, leading to accusations of vulgar Marxism on the one hand and outright fascism on the other. The unsettling presence of fascist tropes in the series reveals the ethical implications of accepting, uncritically, a genre's tropes, themes, and literary modes.

Responding to very different socio-historical circumstances, Zamyatin explores the breakdown of totalising approaches in *Мы*, actively avoiding the kind of generic obligations at play in Asimov's work to create a truly singular text about the irreducible singularity of the Other. The formula-driven and eminently rational philosophy of D-503 cannot withstand the face-to-face encounter with the Other (whether human or animal), which thus becomes the irrational *par excellence*. Realised through a complex and challenging literary form overflowing with interpretive possibilities, *Мы* achieves a more ethical mode of representation while exploring the violence of a society governed by the erasure of alterity. Unknowability has long been a theme of Eastern European SF and, since the 1960s and 1970s, the SF of the West, with today's broad SF field being much more open to writing that challenges the tropes and traditions of earlier, more restrictive notions of genre. Zamyatin's *Мы* demonstrates the enormous potential of SF to open up an unenglobable literary space and critique totalising ideologies of dystopian futures.

Notes

1. Freedman, *Critical Theory and Science Fiction*, xv.
2. Santesso, "Fascism and Science Fiction," 156.
3. Delany, "Science Fiction and 'Literature,'" 110.
4. I am using the term "genre SF" to refer to the Anglophone "hard" SF that originated in the American pulps in the late 1920s and crystallised in John W. Campbell Jr.'s *Astounding Science-Fiction* in the 1940s.
5. Levinas, *Éthique et infini*, 69; Levinas, *Ethics and Infinity*, 75.
6. Levinas, *En découvrant l'existence*, 168; Levinas, *Collected Philosophical Papers*, 50.
7. Levinas, *En découvrant l'existence*, 170–171; Levinas, *Collected Philosophical Papers*, 53.
8. Clute, "Fabulation."
9. Asimov did not return to the *Foundation* series until the 1980s, first with the sequels *Foundation's Edge* (1982) and *Foundation and Earth* (1986), then with the prequels *Prelude to Foundation* (1988) and *Forward the Foundation* (1993). As I am most interested, here, in the generic dispositions of "golden age" SF, the scope of this study is limited to the original trilogy. Although there were minor alterations between the stories' original publications in *Astounding* and their novelised "fixups," including the addition of a new opening story ("The Psychohistorians") in *Foundation*, the novels nonetheless retain pulp SF aesthetics and remain the most influential forms of the stories.
10. Asimov, "Story behind the Foundation."
11. Gunn, *Isaac Asimov*, 28, 44–45.
12. Elkins, "Isaac Asimov's 'Foundation' Novels," 28.
13. Asimov, *Foundation*, 80.
14. Asimov, *Foundation*, 16. Original emphasis.

15. Asimov, *Foundation*, 22.
16. Asimov, *Second Foundation*, 163.
17. Lukács, *Geschichte und Klassenbewusstsein*, 55; Lukács, *History and Class Consciousness*, chap. 2. Original emphasis.
18. Wollheim, *Universe Makers*, 40–41; Milner and Savage, "Pulped Dreams," 38.
19. Elkins, "Isaac Asimov's 'Foundation' Novels," 31.
20. Elkins, "Isaac Asimov's 'Foundation' Novels," 34.
21. Asimov, *Foundation*, 64.
22. As Bill Ashcroft explores in his chapter for this volume, "Postcolonial Science Fiction and the Ethics of Empire," this imperialist drive develops out of the "colonialist orientation" of genre SF.
23. Hassler, "Skepticism, Belief, and Asimov," 3.
24. Langford, "Mystic Star," 540.
25. Santesso, "Fascism and Science Fiction," 139.
26. Asimov, *Foundation and Empire*, 14; Asimov, *Second Foundation*, 89.
27. Asimov, *Second Foundation*, 167.
28. Asimov, *Foundation*, 152.
29. Asimov, *Second Foundation*, 89.
30. Milner and Savage, "Pulped Dreams," 37–38; Asimov, *In Memory Yet Green*, 244.
31. Asimov, "Interview," 40.
32. Wollheim, *The Universe Makers*, 77.
33. Asimov, *In Memory Yet Green*, 276.
34. Asimov, *In Memory Yet Green*, 196.
35. Asimov, for example, was outraged when Campbell added racist passages to his story "Homo Sol" (1940) upon its publication in *Astounding*. Campbell's racist politics come to the fore in his editorials of the 1960s, which Wollheim describes as being "about how there really are superior and inferior people." Asimov, *In Memory Yet Green*, 275; Wollheim, *The Universe Makers*, 78.
36. Moorcock, Interview, 28; Moorcock, "Starship Storm Troopers," 42.
37. Asimov, "Interview," 35.
38. Asimov, *Foundation*, 19.
39. Broderick, *Reading by Starlight*, 28.
40. Levinas, *Totalité et Infini*, 43; Levinas, *Totality and Infinity*, 50–51.
41. Levinas, *Totalité et Infini*, 211; Levinas, *Totality and Infinity*, 194.
42. Levinas, *Totalité et Infini*, 51; Levinas, *Totality and Infinity*, 57.
43. Levinas, *En découvrant l'existence*, 225; Levinas, *Collected Philosophical Papers*, 116.
44. Perpich, *Ethics of Emmanuel Levinas*, 47, 49. Original emphasis.
45. Le Guin, "Science Fiction and Mrs. Brown," 20. This essay is a response to Virginia Woolf's 1924 essay on character, "Mr. Bennett and Mrs. Brown."

46. Jameson, *Archaeologies of the Future*, 107, 108, 116.
47. Le Guin, "Science Fiction and Mrs. Brown," 18. Emphasis mine.
48. Even then, this Russian edition was printed in New York by Chekhov Publishing House. Curtis, *Englishman from Lebedian'*, 4.
49. Walker, "Going after Scientism," 159.
50. Замятин, *Мы*, chap. 6; Zamyatin, *We*, 21.
51. Замятин, *Мы*, chap. 1; Zamyatin, *We*, 3.
52. Замятин, *Мы*, chap. 12; Zamyatin, *We*, 58.
53. Randall, "Introduction," xvi–xvii.
54. Замятин, *Мы*, chap. 2; Zamyatin, *We*, 8.
55. As Randall notes, Zamyatin quite pointedly describes the square root of negative one as *иррациональных* (*irrational*), rather than imaginary, to set it at odds with One State's emphasis on rationality. Randall, "Introduction," xvii.
56. Замятин, *Мы*, chap. 6; Zamyatin, *We*, 25–26.
57. Замятин, *Мы*, chap. 30; Zamyatin, *We*, 153.
58. Замятин, *Мы*, chap. 30; Zamyatin, *We*, 154.
59. Замятин, *Мы*, chap. 8; Zamyatin, *We*, 40.
60. Замятин, *Мы*, chap. 8; Zamyatin, *We*, 39.
61. Замятин, *Мы*, chap. 12; Zamyatin, *We*, 58.
62. Levinas, *Totalité et Infini*, 213; Levinas, *Totality and Infinity*, 195.
63. Замятин, *Мы*, chap. 17; Zamyatin, *We*, 83.
64. Замятин, *Мы*, chap. 17; Zamyatin, *We*, 83.
65. Levinas, *Totalité et Infini*, 237; Levinas, *Totality and Infinity*, 215.
66. Замятин, *Мы*, chap. 39; Zamyatin, *We*, 201.
67. Замятин, *Мы*, chap. 39; Zamyatin, *We*, 201.
68. Randall, "Introduction," xiii.
69. Curtis, *Englishman from Lebedian'*, 13.
70. Stillman, "Rationalism," 161.
71. Asimov, "Story Behind the Foundation."
72. See, for example: Asimov, *In Memory Yet Green*, 197.
73. Gunn, *Isaac Asimov*, ix; Elkins, "Isaac Asimov's 'Foundation' Novels," 26.
74. Barthes, *S/Z*; Asimov, *In Memory Yet Green*, 313.
75. Levinas, "La réalité et son ombre," 786; Levinas, *Levinas Reader*, 141.
76. Levinas, "La réalité et son ombre," 783–784; Levinas, *Levinas Reader*, 139.
77. Asimov, *Foundation and Empire*, 61.
78. Levinas, *Autrement qu'être*, 211; Levinas, *Otherwise Than Being*, 166.
79. Замятин, *Мы*, chap. 1; Zamyatin, *We*, 4.
80. Замятин, *Мы*, chap. 27; Zamyatin, *We*, 134.
81. Замятин, *Мы*, chap. 21; Zamyatin, *We*, 104.
82. Замятин, *Мы*, chap. 40; Zamyatin, *We*, 202.
83. Замятин, "О литературе"; Zamyatin, "On Literature," 112.
84. Замятин, *Мы*, chap. 18; Zamyatin, *We*, 91.

85. Замятин, "О литературе"; Zamyatin, "On Literature," 111.
86. Levinas, *Proper Names*, 100.
87. Levinas, *Sur Maurice Blanchot*, 47; Levinas, *Proper Names*, 151.

Works Cited

Asimov, Isaac. *Foundation*. London: Granada, 1953.

———. *Foundation and Empire*. London: Granada, 1953.

———. *In Memory Yet Green: The Autobiography of Isaac Asimov, 1920–1954*. Garden City, NY: Doubleday, 1979.

———. "An Interview with Isaac Asimov." By James Gunn. In *Conversations with Isaac Asimov*, edited by Carl Freedman. Jackson, MS: University of Mississippi, 2005.

———. *Second Foundation*. London: Granada, 1953.

———. "The Story behind the Foundation." *Isaac Asimov's Science Fiction Magazine* (December 1982). http://www.pannis.com/SFDG/TheFoundationTrilogy/theStoryBehindTheFoundation.html

Barthes, Roland. *S/Z*, translated by Richard Miller. Oxford: Blackwell, 1990.

Broderick, Damien. *Reading by Starlight: Postmodern Science Fiction*. London: Routledge, 1995.

Clute, John. "Fabulation," *The Encyclopedia of Science Fiction*, edited by John Clute, David Langford, Peter Nicholls and Graham Sleight, January 23, 2018. http://www.sf-encyclopedia.com/entry/fabulation

Curtis, J. A. E. *The Englishman from Lebedian': A Life of Evgeny Zamiatin (1884–1937)*. Boston: Academic Studies Press, 2013.

Delany, Samuel R. "Science Fiction and 'Literature'—or, The Conscience of the King." [1979]. In *Speculations on Speculation: Theories of Science Fiction*, edited by James Gunn and Matthew Candelaria, 95–117. Lanham, MD: Scarecrow Press, 2005.

Elkins, Charles. "Isaac Asimov's 'Foundation' Novels: Historical Materialism Distorted into Cyclical Psycho-History." *Science Fiction Studies* 3, no. 1 (1976): 26–36.

Freedman, Carl. *Critical Theory and Science Fiction*. Middletown, CT: Wesleyan University Press, 2000.

Gunn, James. *Isaac Asimov: The Foundations of Science Fiction*. New York: Oxford University Press, 1982.

Hassler, Donald M. "Skepticism, Belief, and Asimov." *Extrapolation* 40, no. 1 (1999): 3–4.

Jameson, Fredric. *Archaeologies of the Future: The Desire Called Utopia and Other Science Fictions*. London: Verso, 2005.

Langford, David. "Mystic Star and Psychohistorian Reborn," *New Scientist*, February 24, 1983, 540–541.

Le Guin, Ursula K. "Science Fiction and Mrs Brown." In *Science Fiction at Large*, edited by Peter Nicholls, 13–34. London: Victor Gollancz, 1976.
Levinas, Emmanuel. *Autrement qu'être ou au-delà de l'essence*. The Hague: Martinus Nijhoff, 1974.
———. *Collected Philosophical Papers*, translated by Alphonso Lingis. Pittsburgh: Duquesne University Press, 1998.
———. *En découvrant l'existence avec Husserl et Heidegger*. 2nd ed. Paris: J. Vrin, 1994.
———. *Ethics and Infinity*, translated by Richard A. Cohen. Pittsburgh: Duquesne University Press, 1985.
———. *Éthique et infini*. Paris: Fayard, 1982.
———. *Otherwise than Being, or, Beyond Essence*, translated by Alphonso Lingis. Dordrecht: Kluwer Academic, 1981.
———. *Proper Names*, translated by Michael B. Smith. Stanford: Stanford University Press, 1996.
———. "La réalité et son ombre," *Les Temps Modernes* 38 (1948): 771–789.
———. *The Levinas Reader*, edited by Seán Hand. Oxford: Basil Blackwell, 1989.
———. *Sur Maurice Blanchot*. Montpellier: Fata Morgana, 1995.
———. *Totalité et Infini*. Paris: Kluwer Academic, 1971.
———. *Totality and Infinity*, translated by Alphonso Lingis. Hague: Martinus Nijhoff, 1979.
Lukács, Georg. *History and Class Consciousness*, translated by Rodney Livingstone. Merlin Press, 1967. https://www.marxists.org/archive/lukacs/works/history/ch02.htm
———. *Geschichte und Klassenbewusstsein*. 1923. https://archive.org/details/GeorgLukacs-GeschichteUndKlassenbewusstsein
Milner, Andrew, and Robert Savage. "Pulped Dreams: Utopia and American Pulp Science Fiction." *Science Fiction Studies* 35, no. 1 (2008): 31–47.
Moorcock, Michael. Interview in: *John W. Campbell's Golden Age of Science Fiction: Text Supplement to the DVD*, by Eric Solstein and Gregory Moosnick (Digital Media Zone, 2002), 28. http://dmznyc.com/pdfs/JWC_Study_Supplement.pdf
———. "Starship Storm Troopers." *Cienfuegos Press Anarchist Review*, no. 4 (1978): 41–44.
Perpich, Diane. *The Ethics of Emmanuel Levinas*. Stanford: Stanford University Press, 2008.
Randall, Natasha. "Introduction: Them." In *We*, by Yevgeny Zamyatin, translated by Natasha Randall, xi–xxi. New York: The Modern Library, 2006.
Santesso, Aaron. "Fascism and Science Fiction." *Science Fiction Studies* 41, no. 1 (2014): 136–162.
Stillman, Peter G. "Rationalism, Revolution, and Utopia in Yevgeny Zamyatin's *We*." In *Critical Insights: Dystopia*, edited by M. Keith Booker, 160–174. Ipswich, MA: Salem Press, 2013.

Walker, Daniel. "Going after Scientism through Science Fiction." *Extrapolation* 48, no. 1 (2006): 152–167.

Wollheim, Donald A. *The Universe Makers*. New York: Harper & Row, 1971.

Zamyatin, Yevgeny. "On Literature, Revolution, Entropy, and Other Matters." In *A Soviet Heretic*, edited and translated by Mirra Ginsburg, 107–112. Chicago: University of Chicago Press, 1970.

———. *We*, translated by Natasha Randall. New York: The Modern Library, 2006.

Замятин, Евгений. *Мы*. Библиотека Максима Мошкова, 2013. http://az.lib.ru/z/zamjatin_e_i/text_0050.shtml

———. "О литературе, революции, энтропии и прочем." Библиотека Максима Мошкова, 2013. http://az.lib.ru/z/zamjatin_e_i/text_1923_o_literature.shtml

CHAPTER 2

Inversion and Prolepsis: Begum Rokeya Sakhawat Hossain's Feminist Utopian Strategies

Sreejata Paul

Within literary representations, utopia generally takes the form of an imaginary past, or an invented present in a geographically distant location, or the future.[1] Lyman Tower Sargent suggests that utopias are presented "not as models of unrealistic perfection, but as alternatives to the familiar, as norms by which to judge existing societies."[2] While this is an interesting and enlightening task, not all utopias posit alternatives to the heteropatriarchal social set-up. This is the task specifically undertaken by feminist utopias. Anglo-American scholars such as Lucy M. Freibert date the beginning of the Anglophone feminist utopian tradition to Charlotte Perkins Gilman's *Herland* (1915), thereby aligning the origin of the genre to the suffragist movement and its demand for equal voting rights.[3] However, some scholars do acknowledge the existence of novels in this genre before 1915, and even before the suffragist era. Carol Farley Kessler, for instance, writes that nearly five dozen utopias by American women had appeared before suffrage

S. Paul (✉)
Indian Institute of Technology Bombay-Monash Research Academy, Mumbai, India

Z. Kendal et al. (eds.), *Ethical Futures and Global Science Fiction*, Studies in Global Science Fiction,
https://doi.org/10.1007/978-3-030-27893-9_2

was granted in 1920 and dates the earliest, Mary Griffith's "Three Hundred Years Hence," to 1836.[4] Kessler's use of the terms "women's utopias" and "feminist utopias" to mean loosely the same thing is not helpful, as not every woman writer of utopias espouses a critical position that can be designated as feminist. Nevertheless, Kessler's focus on early Anglophone feminist utopian writing (and not just women's utopian writing) may reveal how the genre itself emerged from some women writers' impulse to imagine non-heteropatriarchal societies, and is unallied with the claims to gender equality associated with suffrage. Later Anglophone feminist utopian fiction clearly explores more radical ideas. For example, Ursula K. Le Guin's *The Left Hand of Darkness* (1969) is a thought-experiment based on the premise that gender ought not to exist as a binary locus of identity.

Outside of the Anglo-American and European traditions, the genre of the feminist utopia has been deployed for different uses. Given the movement of ideas between metropole and colony, it would be easy to present the non-European or non-Anglophone uses of the genre as merely derivative. However, contextual specificities meant that feminist utopian writers based in South Asia, for instance, were not interested only in presenting a subversion of the sex/gender binary in their work. Instead, it may be argued that some of their utopias went to the heart of the power dynamic inherent in heteropatriarchal domesticity and sought to construct a different vision of that hierarchised world. One such writer is Begum Rokeya Sakhawat Hossain. In this chapter, I examine why Hossain may have chosen to deploy the utopian mode in her novels *Sultana's Dream* (1905) and পদ্মরাগ (*Padmarag*) (1924) to critique Bengali Muslim heteropatriarchal society, how she uses the rhetorical strategies of inversion and prolepsis in these novels, and how, in doing so, she anticipates the ethical stances espoused by radical feminists and care ethics theorists. Like much feminist utopian writing, Hossain's work "deposes the 'wicked father' of the Freudian family romance" and this, according to Cora Kaplan, results in a simultaneous forfeiture of the mother as well.[5] This forfeiture is more deliberate in Hossain's writing, and culminates in an explicit repudiation of motherhood. However, repudiation of motherhood does not equate to repudiation of femininity for Hossain. Coterminous with care ethics, Hossain's writing deconstructs heteropatriarchal fictions such as the myth of female inferiority. Additionally, Hossain's works espouse a valuing of traditionally feminine qualities, and in particular, notions of co-dependency.

Hossain's Life and Works

Hossain (1880–1932) is the most widely studied figure of the Muslim women writers of colonial Bengal. Born into an aristocratic family in the Rangpur district of undivided Bengal, Hossain's father did not allow his daughters to go to school or obtain any kind of formal education. Hossain's brother, however, gave her secret lessons in Bengali at night when the rest of the family was asleep. At a young age, she married a civil servant named Syed Sakhawat Hossain who then encouraged her to read and write in English, and after whom she named the Sakhawat Memorial School that she established after his death.[6] It was highly unusual for a Muslim woman in colonial Bengal to receive this kind of encouragement and support from male authority figures. What must be remembered is that Hossain and some of the other upper-class women of the time were given access to these extra-institutional forms of learning in order to be trained as suitable consorts for a new class of English-educated and sometimes British-employed, salaried men. In the history of Bengali Muslim women's writing, Hossain appears as the pioneering figure, along with Nawab Faizunnesa and Masuda Rahman, and her role in women's education and politics in colonial Bengal deserves further attention.[7] The Sakhawat Memorial School that she founded was the first school aimed primarily at Bengali Muslim girls in Calcutta. The Anjuman-e-Khawateen-e-Islam (Muslim Women's Organisation) that she founded and presided over also did much for women's education and employment in the first half of the twentieth century. It is only in recent years that Hossain's legacy has been institutionally recognised. This is evident from the fact that the School of Women's Studies at Jawaharlal Nehru University, a prestigious Indian university, has named its annual lecture after her. Furthermore, the anniversary of Hossain's death on December 9 is observed as "Rokeya Day" in Bangladesh and, as part of the celebrations, the Bangladesh government confers the "Begum Rokeya Padak" on individual women for their exceptional achievement.

Hossain began her writing career by publishing non-fiction essays in many of the prominent Bengali periodicals of the early twentieth century, such as *Nabaprabha*, *Nabanoor* and *Saogat*, that evolved in response to growing literacy rates, cheap printing technology, and most importantly, to address the intellectual needs of the educated and salaried Hindu and Muslim readership of the time that was composed of both men and, to a smaller extent, of women. Hossain scholarship has concentrated more on her essays than her novels, essays which more often than not took the form of invectives against patriarchal injunctions placed on Muslim women in colonial Bengal. These were severely criticised by the (mostly male) Muslim-reading public in the early twentieth century, who felt that Hossain's apparent attack on Islam and on the Bengali Muslim patriarchy was unfounded. Even Muslim women readers were critical of both the content and the tone of the essays, which were uninhibited in their display of rage and disapproval.[8] Hindu men, and a smaller number of Hindu women, must have been reading these essays as well, but their responses to Hossain's work have not been recorded in the same meticulous fashion. *Sultana's Dream* and পদ্মরাগ were certainly better received by the reading public of her time than her non-fictional essays.[9] This is possibly because they were less directly antagonistic in their portrayal of the power imbalances in colonial Bengali society while simultaneously maintaining the critical mode of Hossain's textual style. *Sultana's Dream*, as her only piece written originally in English, has gone on to attract a considerable amount of scholarly attention; articles on পদ্মরাগ, written originally in Bengali, are more difficult to come by. Following its initial publication in 1924, পদ্মরাগ was read primarily as a tale of adventure and romance. In recent times, it has been lauded for its portrayal of inter-communal harmony and has garnered the attention of feminist critics, such as Barnita Bagchi, who have read the novel as an educational treatise.

Within South Asian feminist circles, scholars have focused on Hossain's critique of the Muslim patriarchal order's attempts to control women's freedom of speech, movement and sexuality; her demand for an identical curriculum, taught in a segregated fashion to Muslim boys and girls in colonial Bengal; and her critique of *ulemas'* interpretation of the Quran and other sacred Islamic texts.[10] Additionally, scholars have evaluated her place within a tradition of Muslim women's writing, comparing her work with that of other women writers, especially of her time, and commenting on her witty and polemical style of writing. Bagchi, for example, argues that Hossain did not indulge in "facile didacticism"; rather, through her humorous and incisive writing, she was able to play with "ideologies of self-formation and processes of gendering."[11] Bharati Ray argues that Hossain called for a restructuring of the institution of family on the basis of gender equality, but not for its dissolution altogether, though I will dispute this with respect to *Sultana's Dream* in this chapter.[12]

Hossain's Rhetorical Strategies in *Sultana's Dream* and পদ্মরাগ

Sultana's Dream first appeared in serial form in *The Indian Ladies' Magazine* in 1905 and was later published as a book by S. K. Lahiri and Company from Calcutta in 1908. The novel begins with the eponymous heroine dozing in an armchair. Upon opening her eyes, she finds herself in an unknown place and is confronted by Sister Sara who tells her she is in Ladyland (an English neologism coined by Hossain) and shows her around. Unlike colonial Bengal, it is the men who are kept in seclusion in Ladyland, after their defeat in military warfare against a neighbouring nation. Ladyland is ruled by a queen with the support of the lady principals of two colleges where the female students have produced innovative technological inventions. There is neither crime nor disease in Ladyland. Trade is conducted in a peaceful manner with other nations where women hold positions of power. Virtue, as a distinctly "feminine" quality lacked by the men who had previously ruled over the territory, reigns supreme in Ladyland. The narrative ends with Sultana waking up to find she has been dreaming all along. What Sultana sees in her dream is almost the polar opposite of the hierarchy of gender relations in her real world, that is, the world of colonial Bengal, where women are confined to the house and exploited for their reproductive abilities. This is what I mean by inversion, and it performs a very important function in Hossain's narrative. The inversions that Hossain describes, especially men's seclusion within the domestic space and women's role in public life and scientific innovation, forces readers of her time and ours to question assumptions regarding culturally specific stereotypes of gendered behaviour. In particular, Hossain prompts us to challenge the equation of women with reproduction and motherhood, as radical feminists have gone on to do.

পদ্মরাগ was written around 1902 and published in the original Bengali in 1924, with Bagchi's English translation appearing in 2005.[13] This novel imagines the model of an organisation named Tarini Bhavan, run by women from Brahmo, Hindu, Muslim and Christian backgrounds; that is, from all the major religious/political/cultural groups in colonial Bengal. These Bengali women come together to educate the girls of the community, to train destitute women in vocational courses and provide them with a means of livelihood, to rehabilitate widows and to treat the sick and the needy of both genders. In this, Hossain anticipates many of the women's cultural associations that, in the 1920s and 1930s, would come to function as spaces where the women of late colonial India (including undivided Bengal) took their first tentative steps outward from the domestic sphere to the larger arena of the public, associations such as the Anjuman-e-Khawateen-e-Islam. This is the trademark of Hossain's proleptic strategy. Prolepsis is the representation or assumption of a future act or development as if presently existing or accomplished. Hossain's vision operates exactly in this manner, looking not ahead, but adjacent, simultaneously representing situations that have not yet come to pass. Using the rhetorical device of prolepsis, Hossain anticipates in পদ্মরাগ many Anglo-American second wave feminist critical stances such as the focus on micropolitics and the intimate sphere, and of the valuing of traditionally feminine virtues seen in care ethics.

The eponymous heroine of পদ্মরাগ, also called by the names Siddika and Zainab at different junctures in her eventful life, arrives at Tarini Bhavan and trains herself in nursing. Through a curious twist of fate, she finds and nurses back to health a man she was once supposed to marry. The two of them develop an intimate relationship, but Siddika chooses not to accept his proposal of marriage. She instead returns to her ancestral estate to right the wrongs committed by an Englishman who had sought to appropriate the property and to bring up her nephew, who will one day run the estate as her father had in the past. Siddika therefore renders her gender-appropriate service to the continuance of the heteropatriarchal tradition of property inheritance, though Hossain's third-person omniscient narrator does not seem to have a sense of what Siddika's return to the estate might imply. In this novel, humour is used in a different manner than in *Sultana's Dream*. It is deployed sparingly and always only in dialogue. Sometimes as the women of Tarini Bhavan exchange stories in a kind of proto-consciousness raising exercise, the mood evoked is either of pathos or defiance. Suspense regarding the identities of Siddika and her lover plays a major role in holding readers' attention, even till the very end. We realise upon reflection that although the characters are didacts and autodidacts, Hossain's tone does not convey any attempt at didacticism.

As a starting point for any discussion on Hossain's utopian writings, it is beneficial to evaluate what kind of social structure she is aiming to critique. In both her non-fictional essays and her utopian novels, Hossain demands the abolition of a specific custom imposed on Hindu and Muslim middle- and upper-class women in colonial Bengal (and also some parts of north India) known as *abarodh*, or seclusion. This term implied that women would be living in secluded living quarters within the *andarmahal*, or inner rooms, that they would use covered transport when stepping out of the house, and that they would wear appropriately covered clothing. Hossain objects to this extreme form of seclusion, but not necessarily to Bengali Muslim women's attempts to maintain their modesty through the use of the veil. This may have been a practical decision that Hossain made so that the Muslim elite of Calcutta would keep sending their daughters to her school. Herself a *purdahnashin* (one who observes *purdah*, or veils her head), *purdah* and *abarodh* are not synonymous for Hossain. In fact, in পদ্মরাগ, Hossain argues that while the practice of *purdah* has the sanction of the Quran and the Hadith,[14] *abarodh* is definitely a man-made custom that has been imposed on women in India to ensure their subservience.[15] The control of women's movement, speech and sexuality in this manner, coupled with the fact that such women had virtually no economic or social role in the public domain, led to the total separation of the public and the private spheres in elite Bengali Muslim society through the practice of *abarodh*.[16]

Hossain's finest indictment of *abarodh* can be found in her collection of non-fictional "reports" titled *Abarodhbasini*, which was first serialised in the Bengali periodical *Mashik Mohammadi* in 1929 and then published as a book in 1931. *Abarodhbasini* was translated as *The Secluded Ones* by noted Hossain scholar Roushan Jahan in her edited volume titled *Sultana's Dream and Selections from The Secluded Ones* (2009). Report Fourteen recounts how a distant relative of Hossain's fell onto the railway tracks after tripping over her voluminous *burqa*[17] while trying to board a train and was eventually crushed to death, as her maid did not allow any of the male porters to lend a hand and she was herself unable to help her mistress return to a standing position. Report Twenty-Three is an autobiographical account, as Hossain herself states in the very first sentence.[18] In this report, Hossain recounts how, at the tender age of five, she had to hide in an attic for four days to avoid the gaze of two Hindu maids brought along by a visitor. Before narrating the incident, Hossain recollects how her mother or aunt would gesture to her about the impending arrival of people and expect her to hide in response. However, as a child, Hossain had no instinctive recognition of such gestures, implying how *abarodh* was in no way a "natural" condition for Muslim women. Rather, the fact that an ordained hiding place was decided upon in the event of the arrival of people outside of the immediate family shows how *abarodh* was, in fact, a form of social conditioning. Failure to retreat into the pre-fixed hiding place meant that a bevy of accusations and insults would come in the way of the *abarodhbasini*, since not maintaining seclusion was equated with drawing inappropriate attention to oneself, allowing the male gaze to view one's body and exercising excessive sexual freedom.

Sultana's Dream and পদ্মরাগ are also subtle but strong indictments of the practice of *abarodh* and aim to illustrate the benefits of living in a society that is free of *abarodh*. Through these novels, Hossain sought to inspire and instigate her women readers to take control of their own destiny and resist seclusion from the outside world. She portrays women having this kind of authority in a proleptic manner in these novels and accords centrality to women educators and leaders within both narratives. The life-trajectories of the women educators and leaders in *Sultana's Dream* and পদ্মরাগ best convey Hossain's anti-*abarodh* stance. Interestingly, the counterintuitive behavioural traits and activities of her protagonists foreground the important features that Hossain's novels share with celebrated novels of the feminist utopian tradition, such as Marge Piercy's *Woman on the Edge of Time* (1976) and Gerd Brantenberg's *Egalia's Daughters: A Satire of the Sexes* (1977). Jean Pfaelzer lists these features in her essay "The Changing of the Avant-Garde: The Feminist Utopia" (1988) and argues that feminist utopian writings expose contemporary social biases by inverting their traditional codification, that they abandon the marriage plot, and that they reverse readers' expectations about the emplotment of gender roles.[19] Both of Hossain's novels provide a catalogue of spectacular inversions.

In *Sultana's Dream*, it is permissible for an *abarodhbasini* to be out at night since no one will be able to view her face at that time. By a curious reversal, night-time, usually thought to be unsafe for women, is rendered safer than daytime for Sultana. Sultana's shyness and timidity is dubbed "mannish." When Sister Sara says that in Ladyland, men are confined to their "proper places" within the domestic sphere,[20] one can see how Hossain's writing is informed by her awareness that the gendering of the public and the private is not natural, but is artificially constructed and forcefully imposed. Sultana laughs at the idea of the *mardana* (the inverted form of the *zenana*) where men are confined indoors and made to engage in housework and childcare, but her laughter stops a second later. Behavioural traits that are generally deemed feminine are here given a counterintuitive spin and revealed to be an easy shorthand for docility and deference to the dominant gender. This both invokes laughter and is the starting point of an introspective questioning of gender norms, for Hossain makes them so easily substitutable. Sister Sara tells Sultana that the men of Ladyland have had to retire to the *mardana* "for the sake of honor and liberty."[21] Hossain thus critiques the skewed logic of how confinement in the domestic space can translate to retention of "honor," a distinctly feminine virtue within the discourse of Bengali upper-class domesticity, and even more ludicrously to "liberty," the exact opposite of confinement. The question that begs to be asked is, whose honour and liberty are at stake here? Hossain exposes how one-half of the population is able to live free of all responsibilities and are bestowed with honour and commendations for their work, based on the hidden labour of the other half. Thus, Hossain's critique is primarily aimed at this unequal distribution of assets and liabilities in the heteropatriarchal social structure.

With the confinement of the men in the *mardana*, women naturally come to occupy all positions of power in Ladyland. The queen puts an end to the war with a neighbouring nation that had attacked Ladyland, after the men fail to do so using their military prowess. The inventions of the Ladyland students include a balloon to capture rainwater and redistribute it as and when necessary, thereby controlling the weather; and a method of capturing the sun's intense rays and using this excruciating heat to re-route the enemy. Thus, as Fayeza Hasanat notes, the most impactful reversal that occurs in *Sultana's Dream* is that women defy "the masculine notion of power by gaining control over both man and nature."[22] Here it is significant that nature, usually rendered feminine, is tamed but is not blatantly exploited. Natural resources are conserved and re-used for the sustenance of Ladyland's inhabitants, rather than being harnessed for profit-making purposes. As a corollary, phallogocentric definitions of "strength," "power" and so on undergo a revision in this novel. The equation of physical strength with dominance is undermined when Sister Sara tells Sultana that the lion is "stronger" than man but cannot subjugate the human race for that reason.[23] In fact, Sister Sara makes it clear that real strength lies not in the body, but in the mind. Brain trumps over body, and deed over word, and the former terms are both associated with women here. The notion of power is also redefined in *Sultana's Dream*. The queen, though a towering figure, receives Sultana "cordially without any ceremony."[24] Power is not for display, but is secure in its existence. Moreover, the queen says that she does not covet the *Koh-i-noor* diamond or the peacock throne. While it is possible to argue that this is an anti-colonialist stance, an alternate reading is also possible whereby such ornamentation is deemed unnecessary by the queen for she values only knowledge. In any case, the visible signs of power are disdained in Ladyland, though power itself is not. If all claims to power were renounced by the queen, then Hossain's strategy of inversion would not work.

Hossain's aim is not to envision a world where the notion of centralised control no longer exists. Rather she portrays a world where the figurehead of this centralised control does not use her power to intimidate the uninitiated; rather, she welcomes Sultana into her kingdom and is happy to share with her and teach her the alternative way of life she has cultivated there.

In পদ্মরাগ, too, reversals constitute a major mode of storytelling. Md. Rezaul Haque argues that there is a structural divergence of পদ্মরাগ from nationalist narratives in the way in which it deploys the trope of rescue through gender reversal.[25] Nationalist narratives always depict the male protagonist saving the mother/land and her children, especially daughters, from all disasters, natural or socio-political. However, in the two rescue episodes in পদ্মরাগ, the rescuers are always women, the "poor sisters" of Tarini Bhavan who have devoted their lives to the service of the ailing and the needy. These women first rescue Latif Almas, whom Siddika was supposed to marry, and then the Englishman, Robinson, who had appropriated Siddika's father's property. In marked contrast, the character Usha is abandoned to the mercy of dacoits by her cowardly husband when they break into her in-laws' house. This kind of gender reversal allows Hossain to deconstruct the notion of women being the weaker sex. Women in this narrative are weak neither in body nor in mind. In contrast to the cowardice of men like Usha's husband, the courage of the "poor sisters" leads them to search for and find Almas and Robinson in the midst of inhospitable terrain in the foothills of the Himalayas in north Bengal. In addition to this, the sisters' forgiveness and compassion leads them to care for both men despite the fact that they had both done Siddika wrong in the past.

Usha's story is one of a series of personal narratives in পদ্মরাগ which expose the familial and marital oppression suffered by various inhabitants of Tarini Bhavan. A more striking indictment of the same occurs when Hossain refuses to define the telos of her protagonist's life through marriage and heteropatriarchal domesticity. Here Hossain once again critiques the equation of women's lives with childbirth, childrearing and tending to the household. Hossain's narratives, then, anticipate the contemporary repudiation of the compulsory reproductive economy of heteropatriarchy. The women of Ladyland are happy to leave the childrearing to men, none of whom are ever referred to as husbands, by which it would be safe to assume that there is no concept of marriage in Ladyland (though Hossain never says so explicitly within the novel). The poor sisters of Tarini Bhavan, in contrast, are content to serve anyone in need of their care and do not find it necessary to channel this care through husbands or children of their own. In fact, some of these women have voluntarily chosen to live away from husbands who have proven a burden to them in various ways. In both novels, men are sidelined as an avoidable evil. In this context, I argue that *Sultana's Dream* and পদ্মরাগ, like many feminist utopias, participate in the idealisation of "femininity" as a haven or a beacon for the rescue of a masculinised society from amorality and degeneration.[26] As a consequence, these novels uphold a "femininity" that is "essentially charitable, benevolent and sociable," but also asexual.[27] This unwillingness—or perhaps inability—to portray women as sexual beings is the price that Hossain's utopian visions must pay in order to effectively critique marital domesticity.

One of Hossain's clever wordplays captures her critique of marital oppression and her emphasis on women's education simultaneously. In পদ্মরাগ, Hossain plays on the double meaning of the Bengali word *biye*, meaning marriage, and the Bachelor of Arts degree, shortened as BA and pronounced similarly bee-yay. Bagchi points out that Hossain's phrase "*biye* fail" connotes both the "pathos of a failed marriage—the most common vocation of women in those days—as well as the failure to obtain a BA, a college degree."[28] This is best exemplified through the character of Usha within the narrative of পদ্মরাগ. After being abducted by dacoits, Usha is refused entry into her in-laws' home under the pretext that she has spent the night with multiple men, and thus her marriage fails. She takes shelter at Tarini Bhavan and its proprietor, Dina-Tarini, pays for Usha's education, although she fails her BA examination by just three marks. However, Hossain shows that life is both possible and productive after both kinds of "*biye* fail." Usha stays on as a teacher-in-training at Tarini School and eventually becomes the head-teacher at the school. Hossain was acutely aware of the importance of women's education in a day and age when it was not looked upon favourably. Traditionalists in the Bengali society of Hossain's time viewed education as a vehicle of moral corruption and sexual laxity, and every word and deed of an educated woman was scrutinised and criticised. Hossain viewed education as a tool of self-improvement. She realised that it was not enough to bring women out of seclusion without making access to education and gainful employment available to them. Hossain was also aware that early marriage and the burden of housework often became an impediment to the education of girls. Hence, neither marriage nor motherhood is part of women's lives in Ladyland. In fact, Hossain's emphasis on scientific innovations in *Sultana's Dream* should also be read as a critique of various *zenana* education schemes in colonial Bengal that focused solely on hygiene and needlework and did not give women access to knowledge of subjects such as science and mathematics.

This kind of curricular critique is also found in পদ্মরাগ where Siddika realises, soon after coming to Tarini Bhavan, that her elite education which had concentrated on language skills and fine embroidery, is no good for income-generation. She goes on to educate herself and acquire useful skills (most notably, nursing) in order to make herself fit for service like the other women at Tarini Bhavan. In fact, Hossain presents us with a compiled list of marketable skills for women in the nineteenth and twentieth centuries through Siddika and the other "poor sisters." In the workshop at Tarini Bhavan, single, married, separated and widowed women are taught book-binding, spinning, sewing, typing, nursing and sweet-making—all aimed at equipping them to earn their own living. পদ্মরাগ can, in fact, be read as a treatise on women's education. After all, the novel catalogues the problems faced by teachers and principals of a girls' school, proleptically drawing on Hossain's future responsibilities and challenges in the context of the Sakhawat Memorial School. It also foregrounds the heuristic process by which the "poor sisters" of Tarini Bhavan become autodidacts in order to work with and educate others. Finally, it highlights the mutual education of women as they exchange personal narratives in a fashion somewhat anticipating the Anglo-American feminist second wave's focus on micropolitics. In doing so, পদ্মরাগ focuses strongly on how education can enable women to become a support system for one another, and to breach the gap between the private and the public spheres. The characters who are made most compelling in Hossain's utopian works are female educators. This is more evident in *Sultana's Dream* where Sister Sara, the queen and the lady principals of the universities of Ladyland emerge more prominently as inspirations rather than the learner Sultana. In the case of পদ্মরাগ, while Siddika occupies more narrative space than Dina-Tarini (the life-force behind Tarini Bhavan), it is the latter figure that is the very embodiment of the unsectarian welfare work and the highly syncretic spiritualism inherent in the sisterhood that runs Tarini Bhavan.

Through *Sultana's Dream* and পদ্মরাগ, Hossain issues some bold challenges to the Bengali Muslim patriarchy that had previously refused to pay heed to the grievances and demands she had made through her non-fictional essays. She does so through inversion of traditional gender roles as shown above, and by proleptically envisioning how the personal can become the basis of a political demand for women's vocational training and future employment in the paid workforce. For Hossain, then, rhetorical strategies become the lens through which she questions the machinations of the hierarchised society of colonial Bengal, and the genre of the utopia becomes the tool through which she envisions the reformation and restructuring of this unjust hierarchy. The innovative solutions that Hossain proposes to the power imbalances in colonial Bengali society can best be encapsulated using two conceptual paradigms within the contemporary field of feminist ethics: radical feminism and care ethics.

Hossain's Dual Vision of Feminist Ethics

While both *Sultana's Dream* and পদ্মরাগ portray only women occupying positions of authority and responsibility, men are not entirely absent in either narrative. Ladyland does not have a gynocratic society, and the women living there do not take recourse to parthenogenic reproduction. Sexual reproduction continues to be the way of life in Ladyland, but men are confined to the *mardana* after doing their part for the perpetuation of the species. In this, *Sultana's Dream* participates in the suffragist fantasy of reducing the power of man and then completely cutting him off after a short functional use,[29] seen in such notable suffragist science fictional titles as Elizabeth Burgoyne Corbett's *New Amazonia: A Foretaste of the Future* (1889).[30] *Sultana's Dream*, then, may be seen to be espousing a radical feminist ethics that believes that the patriarchal system cannot be reformed and must be eliminated altogether. Radical feminism locates the source of women's oppression in women's biological capacity for motherhood. This is seen in the works of many theorists such as Shulamith Firestone, Kate Millett, Mary Daly and Susan Griffin. As Rosemarie Tong explains in *Feminine and Feminist Ethics* (1993), according to radical feminists, neither women's economic position nor women's lack of equal rights provides an adequate explanation for women's oppression by men. Rather, it is women's reproductive roles and responsibilities that are "the fundamental causes of female subordination and male domination."[31] As a result, radical feminists believe that liberating women from these roles and responsibilities can effectively overturn patriarchy. This, arguably, is what Hossain was hoping to achieve in *Sultana's Dream*. There is no mention of how babies are born in Ladyland, but it is safe to assume that it is nothing out of the ordinary. However, as soon as babies are born, they are left to the men to be fed, cleaned and taken care of. Women are free to pursue other kinds of interests, such as scientific innovation. Hossain's point is not that childrearing should be left to a particular demographic of the population (the broad category of "men" in Ladyland, and the broad category of "women" in colonial Bengal) so that everyone else can shrug this responsibility off their shoulders to focus on more "noble" endeavours. In this, Hossain seems to anticipate some of the backlash that Anglo-American second wave feminists have received since the 1980s onwards. Third wave feminists have argued that most middle- and upper-class women, both of Anglo-American background and those from developing economies, are able to pursue careers outside the home only because of the domestic service rendered by maids of lower socio-economic circumstances inside the home. Hossain's point is that motherhood cannot be designated as colonial Bengali women's sole occupation in life. These women must be given education on a par with men, and then given entry to professions outside the home if they are to be valued in society at large.

In much the same way I have drawn upon radical feminist ethics to read *Sultana's Dream* as representing a repudiation of the compulsory reproductive economy, I read the activities of the sisterhood that runs Tarini Bhavan in পদ্মরাগ as being rooted in a feminine ethics of care. A prominent branch of care ethics is maternal ethics, as seen in the work of Virginia Held, Sara Ruddick and Caroline Whitbeck. However, Hossain does not espouse any form of maternal ethics. Instead, her alignment with the contemporary understanding of care ethics is more fundamentally connected with women's personal experiences and their ability to move away from the capitalism-driven narrative of individual merit and ascendancy. Scholars writing on care ethics have argued that negative masculine psychological traits such as antagonism have set the standard for behaviour in the public world, and that positive feminine psychological traits such as mutual co-dependency have been denigrated.[32] Care ethics has consistently been criticised for being overtly essentialist in nature, but its defenders have tried to focus attention on the belief that "women's distinctive social experiences make them especially perceptive regarding the male bias."[33] This is what I believe Hossain tries to portray in পদ্মরাগ. Much of the narrative revolves around the personal accounts of the "poor sisters" of Tarini Bhavan which acts as their means of education in the school of life. These experiences teach them how all scales in the heteropatriarchal social system are tipped against women, especially those wanting to come out of seclusion and enter into the economy of paid labour outside the home. Having learned their lesson, these women distance themselves from the heteropatriarchal unit of the family, but develop a keen sense of compassion towards anyone who suffers in the hands of the same larger social system that rests on the foundation laid through exploitation of the labour of society's most vulnerable sections. Their specific knowledge of how the phallocentric economy marginalises all identities that do not fit into the mould of the dominant (male, upper class, upper caste, heteronormative) identity makes them especially skilled at identifying such patterns of exclusion and ensuring they are not replicated within the space they inhabit. These women wear saffron or blue uniforms and live together in a space akin to that of a nunnery (in the Christian imagination) or an *ashram* (in the Hindu imagination). While they are constantly in touch with the outside world through their work, they are firmly removed from domesticity, and therefore represent a relation between the public and the private that is diametrically opposed to the total separation of the spheres in colonial Bengali society. Their agency is contained in the ideals of service, renunciation and empathy. In this, Hossain envisions a valuing not just of women, but of qualities deemed feminine and mutually nourishing. My reading suggests that these are the same qualities that form the basis of care ethics theory.

Though the ethical stand that Hossain advocates in *Sultana's Dream* seems to be far apart from the one she advocates in পদ্মরাগ, it is important to note that both novels problematise culturally specific notions of masculinity and femininity, most significantly through instances of inversion in the respective plots. They repudiate the equation of men with political domination, charitable paternalism and phallogocentric definitions of strength (both mental and physical). They also repudiate the equation of women with motherhood, family, childcare and domesticity. Significantly, then, both novels move away from the ideal of the "feminine woman" who lives only for her children and/or men. Instead, they showcase in a proleptic manner the ideal of the "feminist woman" who has loving and nurturing relationships with other women, keeping in mind that both of these are descriptive rather than prescriptive categories.[34] Hossain's elision of men in this manner in her utopian vision(s) may be what her husband alludes to upon reading her first draft of *Sultana's Dream*. Hossain recounted this incident as follows to her first biographer Shamsunnahar Mahmud, who included it in the biography she wrote titled *Rokeya Jibani* (Life of Rokeya) (1937):

> সে বহু দিনের কথা (১৯০৫খৃ:)।আমরা তখন ভাগলপুরের বাঁকা সাবডিবিসনে ছিলাম। আমার পূজনীয় স্বামী '**টুর**'এ গিয়াছিলেন; আমি বাসায় সম্পূর্ণ একাকী ছিলাম। সময় যাপনের নিমিত্ত কিছু একটা লিখিলাম। তিনি দুই দিন পরে ফিরিয়া আসিয়া জিজ্ঞাসা করিলেন, এই দুই দিন আমি কি করিতে ছিলাম। তদুত্তরে আমি তাঁহাকে খসড়া লেখা, "Sultana's Dream" দেখাইলাম। তিনি দাঁড়াইয়া সমস্ত পাঠ করিয়া বলিয়া উঠিলেন – "A terrible revenge!"

(It was long ago (in 1905). We lived in the Banka subdivision of Bhagalpur then. My honourable husband had gone on a "tour"; I was all alone at home. I wrote something to pass the time. He returned two days later and asked what I was doing in the interim. In reply, I showed him a rough draft of *Sultana's Dream*. He kept standing, read it all in one go and then declared—"A terrible revenge!")[35]

Hossain's husband, Syed Sakhawat Hossain, tells her that she has exacted a terrible revenge upon men through the writing of this short novel. Here it is important to emphasise that his tone is appreciative rather than hostile. For the revenge that Hossain wrecks upon men is not to kill them off, but to declare them merely incidental to her narrative and to the lives of women in Ladyland and Tarini Bhavan. In *Sultana's Dream*, women are doing a better job of ruling over Ladyland than men have ever done. Men sometimes render the service required of them in the context of sexual reproduction and then are made to take responsibility for the lives they have supposedly engendered, which is also the only point in the story where men appear in the flesh. In পদ্মরাগ, women are the ones whose forbearance of difficult living conditions is rewarded, and whose labour enriches countless human lives. The only men who occupy any space in the narrative are the character of Siddika's brother, revealed to be Siddika herself in disguise, and Siddika's one-time fiancé whom she rescues, but does not allow to re-enter her life in the way he desires, that is, in marriage. Hossain may seem ruthless in the manner in which she summarily dismisses men in these narratives, but she does so in order to emphasise how it is possible for women to build mutually supportive relationships for the benefit not just of themselves, but of the heteropatriarchal society that had previously devalued and denigrated them. This is Hossain's most significant contribution to the feminist utopian genre: to show how women-centric societies can exist, not in the absence of male members, but despite the presence of male members; and how such societies can be built on models of co-dependency and mechanisms of mutual support.

Notes

1. Greene, "Utopia/Dystopia," 2.
2. Sargent, "Three Faces," 5.
3. Freibert, "World Views," 49.
4. Kessler, "Fables," 190.
5. Kaplan, *Sea Changes*, 210.

6. Ray, *Early Feminists*, 22.
7. Hasan, "Commemorating," 47.
8. Haque, "Educating Women," 102.
9. Haque, "Educating Women," 103.
10. An *ulema* is a body of Muslim scholars who are recognised as having specialist knowledge of Islamic sacred law and theology.
11. Bagchi, "Fruits of Knowledge," 127.
12. Ray, "Feminist Critique," 21.
13. There are recorded conversations in Shamsunnahar's biography of Hossain to indicate that Hossain did not believe that she was skilled enough in English to write a full-length novel in the language. Perhaps she was also aware that she would be able to reach a wider and more class-differentiated audience by writing in Bengali, though I have no direct evidence to support this. From a reading of her other works, primarily her essays, which were clearly directed towards Bengali Muslim women of her time, it seems that she does believe she has a didactic duty to her readers, and writing in Bengali is a means to that end as well, since she can clearly express her thoughts in that language and reach a (female) audience unfamiliar with any other language to a considerable extent.
14. Didactic tales based on the life of the Prophet.
15. Hossain, *Sultana's Dream and Padmarag*, 104.
16. Basu, "Nation and Selfhood," 40.
17. A black full-length garment covering the woman's body from her neck to her feet, and usually coupled with the *hijab*, a veil covering the entirety of a woman's face.
18. Hossain, *Sultana's Dream: A Feminist Utopia*, 29.
19. Pfaelzer, "Changing," 282.
20. Hossain, *Sultana's Dream and Padmarag*, 4.
21. Hossain, *Sultana's Dream and Padmarag*, 14.
22. Hasanat, "Sultana's Utopian Awakening," 115.
23. Hossain, *Sultana's Dream and Padmarag*, 4.
24. Hossain, *Sultana's Dream and Padmarag*, 17.
25. Haque, "Educating Women," 111.
26. Pohl and Tooley, "Introduction," 12.
27. Pohl and Tooley, "Introduction," 13.
28. Bagchi, "Introduction," xiv–xv.
29. Hasanat, "Sultana's Utopian Awakening," 119.
30. I cannot claim to know whether Hossain was familiar with this particular suffragist text, but there is evidence to suggest that the work of the suffragists was much lauded at meetings of the Bengal chapter of the Anjuman-e-Khawateen-e-Islam (Islamic Women's Association), the chapter that Hossain founded.

31. Tong, *Feminine*, 8.
32. Tong, *Feminine*, 40.
33. Jaggar, "Feminist Ethics," 91.
34. Tong, *Feminine*, 8.
35. Mahmud, *Rokeya Jibani*, 55. English translation mine.

Works Cited

Bagchi, Barnita. "Fruits of Knowledge: Polemics, Humour and Moral Education in the Writings of Rokeya Sakhawat Hossain, Lila Majumdar and Nabaneeta Dev Sen." *Asiatic* 7, no. 2 (2013): 126–138.

———. "Introduction." In *Sultana's Dream and Padmarag: Two Feminist Utopias*, by Rokeya Sakhawat Hossain, vii–xxvi. New Delhi: Penguin Books India, 2005.

Basu, Shamita. "Nation and Selfhood: Memoirs of Bengali Muslim Women." In *Muslim Women in War and Crisis: Representation and Reality*, edited by Faegheh Shirazi, 37–51. Austin: University of Texas Press, 2010.

Corbett, Elizabeth Burgoyne. *New Amazonia: A Foretaste of the Future*. London: Tower Publishing, 1889.

Freibert, Lucy M. "World Views in Utopian Novels by Women." *Journal of Popular Culture* 17, no. 1 (1983): 49–60.

Gilman, Charlotte Perkins. *Herland*. New York: Pantheon Books, 1979.

Greene, Vivien. "Utopia/Dystopia." *American Art* 25, no. 2 (2011): 2–7.

Haque, Md. Rezaul. "Educating Women, (Not) Serving the Nation: The Interface of Feminism and Nationalism in the Works of Rokeya Sakhawat Hossain." *Asiatic* 7, no. 2 (2013): 95–113.

Hasan, Md. Mahmudul. "Commemorating Rokeya Sakhawat Hossain and Contextualising Her Work in South Asian Muslim Feminism." *Asiatic* 7, no. 2 (2013): 39–59.

Hasanat, Fayeza. "Sultana's Utopian Awakening: An Ecocritical Reading of Rokeya Sakhawat Hossain's *Sultana's Dream*." *Asiatic* 7, no. 2 (2013): 114–125.

Hossain, Rokeya Sakhawat. *Sultana's Dream and Padmarag: Two Feminist Utopias*. Translated by Barnita Bagchi. New Delhi: Penguin Books India, 2005.

———. *Sultana's Dream: A Feminist Utopia and Selections from the Secluded Ones*. Edited and translated by Roushan Jahan. New York: The Feminist Press at CUNY, 2009.

Jaggar, Alison M. "Feminist Ethics: Projects, Problems, Prospects." In *Feminist Ethics*, edited by Claudia Card, 78–104. Lawrence, KS: University Press of Kansas, 1991.

Kaplan, Cora. *Sea Changes: Culture and Feminism*. London: Verso, 1986.

Kessler, Carol Farley. "Fables toward Our Future: Two Studies in Women's Utopian Fiction." *The Journal of General Education* 37, no. 3 (1985): 189–202.

LeGuin, Ursula K. *The Left Hand of Darkness*. New York: Ace Books, 1969.

Mahmud, Shamsunnahar. *Rokeya Jibani*. Kolkata: Vishwakosh Parishad, 2005.

Pfaezler, Jean. "The Changing of the Avant-Garde: The Feminist Utopia." *Science Fiction Studies* 15, no. 3 (1988): 282–294.

Pohl, Nicole, and Brenda Tooley. "Introduction." In *Gender and Utopia in the Eighteenth Century: Essays in English and French Utopian Writing*, edited by Nicole Pohl and Brenda Tooley, 1–16. Farnham: Ashgate Publishing Company, 2007.

Ray, Bharati. "A Feminist Critique of Patriarchy: Rokeya Sakhawat Hossain (1880–1932)." *Asiatic* 7, no. 2 (2013): 60–81.

———. *Early Feminists of Colonial India: Sarala Devi Chaudhurani and Rokeya Sakhawat Hossain*. New Delhi: Oxford University Press, 2012.

Sargent, Lyman Tower. "The Three Faces of Utopianism Revisited." *Utopian Studies* 5, no. 1 (1994): 1–37.

Tong, Rosemarie. *Feminine and Feminist Ethics*. Belmont, CA: Wadsworth Publishing Company, 1993.

CHAPTER 3

Better Societies for the Ethical Treatment of Animals: Vegetarianism and the Utopian Tradition

Joshua Bulleid

Introduction

Vegetarianism is frequently presented as a marker of idealised societies throughout utopia's literary history. As Fredric Jameson observes in *Archaeologies of the Future* (2005), "beyond the merely aesthetic, it is certain that the issue of the kitchen and the dining room is a central feature of the Utopian text from More to Bellamy and down to our own time."[1] Jameson examines these utopian spaces only insofar as their arrangement is often allegorical of equality between the sexes. Yet his statement is also true of the more essential aspects of utopian-eating habits. As Brett Cooke has stated, "an easy way to detect the true sympathies of the fantasist is to peek at what [they put] on the communal table."[2] The following survey will show, not only that the diets of Western utopias are overwhelmingly averse to the consumption of animal flesh, but that their vegetarianism is almost inextricably linked to, and informed by, broader ethical concerns.

J. Bulleid (✉)
Monash University, Clayton, VIC, Australia

Z. Kendal et al. (eds.), *Ethical Futures and Global Science Fiction*,
Studies in Global Science Fiction,
https://doi.org/10.1007/978-3-030-27893-9_3

Utopia's preoccupation with vegetarianism has not gone unnoticed. In a 2015 survey of utopian food production and eating habits, Lyman Tower Sargent identified vegetarianism as a "regular theme" of utopian literature, which he argued predominantly "reflected a concern with health," and was "often connected with the prohibition of alcohol, coffee, tea, and tobacco."[3] However, this survey reveals that depictions of vegetarianism within the genre are almost invariably paired with concerns and motivations surrounding animal ethics and environmental stability, as opposed to issues of prohibition or physical well-being. By promoting a greater ethical consideration towards non-human "others," utopian representations of vegetarianism encourage both a greater, intrinsic human morality, as well as the acknowledgment of non-human animals as beings worthy of their own inherent ethical consideration.

Unlike the other chapters in this collection, this survey focusses primarily on Anglo-American and English literature. However, examples from outside the Western canon are occasionally provided as points of comparison, and will hopefully inspire further investigation into other, global traditions. Although the survey focusses foremost on utopian literature, consideration must also be given to works more commonly considered as science fiction. Utopia and science fiction have long been treated as interconnected or even identical genres by some of the field's most eminent scholars—most notably Jameson and Darko Suvin.[4] The arguments for and against their identification are too intricate to detail here, save that they each present an alternate reality to the author's and reader's own, while remaining (or at least claiming to be) rooted in the rules which govern that base reality. Regardless of whether one accepts the identification of the two genres, their entanglement has only grown more complex since the turn of the twentieth century, with utopian authors ever more frequently turning to technological and temporal nova as a means to assert their estranged worldviews. Works generally considered as science fiction will therefore become more prominent as the survey progresses.

The first section of this chapter examines the common connections between carnivorousness and violence among early, foundational Western utopias. The second section examines the prominent positive portrayals of vegetarianism within early science-fictional and utopian works during the nineteenth century. The third section is devoted to H. G. Wells, who proved central to both the utopian and science fiction traditions—moving into the twentieth century—and who maintained a critical, although ultimately ambiguous, relationship with vegetarianism throughout his many influential works. The fourth section focusses on early modern feminist utopias and the "critical" utopias of the 1970s, within which positive portrayals of vegetarianism remain prevalent. Finally, consideration will be given to the presence of vegetarianism within the ecological and climate-driven works of utopia and science fiction that have dominated the genres during the first part of the twenty-first century.

Formative Utopias

Idealised endorsements of vegetarianism can be dated back to the beginning of the utopian literary tradition. Sargent dates the emergence of the utopian vegetarian trend to the anonymous *Voyage to the Centre of the Earth* (1755).[5] However, as Sargent acknowledges in a footnote to his claim, idealised illustrations of vegetarianism date back to at least ancient Greece.[6] The Hellenistic period also contains an important, though often overlooked, early utopian endorsement of vegetarianism. As Daniel Dombrowski claims—it is "one of the best kept secrets in the history of philosophy" that Plato's prototypical *Πολιτεία* (*The Republic*) (ca. 308 BCE) was "to be a vegetarian city."[7] The soldiers of Plato's idealised city state are to be fed roast meat, primarily out of practicality, but its other citizens are sustained on a vegetarian diet of cakes, loaves, figs, berries and other luxurious vegetables.[8] A vegetarian diet is adopted in order that the Republic's citizens lead a "peaceful and healthy life" and, while their meat-free regimen also promises "a ripe old age," its moral precedent is emphasised by the following condemnation of carnivorousness as part of an "unjust" society, which would inevitably "lead to war."[9] Similar connections between violence and carnivorousness continue all the way through to the modern era. Along with its endorsement of the "greater healthfulness of a fruit, nut, and vegetable diet," Horace N. Fowler and Samuel T. Fowler's *The Industrial Public* (1941), for example, warns that "the murdering of animals and using them for food, is one of the things that keeps alive the spirit of war, destruction and murder within us."[10] Plato also provides additional support for vegetarianism in his *Νόμοι* (*Laws*) and *'Επινομίς* (*Epinomis*) (ca. 348 BCE), and further associations between meat-eating and violence are repeated throughout many foundational works of the utopian tradition.[11]

Associations between carnivorousness and violent behaviour play a prominent role in Thomas More's eponymous *Utopia* (1516). More's Utopians "suffer none of their citizens to kill their cattle," out of a concern that the butchering of animals will lead their natural sense of "pity and good nature" to become "much impaired."[12] The Utopians also consider hunting to be a "foolish thing," which ought to "stir pity" rather than inspire pleasure. The only Utopian slaughter that takes place is, therefore, carried out only by slaves outside of the Utopian's towns, with procurative hunting looked upon as one of the "basest parts" of their work.[13] The unsettling ethical dynamic established here—whereby the slaughter of non-human animals is perceived as morally degrading but slavery is accepted—exposes the colonial underpinnings of More's *Utopia*, which has lead critics such as Jameson to characterise it as a "forerunner of modern imperialism."[14] Yet, its colonialist undertones also emphasise its uncomfortable relationship with carnivorousness. In order for More's Utopians to continue to eat meat, they must outsource any potential moral degradation to their own dehumanised social classes. As Sargent notes, *Utopia* encapsulates the notion that "who gets to consume and what they consume versus who does the work, says a lot about the structure of the society," and *Utopia*'s classist and colonialist moral imbalances are never more evident than in the matter of its food supply.[15] Similar concerns surrounding carnivorousness are contained within Tommaso Campanella's *La Città del Sole* (*The City of the Sun*) (1602), whose utopian inhabitants are unwilling to slay their animals for meat because it "seemed cruel" and only finally accept their slaughter as "an unjustifiable action for the sake of justifiable ones."[16] Although Campanella and More's utopias are not vegetarian, each of these foundational works shows a clear reluctance on the part of its citizens to engage with carnivorousness, due to concerns for the welfare of both human and non-human animals.

Active endorsements of carnivorousness are tellingly rare among early structural utopias. Meat-eating is rampant and glorified among what Sargent would call "fantasies of abundance," such as the medieval myth of Cockaigne.[17] The almost miraculous abundance of fowl available to the shipwrecked settlers of Henry Neville's *The Isle of Pines* (1668) would also seem more suited to such "fantasies" than those of traditional utopias.[18] Yet the only major example of a classic, programmatic utopia that actively glorifies meat consumption appears to be Francis Bacon's "New Atlantis" (1624). The inhabitants of Bacon's utopian isle of Bensalem are privy to meats whose consumption greatly increases their strength and longevity.[19] Even so, his utopians' reverence for flesh food remains at odds with utopian literature's frequent endorsement of vegetarianism and aversions to flesh-eating. It is also rather ironic that Bacon himself perished of pneumonia, supposedly caught while conducting an experiment into the preservative effects of freezing meat.[20] Where present, however, utopian endorsements of vegetarianism appear to be overwhelmingly driven by ethical rather than hygienic concerns, especially during the nineteenth century.

Nineteenth-Century Utopias

Along with continuing aversions towards carnivorousness, overt endorsements of vegetarianism became prominent among nineteenth-century utopias. In fact, vegetarianism and its connections to animal rights became such a cliché of utopian literature by the end of the nineteenth century that Samuel Butler saw fit to amend two chapters lampooning vegetarian and animal rights advocates to the 1901 edition of his seminal utopian satire *Erewhon* (1872). The first of these chapters concerns a prophet who promotes a meat-free lifestyle in order to increase the Erewhonian's happiness and "prosperity."[21] However, increasing legal and religious sanctions against the consumption of further animal foods—what modern readers would recognise as "veganism"—drives many of the Erewhonians to insanity and self-harm. The second chapter sees a botanist philosopher (secretly a "great meat-eater" himself) seek to expose the "absurdity" of the newly established Puritan Party by extending such restrictions to vegetable as well as animal life.[22] Yet while he portrays vegetarian advocates as irrational extremists, Butler also acknowledges that "even in flesh-eating countries … the poor seldom see meat from year's end to year's end," and that many "law-abiding people" likewise go without animal food and seem "none the worse" for it.[23] Butler's critique is therefore more comical than it is scathing and, despite such satires, as Warren Belasco notes in his 2006 survey of science-fictional and utopian eating habits, the claims of nineteenth-century vegetarianism were often "aired more freely and fairly in utopian literature than elsewhere."[24]

Perhaps the most influential vegetarian writer of the century—and certainly one of the most militant—was Percy Shelley. In his philosophical poem *Queen Mab* (1813), Shelley depicts a utopia wherein immortal "man" no longer "slays the lamb that looks him in the face, /and horribly devours his mangled flesh."[25] Attached to the poem was Shelley's profoundly influential essay "A Vindication of Natural Diet," wherein he charges carnivorousness with both humanity's moral and physical degradation.[26] Both *Queen Mab* and the "Vindication" had a lasting effect on the vegetarian movement of the nineteenth century, with the Irish playwright and noted vegetarian advocate George Bernard Shaw, recording being told by an old Chartist that the early vegetarian sect considered *Queen Mab* to be their "Bible."[27] A vegetarian utopia is also established, via political revolution, in Shelley's later poem *Laon and Cythna; or, The Revolution of the Golden City* (1817).[28]

Mary Shelley depicts her own vegetarian utopia in *The Last Man* (1826). In the novel's vision of the future, England is rendered as a "scene of fertility and magnificence," following a republican revolution, wherein "machines [exist] to supply with facility every want of the population."[29] Adrian, one of its leaders (and a transparent caricature of Percy Shelley) praises the "fruits of the field" that provide its sustenance, while Verney—the "Last Man" of the book's title (and a potential stand-in for Mary Shelley herself)—resists his carnivorous temptations, respecting the distinctions between wild omnivorousness and civilised, utopian vegetarianism set up in the novel.[30] Her earlier *Frankenstein* (1818)—which Brian Aldiss (among others) has influentially argued for as the original work of science fiction[31]—also sees its famous creature offering his creator a utopian vision, during which he describes his vegetarianism as being both "peaceful and human."[32] A moral basis for vegetarianism has, therefore, also been embedded within the science fiction tradition, as much as the utopian tradition, from its very beginning.

Morality and animal welfare are also driving forces behind one of the most explicit and adamant of all nineteenth-century utopian endorsements of vegetarianism. The utopian race of Vril-ya in Edward Bulwer-Lytton's *The Coming Race* (1871) practice "abstinence from [all] other animal food than milk." Their imposing physical forms "suffice to show that … meat is not required for superior production of muscle fibre."[33] Yet ethical, rather than health concerns continue to be the driving force behind their vegetarianism. The Vril-ya are said to have evolved from humans who sought refuge underground during the Biblical deluge, and are therefore primarily motivated by the "contempt and horror" inspired by their forefathers, who "degraded" their lives by eating the flesh of the animals who sought shelter alongside them.[34] Ironically, Lytton's tale went on to inspire the name of the British meat spread Bovril, whose name combines "bovine" with "vril" (the pseudo-electromagnetic force from which the Vril-ya take their name) to suggest physical enhancement via the consumption of meat. Nevertheless, the utopians of Bulwer-Lytton's text itself remain adamant that carnivorousness is both unnecessary and morally reprehensible.

Further examples of ethically driven vegetarianism abound throughout nineteenth-century utopian and science-fictional literature. The utopians of Henry Olerich's *Cityless and Countryless World* (1893) find carnivorousness to be both "antiquated" and "repugnant."[35] As the novel's utopian ambassador explains: "In our opinion, a flesh diet is degenerating, as well as unwholesome. May it not be possible that a human body, built up on the flesh and blood of a carnivorous brute, cannot be expected to contain within itself genuine purity, love and kindness toward others?"[36] Connections to the prohibition of caffeine, tobacco and alcohol are made on similar grounds. However, issues of physical health are never broached in any of the passages that discuss its vegetarian diet. Camille Flammarion's 1894 novel *La Fin du Monde* (*Omega: The Last Days of the World*) features a future utopian society "freed from the vulgar necessity of masticating meats" via technological and evolutionary advancements.[37] Women's suffrage leads to an earthly, vegetarian utopia in John F. McCoy's *A Prophetic Romance* (1896), wherein "murdering to satisfy the appetite" is considered a kind of madness.[38] Fruits and vegetables are also able to be chemically concocted from elements in the air in McCoy's utopia, leading to a more sustainable and moral existence.[39]

Discussions of vegetarianism—or indeed any diet—are, however, conspicuously absent from what are perhaps the nineteenth century's two most notable utopian examples: Edward Bellamy's *Looking Backward* (1887) and William Morris's *News from Nowhere* (1890). Descriptions of food production (along with its consumption) are all but ignored by Bellamy in *Looking Backward*, while Morris's utopia can be assumed to be at least pescatarian due to the presence of salmon farmers.[40] A further curious exception is James Silk Buckingham's *National Evils and Practical Remedies* (1849), which Sargent cites as a primary example of utopian vegetarianism's connections to health and prohibition.[41] Buckingham's utopia includes prohibitions against the introduction of intoxicating substances, including alcohol, opiates and tobacco. However, no mention of vegetarianism is made throughout the cited passage, and the same section goes on to describe regulations for the placement of "cattle-markets" and "slaughter-houses for butchers"—implying both the presence and acceptance of carnivorousness.[42] Even so, these notable carnivorous examples remain exceptions, rather than the rule, when it comes to nineteenth-century utopian eating habits.

A vegetarian diet *is*, however, overtly endorsed by Bellamy in *Looking Backward*'s sequel, *Equality* (1897). In a lengthy chapter titled "Several

Important Matters Overlooked," it is revealed that Bellamy's future utopians "don't eat the flesh of animals any more" and hold a "decided revulsion in sentiment against the former [carnivorous] practices."[43] Once again, ethical and sympathetic reasons are given for the vegetarianism of Bellamy's utopians. Although it is conceded that their vegetarianism has undoubtedly has something to do with the "great physical improvement of the [human] race," their vegetarianism is foremost taken up out of a "great wave of humane feeling … and compunction for all suffering."[44] This same outburst of sympathy is further credited with inspiring the utopian revolution detailed in *Looking Backward*, with Bellamy also explicitly positioning the vegetarianism of his working class revolutionaries in direct opposition to the pre-revolutionary, capitalist, upper classes who "lived chiefly on flesh."[45] While *Equality* proved less influential than its predecessor, it was nonetheless immensely popular—selling out within 36 hours of its original publication—and is yet another significant example where a utopian text has endorsed vegetarianism for overwhelmingly ethical reasons.[46]

H. G. Wells

As noted earlier, the late nineteenth century and early twentieth century saw the beginning of increased interconnectivity between the traditions of utopia and science fiction. The same period also saw the emergence of H. G. Wells, who proved to be one of the most pivotal writers in both traditions, particularly among their Western variants.

Wells held a lifelong fixation with depictions of diet, of which vegetarianism was a frequent feature.[47] His early, influential "scientific romances" regularly contained ambiguous criticisms of vegetarianism. In *The Time Machine* (1895), the peaceful existence of the vegetarian Eloi is supposed to have led to the deterioration of their intellect, and they survive only as cattle—raised by the carnivorous (and potentially cannibalistic) subterranean Morlocks. Similar revelations of carnivorousness are repeated to shocking effect in *First Men in the Moon* (1901), which concludes with a newly discovered, hyper-intelligent lunar species foregoing further communication with humanity, following the proclamation of their vegetarian visitor, Mr. Cavor, that humans do little more than "run about over the surface of [their] world … killing one another for beasts to eat."[48] A similar scenario occurs at the conclusion of *The War in the Air* (1908), which finds the complex economic and nationalistic motivations of its central conflict reduced to people having had "too much meat and drink."[49] Likewise, the Beast People of *The Island of Doctor Moreau* (1896) maintain their humanity by adhering to a quasi-religious "Law," which forbids the consumption of "Fish or Flesh," while the novel's human characters are constantly compromised by the very carnivorous tendencies they seek to suppress.[50] However, the ambiguity of his treatments meant that Wells's condemnation of vegetarianism was not always as effective as he might have intended.

Although Wells might not have seen vegetarianism as humanity's salvation, he also consistently and distinctly associated carnivorousness with its lesser qualities. Wells's vegetarian characters also prove far more sympathetic and effective than his carnivorous critiques. The Eloi's vegetarianism in *The Time Machine* is as much an indication of their feebleness as the Morlocks' carnivorousness is of their violence and inhumanity. Wells also concludes his anti-utopia ironically, by having his narrator commend the Eloi as proof that "even when mind and strength had gone, gratitude and a mutual tenderness still lived on in the heart of man."[51] The narrator's sympathetic presentation is at odds with the Time Traveller's own disapproving depiction of the Eloi's child-like naïveté during his narration of the tale. Yet, it is the narrator's romantic reverence, rather than the Time Traveller's condemnation, that echoes throughout adaptations such as Stephen Baxter's authorised and heavily awarded sequel *The Time Ships* (1995); as well as the Indian-born, Australian author David J. Lake's continuations, *The Man Who Loved Morlocks* (1981) and "The Truth about Weena" (1998). Both Lake and Baxter's sequels are motivated by the Time Traveller going back to the future in an attempt to rescue Weena, his Eloi companion during Well's original narrative. However, as Lake later realised, both his and Baxter's continuations constitute "creative misreadings" of Wells, who "never intended his Traveler … to try and rescue Weena."[52] Likewise, as Aldiss has pointed out, it is the Beast people in *The Island of Doctor Moreau* with whom readers typically sympathise, rather than its ravenous human characters.[53]

The misanthropic sentiments of *The War of the Worlds* (1898) also seem to have been more sympathetic than Wells intended. In what is undoubtedly one of his most influential works, along with *The Time Machine*, Wells pessimistically casts humanity in the role of supposedly "lesser animals" as they face an invasion of hematophagous Martians, in an attempt to deflate his readers' sense of human superiority. Yet, he also asks them to consider "how repulsive our carnivorous habits would seem to an intelligent rabbit."[54] Such a display of animal empathy led to the novel being praised within vegetarian publications such as the international animal rights periodical *The Herald of the Golden Age*; as well as being recommended in the pages of *The Vegetarian* as "a good story for vegetarians to get their friends to read."[55] The same page of *The Vegetarian* also praises George du Maurier's *The Martian* (1897), whose extra-terrestrials, in contrast to Wells's blood-thirsty invaders, possess a "moral sense … so far in advance of ours that we haven't even a terminology for it" and who "feed exclusively on edible moss and roots and submarine seaweed."[56] Although Wells may have intended *The War of the Worlds* to suggest the insignificance of humanity, rather than the significance of the creatures with whom they share the earth—as with *The Time Machine* and *The Island of Doctor Moreau*—his sympathetic stylings proved far more resonant than his satirical ones. However, such animal sympathies were not entirely out of character for Wells, who later advocated for the "urgent need of international game laws and a supernational game-keeper" to protect and preserve endangered species.[57]

Wells was far more direct in his criticisms of vegetarianism in his later, utopian writings. Vegetarians and anti-vivisectionists are criticised as being "impracticable and unconvincing" in *The Shape of Things to Come* (1933).[58] Likewise, Wells's real-life utopian efforts saw him dismissing those who wanted to establish the legal rights of non-human animals as mere representatives of "some particular fad" when seeking to establish a universal Declaration of Human Rights.[59] His most scathing assessment, however, comes in his later utopian effort *Men Like Gods* (1923). Meat and its consumption are of continuous concern to the novel's band of protagonists, who find themselves transported to the parallel realm of Utopia. Upon their arrival, they are hastily reassured that its inhabitants are "*not* vegetarians!" and promptly treated to a "pleasant meat pate."[60] They also encounter a biologist who is working on creating a new kind of fowl that will hopefully combine all the best qualities of beefsteak and chicken breast.[61] Yet, even here, Wells's critique of vegetarianism is ambiguous. It is later revealed that "eating bacon has gone out of fashion" in Utopia and, when one of its cooks must kill a pig in order to satisfy his visitor's appetites, he condemns their diet as being "rather destructive."[62] The Utopians are also discovered to have abolished hunting and to have rendered all remaining non-human predators vegetarian through genetic manipulation. A "greatly increased" intelligence is observed among some of the now-vegetarian species, at the supposed loss of "nothing worth having," suggesting psychological and intellectual benefits to vegetarianism on top of the moral and physiological benefits endorsed elsewhere.[63]

Another landmark work of utopian science fiction, written during the early twentieth century, sees a similar correlation between vegetarianism and increased intelligence. In Olaf Stapledon's *Last and First Men* (1930), the perfected race of Last Men find vegetables to be "very beneficial to the [human] race psychologically," while meat is eaten only on very "rare and sacred occasions."[64] Stapledon also endorses a diet of synthetic food concocted entirely from vegetable matter in the utopian section of his 1942 novel *Darkness and the Light*.[65] The development of artificial food results in a similar state of proxy-vegetarianism in John Macnie's *The Diothas* (1883), whose utopians derive "all they need" to produce their artificial sustenance from the "vegetable world," with eggs and fish being the "only animal products used as food."[66] Wells's earlier, prophetic utopia, *The World Set Free* (1914), likewise, sees animal agriculture passing out of human experience due to advances in the production of synthetic food.[67] However, it is also during the early stages of the twentieth century that the dystopian mode—of which Well's *When the Sleeper Wakes* (1898–1899)[68] arguably constitutes the earliest, modern example—becomes truly established and begins to dominate both utopian and science fiction literature. As Belasco has observed, synthetic meats have their own connotations within the dystopian tradition, often serving as "a culinary symbol of the worst that humans can do to each other."[69] It is not within the scope of this survey to go into the dystopian representation of vegetarianism in any detail. Yet, while Belasco also notices "a strong dose of vegetarianism" among the heroic survivors and rebuilders of dystopian literature, he gives no specific examples, and none appear to be forthcoming within the subgenre's early, foundational texts.[70]

Despite his earlier ambiguities and later hostilities towards vegetarianism, however, Wells's earliest and most influential utopian effort—written towards the end of his original cycle of scientific romances—contains what appears to be a genuine endorsement of a meat-free diet. The world of Wells's *Modern Utopia* (1905) is one in which "no meat figures."[71] As the novel's utopian chaperone explains:

> In all the round world of Utopia there is no meat. There used to be. But now we cannot stand the thought of slaughter-houses. And, in a population that is all educated, and at about the same level of physical refinement, it is practically impossible to find anyone who will hew a dead ox or pig. We never

> settled the hygienic question of meat-eating at all. This other aspect decided us. I can still remember, as a boy, the rejoicings over the closing of the last slaughter-house.[72]

A Modern Utopia is perhaps the only truly programmatic of Wells's utopias, and many real-life—and often vegetarian—utopian communities (including notable pockets within the early Fabian Society) attempted to imitate it during the early decades of the twentieth century.[73] Furthermore, the fact that Wells's Modern Utopians are physically refined is cause for them to abstain from, rather than continue, eating meat, and the above proclamation also causes the novel's narrator to reflect upon the carnivorous practices of twentieth-century English society as being "barbaric."[74] For all his criticisms and satirising of vegetarianism, Well's most endearing and influential utopian effort is also the one which contains his strongest ethical endorsement of a meat-free lifestyle.

Feminist and Critical Utopias

Vegetarianism is particularly common among feminist utopias. A meat-free diet is a primary feature of Charlotte Perkins Gilman's *Herland* (1915), which is perhaps the most significant utopian work to emerge during the early twentieth century outside of Wells. One of the first observations made by the visitors to Gilman's utopia is that it is entirely devoid of cattle.[75] It is later confirmed that the women of the novel's titular enclave live entirely off the fruit of their forest, which has been cultivated to achieve maximum efficiency and environmental sustainability. Along with its endorsement of vegetarianism, Gilman's utopia also critiques the carnivorous attitudes and practices of her contemporary society. The visiting men regularly struggle to conceal the conditions of their outside culture in order to avoid confronting the abundant and readily apparent inequalities inherent within it, and this embarrassment is "nowhere better shown than in the matter of the food supply."[76] The Herlanders are horrified to discover that dairy farming "robs [both] the cow of her calf, and the calf of its true food"; as further workings of the "meat business" are revealed, they turn "very white" and beg to be excused.[77] Such a horrified reaction to descriptions of contemporary farming procedures adds an element of animal empathy to the Herlanders' environmental endorsement of vegetarianism. Likewise, although a vegetable diet was not yet absolute in Gilman's precursory utopian effort, *Moving the Mountain* (1911), meat eating is said to be "decreasing every day," while both zoos and hunting have been completely abolished.[78]

Gilman's emphasis on vegetarianism is echoed among previous works of feminist utopianism. A notable example from outside the Anglo-American tradition is Bengali author Rokeya Sakhawat Hossain's *Sultana's Dream* (1905) (discussed in the previous chapter), whose utopian women take fruit to be their "chief food" and "do not take pleasure in killing [any] creature of God."[79] The women of the English writer Elizabeth Burgoyne Corbett's earlier feminist utopia *New Amazonia* (1889) display a similar "repugnance or inability" to slaughter animals for food, such that "the trade in meat carcases ceased entirely, to the ultimate permanent advantage of [their] nation."[80] Meat is noticeably absent from the meals of American authors Alice Ilgenfritz Jones and Ella Merchant's *Unveiling a Parallel* (1893)[81]; and it is, again, fruit that forms the "principle part" of the utopian diet in fellow-American Mary E. Bradley Lane's *Mizora* (1880–1881/1890).[82] Lane's utopian cuisine is further supplemented by chemically prepared, artificial meats, and her utopians notably "keep no cattle, nor animals of any kind for food or labor."[83] The Mizorans are no animal lovers, however, and are later revealed to have exterminated from their society all non-human species, whose association they find "degrading."[84] Lane's novel is instead representative of an uncomfortable eugenic undertone that informed many early twentieth-century utopias; the supposed "elevated moral character" of her utopian women is framed as stemming as much from the elimination of those with "dark complexions" from their society as from the increased health and eradication of poverty brought about by their synthetic diet.[85] Similar eugenic undertones are implied in Gilman's *Herland*—becoming overt in its sequel, *With Her in Our Land* (1915). However, while eugenics have not disappeared completely from utopian literature—they are driving force behind the plot of Margaret Atwood's *MaddAddam* trilogy (2003–2013) (discussed below) to name just one prominent example—it is more often the vegetarian aspect of these early feminist utopias that have been repeated throughout, and which continue to inform, both the general and feminist-focused utopian traditions of the twentieth century.

Utopian endorsements of vegetarianism appear again, most notably, in Marge Piercy's *Woman on the Edge of Time* (1976). Much like her turn-of-the-century predecessors, Piercy makes many connections between the treatment of non-human animals and treatment of women in her feminist utopian opus. Her protagonist, Connie Ramos, remarks early on that prostitution—the selling of a person's "flesh"—is not dissimilar to the consumption of pigs and even "cannibalism."[86] The fetishisation of cannibalisation and consumption provides a recurring link between human violence and animal abuse throughout the narrative. The connection is made truly damning when Connie is accidentally projected into an alternate, dystopian future. Here she is confronted by a "holigraph" entertainment titled "Good Enough to Eat," which promises such experiential delights as "Mass rapes, torture" and the "Ultimate cannibal scene"—all in extreme close-up. In a surprising twist, however, it is also revealed that the less-privileged inhabitants of this dystopian future are themselves vegetarians—inasmuch as they are forced to eat "mined" food, made from "coal and algae and wood by-products"—while the nefarious upper classes continue to feed on a "sexy" diet of animal tissue.[87]

The novel's carnivorous dystopia is contrasted with the largely vegetarian utopia of Mattapoisett. Again, the driving force behind this vegetarian impulse is one of empathy. In contrast to Wells's carnivorous Utopians in *Men Like Gods*—who, though telepathic, are unable to communicate with non-human animals—Piercy's Mattapoiseans credit their ability to communicate with non-human species using sign language with changing their diets. Piercy's utopia is also home to both dedicated Environmental and Animal Advocates; there is also a political movement in place which aims to outlaw hunting; and they also recognise the cultivation of meat as an "inefficient use of grains."[88] Yet, despite these ethical and environmental concerns, the Mattapoiseans are not entirely vegetarian. They admit to eating meat on holidays as a way of "culling the herd," and Connie is later introduced to visitors from cultures where they continue to "eat plenty of meat."[89] Nevertheless, Piercy's utopia shows an aspiration towards a completely vegetarian state, even if it has not yet been achieved, with imperfection and progress being key elements of the "critical" utopian model, of which *Woman on the Edge of Time* constitutes a pre-eminent example.

Tom Moylan has identified *Woman on the Edge of Time* as one of four key "critical utopias," alongside Joanna Russ's *The Female Man* (1975), Ursula Le Guin's *The Dispossessed* (1974) and Samuel R. Delaney's *Triton* (1977), Moylan argues that, rather than presenting a blueprint for an alternate society, critical utopias focus on "the continuing presence of difference and imperfection" within utopia, in order to create new spaces which allow for "fundamental social change."[90] As Andrew Milner states, these four novels have "become something like a canon for American SF studies," of which—although less overtly than in earlier, structural utopias—vegetarianism continues to be a common feature.[91] The one exception is *The Female Man*, which begins with its primary utopian character describing how she single-handedly stalked and killed a wolf at age 13.[92] The scene can arguably be interpreted as being suggestive of an uncomfortable underlying violence within her utopian society, although no contrasting vegetarian endorsement is provided either.

Yet, despite Russ's exception, the other three entries in the critical canon continue to espouse a familiar, vegetarian ethic. Shevek, the hero of *The Dispossessed*, is a vegetarian.[93] His diet primarily stems from a lack of animal life on his home planet; although he goes on to recognise his hosts' pet otter as his "brother," suggesting a utopian kinship with non-human animals.[94] A similar animal kinship is also emphasised in Le Guin's later utopia *Always Coming Home* (1985), in which meat-eating is abundant. Nevertheless, while the later novel's utopians continue to raise animals such as goats, chickens, rabbits and quail for meat; the "factual" section in detailing their eating habits emphasises that hunting is "of very little real importance" to their food supply, and is primarily preserved as a sport for children, who also consume most of its spoils. Conversely, the gathering of vegetable foods, such as acorns, greens, roots, herbs, seeds and berries, is said to be a "major source of food"—suggesting a far less meat-intensive existence than its narrative implies.[95] Bron Helstrom, the lead character in Samuel R. Delany's *Triton* (1977), is not a vegetarian. However, they also play a more antagonistic role in the novel, and their carnivorous eating habits are further called into question via their juxtaposition with those of their idealised, vegetarian co-worker Mirriam.[96] Bron is also made explicitly "uncomfortable" by a bloodstained butcher encountered while visiting Earth, which is further suggested to be the "most conservative" of the solar system's planets.[97] Later, during an attempt to fit in with Triton's more progressive, utopian culture, Bron orders a vegetarian meal, which they then declare "pleasanter than any" they had had while abroad.[98] Although Delaney's treatment of the vegetarian trope is—like all his renderings—complex and implicit, he nevertheless appears to both acknowledge and utilise the moral continuum established between vegetarianism and carnivorousness throughout utopia's history.

Associations between utopia and vegetarianism are not absolute in the critical era. Alongside Russ's counterexample may be added Ernest Callenbach's controversial *Ecotopia* (1975), which Jameson has declared "the most important Utopia to have emerged from the North American 1960s."[99] As with Le Guin's *Always Coming Home*, Callenbach's utopia is, at heart, an idealisation of Californian Native American cultures. As such, meat is valued in Ecotopia "for its spiritual qualities"; its children are taken on hunting "field trips"; and kills are paraded about its town square.[100] Nevertheless, Callenbach's attitude towards industrialised carnivorousness remains somewhat critical, if not suggestively ambiguous. His protagonist, the American journalist William Weston, is initially "shocked" and "disgust[ed]" by the "ghoulish" scene in the town square and describes the recent introduction of synthetic meat as one of the "great achievements" of American industry.[101] Weston's declaration primarily functions as a representation of the industrial Americanism he is eventually convinced, by the Ecotopians, to leave behind. Yet, later, when visiting an abandoned whaling station, he is prompted to reflect upon the "tragic and irreversible" number of extinct species that have been "gobbled … up" as part of humanity's "relentless increase."[102] Callenbach, therefore, proposes hunting as a more sustainable form continued carnivorousness than industrialised farming. However, Callenbach's prequel novel, *Ecotopia Emerging* (1981), reveals that many members of the Survivalist party, who helped engineer Ecotopia's creation, "were either vegetarians or followed diets in which … meat and fish were used sparingly."[103] The Survivalists also regularly emphasise the economic and environmental advantages of a less meat-intensive diet during their political campaign.[104] Moreover, the masculinist ethic from which *Ecotopia*'s carnivorous endorsement derives is used to justify murderous war games and a vicious sexual encounter Weston himself recognises as being "more or less rape"—suggesting an inherent connection between carnivorousness and aggressive (male) behaviour.[105] Carnivorousness in Callenbach's Ecotopia is therefore emblematic of a violence and cruelty that stands in opposition to the more humane and peaceful vegetarian options presented in Delany, Le Guin and Piercy's more prominent, critical utopian works.

Ecological Utopias

The twenty-first century, and the later decades of the 1900s, have largely seen a shift in focus towards issues of ecological and environmentally sustainability within science fiction and utopian literature. Yet, despite the precedent set by earlier authors, depictions—or even considerations—of vegetarianism appear to have become less frequent among notable genre works. The sudden scarcity of vegetarian narratives seems counter-intuitive, given the environmental benefits of a move towards meat-free diets stressed by such organisations as the United Nations Environmental Programme (UNEP) and the Intergovernmental Panel on Climate Change (IPCC).[106] Minor examples persist. Alan Marshall's *Ecotopia 2121* (2016), for example, sees San Francisco transformed into an organic "Growhemia," in opposition to the many twenty-second-century supermarkets that seem to "cater only to rich carnivores," while the Belgian municipality of Leuven—where the humble cabbage is considered "queen," since its production "does not require heated glass greenhouses"—promotes itself as the "Winter Vegetarian Capital of the World."[107] However, major modern examples of sustainability-driven vegetarianism within utopian literature appear few and far between.

Despite his considerable contribution to the field, the specifics of diet and food sustainability are barely addressed in any of Kim Stanley Robinson's many works of climate fiction and ecological utopia. In *2312* (2012), non-human animals are re-introduced to an ailing Earth in an attempt to restore its ecological balance. The novel's protagonist, Swan, laments that "our horizontal brothers and sisters" should be "enslaved as living meat." However, her earlier poaching activities show she is no vegetarian.[108] Robinson's only real engagement with vegetarianism is contained within his recent *New York: 2140* (2017), in which a return to localised farming has caused meat to largely become a speciality. The reduction in available farmland due to rising sea levels means that animals must be raised on dedicated "farm floors," within the submerged towers of its titular future city. It is therefore left up to the building's inhabitants to slaughter their own meat, with many finding it "easier to eat fake meat or become vegetarian" than to raise and kill their own meat, despite the supposedly "super-humane zappers" provided to them.[109] Although Robinson's treatment of vegetarianism is brief, it suggests that, even in the ecologically devastated, post-global-warming era of the novel's future, ethical concerns remain more effective at converting people to vegetarianism than the diet's many ecological benefits.

Vegetarianism is a central feature of Margaret Atwood's *MaddAddam* trilogy, which has probably become the most critically examined work of twenty-first-century utopian science fiction. The first novel in the series, *Oryx and Crake* (2003), concerns the development of a post-human species called "Crakers," who are rendered both more environmentally and ethically friendly via their vegetarianism. Since the Crakers are neither hunters nor agriculturalists, they are not required to mark out territory or claim possessions, therefore resulting in a species that is neither "violent" nor "given to bloodthirsty acts of retribution."[110] The Crakers' peaceful vegetarianism is contrasted against a terrifying, carnivorous vision of humanity: "They were killing other people all the time. And they were eating up all the Children of Oryx [animals]. … Every day they were eating them up. They were killing them and killing them, and eating them and eating them. They ate them even when they weren't hungry."[111] Vegetarianism is therefore set up as a clear indication of the moral superiority of the utopian Crakers over an amoral and environmentally devastating humanity. Yet, As Jovian Parry points out, "Atwood's novels make clear that she is no supporter of vegetarianism."[112]

As with Wells, Atwood's treatment of vegetarianism remains highly ambiguous. Like the Eloi in *The Time Machine*, the Crakers are initially perceived to be without culture or creativity—calling into question whether they are a truly "superior" option to a carnivorous and consumptive humanity. As the series progresses, it comes to focus on a band of post-plague, human survivors, who are all ex-members of a vegetarian religious cult, called the God's Gardeners. Atwood has stated that she intended both the God's Gardeners and the Crakers as genuine utopian elements within the overbearing dystopia of both *Oryx and Crake* and *The Year of the Flood* (2009). However, the Gardeners' behaviour is so overblown as to be almost satirical, and the trilogy's final volume, *MaddAddam* (2013), sees the survivors rejecting many of their strict impositions—including vegetarianism. Although they end up forming a pact with the intelligent "pigoons" (genetically modified pigs) that prohibits the eating of each species by the other, the humans continue to farm perceivably less intelligent animals, such as deer, for meat, and even (incidentally) introduce the mastication (without swallowing) of fish as a sacred part of a now-burgeoning Craker culture.[113] The *MaddAddam* trilogy therefore concludes with the vegetarian aspect of its utopian vision almost entirely undermined.

Although often acknowledged throughout utopia's history, it appears that the environmental benefits of vegetarianism are yet to be widely or fully realised within the current ecological era of utopian fiction.

Conclusion

The survey presented here is by no means exhaustive, nor does it find utopian endorsements of vegetarianism to be absolute. More investigation into the representation of vegetarianism within modern, ecologically driven science fiction and utopian literature also needs to be undertaken as the genre develops. Nevertheless, whenever vegetarianism has been implemented in any of the landmark utopias discussed in this survey, it has been principally due to concerns of animal ethics, human morality, environmental sustainability or a combination thereof. Although vegetarian utopias such as those of Bellamy and Bulwer-Lytton provide reassurance that their vegetarian diets are perfectly nutritious, health is never the primary motivation for adopting a meat-free diet in any of the classic utopian texts examined in this survey. Of all the utopian texts examined here, it is only in Bacon's carnivorous "New Atlantis" that health benefits are appealed to as the primary justification for a supposedly utopian diet. Even other early non-vegetarian utopias, such as those of More and Campanella, express considerable concern over the moral detriment of carnivorousness and attempt to minimise its impact. Even so, these reluctantly carnivorous early examples remain an exception within the utopian tradition—being both pre-empted by Plato and overwhelmed by the multitude of decidedly vegetarian utopias that followed in their wake. Most significant, however, is not the frequency by which vegetarianism has been featured within utopian literature, but the degree to which it has remained central to the ethical and environmental ethos of these idealised societies.

Notes

1. Jameson, *Archaeologies*, 51.
2. Cooke, "Utopia," 188.
3. Sargent, "Everyday Life," 22.
4. See Milner, *Locating Science Fiction*, 90–96.
5. Sargent, "Everyday Life," 22.
6. Sargent, "Everyday Life," 28n15. See also Spencer, *Vegetarianism*, 38–69; Williams, *Ethics of Diet*, 1–22.
7. Dombrowski, *Philosophy of Vegetarianism*, 62.
8. Plato, *Republic*, 59, 103.
9. Plato, *Republic*, 59–61.
10. Fowler and Fowler, *Industrial Public*, 49.
11. Spencer, *Vegetarianism*, 85–87; Dombrowski, *Philosophy*, 61–63.
12. More, *Utopia*, trans. Burnet, 62. The language condemning butchery in Burnet and other earlier English translations is considerably harsher than that of modern editions. To what extent the exact wording of the citation can be attributed to Thomas More is debatable. However, its sentiment remains consistent across the earlier translations. Ralph Robinson's

original 1556 English translation, for example, gives the cited passage as: "they permitte not their frie citezens to accuftome themfelfes to the killing of beaftes, through the vfe whereof they thinke, clemencye the gentelefte affection of oure nature by lytle and lytle to decaye and peryfhe" (More, *Utopia*, trans. Robinson, 91). Burnet's popular translation has been used here for the sake of clarity and would also have been available to, and likely influential upon, the nineteenth-century utopian authors discussed in the following section.

13. More, *Utopia*, trans. Burnet, 83–84.
14. Jameson, *Archaeologies*, 205.
15. Sargent, "Everyday Life," 19.
16. Campanella, *City of the Sun*, 52.
17. Sargent, "Everyday Life," 18.
18. Neville, *Isle of Pines*, 196.
19. Bacon, "New Atlantis," 180.
20. Bevan, *Real Francis Bacon*, 295–296.
21. Butler. *Erewhon and Erewhon Revisited*, 263.
22. Butler, *Erewhon and Erewhon Revisited*, 272.
23. Butler, *Erewhon and Erewhon Revisited*, 269, 273.
24. Belasco, *Meals to Come*, 112.
25. P. Shelley, *Complete Works*, I, 126, lines 8.211–8.213.
26. P. Shelley, *Complete Works*, VI, 3–20, lines 338–344.
27. Holmes, *Shelley*, 208. However, as Andrew Milner has pointed out, the Chartists were a later political movement, and is rather the Luddites who provide the political backdrop to Shelley's early nineteenth-century writings. Milner, *Literature*, 228.
28. Later published as *The Revolt of Islam*. P. Shelley, *Complete Works*, I, 325, lines 526–527.
29. M. Shelley, *Last Man*, 84.
30. M. Shelley, *Last Man*, 59, 61.
31. Aldiss, "On the Origin," 25–52.
32. M. Shelley, *Frankenstein*, 120.
33. Bulwer-Lytton, *Coming Race*, 119.
34. Bulwer-Lytton, *Coming Race*, 33. Indeed, the eating of animal flesh is never explicitly ordained by God in Genesis until after the flood, which perhaps suggests that, according to Abrahamic legend, humans lived a vegetarian existence while in the Garden of Eden (Gen. 9:2; see also: Spencer, *Vegetarianism*, 112–113).
35. Olerich, *Cityless and Countryless World*, 85.
36. Olerich, *Cityless and Countryless World*, 333.
37. Flammarion, *Omega*, 198.
38. McCoy, *Prophetic Romance*, 126–127.

39. McCoy, *Prophetic Romance*, 209.
40. Morris, *News from Nowhere*, 2. Morris himself appears to have been a decidedly non-vegetarian figure, with George Bernard Shaw recording having once being served a pudding made with suet by Morris's wife, who "couldn't conceal her contempt at [Shaw's vegetarian] folly." See Shaw, *Collected Letters*, 106.
41. Sargent, "Everyday Life," 22.
42. Buckingham, *National Evils*, 144–143, 151.
43. Bellamy, *Equality*, 285.
44. Bellamy, *Equality*, 85–86.
45. Bellamy, *Equality*, 87.
46. "Books and Authors," *New York Times*, July 3, 1897.
47. Kemp, "Edible Predator," 15–18.
48. Wells, *First Men*, 199. (See also page 14, regarding Cavor's vegetarianism.)
49. Wells, *War in the Air*, 225.
50. Wells, *Island of Doctor Moreau*, 59.
51. Wells, *Time Machine*, 91.
52. Lake, "Truth About Weena," 193.
53. Brian Aldiss, introduction, xxxvi.
54. Wells, *War of The Worlds*, 125.
55. "Man under Martian," *The Herald of the Golden Age*, August 15, 1898, 94; Harpur, "Martians and Sportsmen," 540.
56. Du Maurier, *Martian*, 366. Du Maurier's human protagonist also frequently expresses his "loathing for meat" (40, 233, 242).
57. Wells, *Year of Prophesying*, 308.
58. Wells, *Shape of Things to Come*, 276.
59. Wells, *Guide to the New World*, 49.
60. Wells, *Men Like Gods*, 35, 4. Emphasis in original.
61. Wells, *Men Like Gods*, 169.
62. Wells, *Men Like Gods*, 108.
63. Wells, *Men Like Gods*, 73–74, 81. Such occasions include the ceremonial cannibalism of the deceased by their friends and family.
64. Stapledon, *Last and First Men*, 257.
65. Stapledon, *Darkness and the Light*, 146.
66. Macnie, *Diothas*, 86.
67. Wells, *World Set Free*, 147–151.
68. Later revised as *The Sleeper Awakes* (1910).
69. Belasco, *Meals to Come*, 100.
70. Belasco, *Meals to Come*, 100.
71. Wells, *Modern Utopia*, 42.
72. Wells, *Modern Utopia*, 192. "Cattle men" still figure among the wooden toys the Modern Utopians make for their children, however (Wells, *Modern Utopia*, 150).

73. Smith, *H. G. Wells*, 101.
74. Wells, *Modern Utopia*, 192.
75. Gilman, *Herland*, 11.
76. Gilman, *Herland*, 79.
77. Gilman, *Herland*, 47–48.
78. Gilman, *Moving the Mountain*, 144, 202.
79. Hossain, *Sultana's Dream and Padmarag*, 12–3.
80. Corbett, *New Amazonia*, 52.
81. Jones and Merchant, *Unveiling a Parallel*, 19.
82. Lane, *Mizora*, 18.
83. Lane, *Mizora*, 20.
84. Lane, *Mizora*, 113.
85. Lane, *Mizora*, 27, 92.
86. Piercy, *Woman on the Edge*, 39.
87. Piercy, *Woman on the Edge*, 293–296.
88. Piercy, *Woman on the Edge*, 100, 151.
89. Piercy, *Woman on the Edge*, 100, 210.
90. Moylan, *Demand the Impossible*, 39.
91. Milner, *Locating Science Fiction*, 100.
92. Russ, *Female Man*, 1.
93. Le Guin, *Dispossessed*, 72.
94. Le Guin, *Dispossessed*, 131.
95. Le Guin, *Always Coming Home*, 437. The novel's utopians also possess a strict, though seemingly unexplained, aversion towards beef (see 366, 415, 421).
96. Delany, *Triton*, 67.
97. Delany, *Triton*, 168, 172.
98. Delany, *Triton*, 278.
99. Jameson, *Archaeologies*, 13n5. Jameson's claim seems curious, however, given both then novel's publication date and his own later declarations about *Woman on the Edge of Time* (see Jameson, *Archaeologies*, 233).
100. Callenbach, *Ecotopia*, 15, 35.
101. Callenbach, *Ecotopia*, 15–18.
102. Callenbach, *Ecotopia*, 69–70.
103. Callenbach, *Ecotopia Emerging*, 166.
104. Callenbach, *Ecotopia Emerging*, 93, 186.
105. Callenbach, *Ecotopia*, 36, 76.
106. United Nations Environment Programme, *Assessing the Environmental Impacts*, 82; Intergovernmental Panel on Climate Change, "IPCC Special Report," 12.
107. Marshall, *Ecotopia 2121*, "San Francisco 2121" and "Leuven 2121." (No pagination; references given as section titles.)

108. Robinson, *2312*, 416, 54.
109. Robinson, *New York: 2140*, 132.
110. Atwood, *Oryx and Crake*, 120.
111. Atwood, *Oryx and Crake*, 119.
112. Parry, "*Oryx and Crake*," 254.
113. Atwood, *MaddAddam*, 458, 435–436.

Works Cited

Aldiss, Brian. Introduction to *The Island of Doctor Moreau*, by H. G. Wells, xxix–xxxvi. London: Orion, 1993.

———. "On the Origin of Species: Mary Shelley." In *Trillion Year Spree*, by Brian W. Aldiss and David Wingrove, 25–52. London: Victor Gollancz, 1986.

Atwood, Margaret. *MaddAddam*. London: Virago, 2014.

———. *Oryx and Crake*. London: Virago, 2009.

Bacon, Francis. "The New Atlantis." In *Three Early Modern Utopias*, edited by Susan Bruce, 149–186. Oxford: Oxford University Press, 2010.

Baxter, Stephen. *The Time Ships*. London: Voyager, 1995.

Belasco, Warren. *Meals to Come: A History of the Future of Food*. Berkeley: University of California Press, 2006.

Bellamy, Edward. *Equality*. Toronto: George N. Morang, 1897.

Bevan, Bryan. *The Real Francis Bacon: A Biography*. London: Centaur Press, 1960.

Buckingham, James S. *National Evils and Practical Remedies*. London: P. Jackson, 1849.

Bulwer-Lytton, Edward. *The Coming Race*. New York: Henry L. Hinton, 1873.

Butler, Samuel. *Erewhon and Erewhon Revisited*. New York: The Modern Library, 1922.

Callenbach, Ernest. *Ecotopia*. Berkeley: Banyan Tree Books, 2004.

———. *Ecotopia Emerging*. New York: Bantam, 1982.

Campanella, Tommaso. *The City of the Sun*. Auckland: Floating Press, 2009.

Cooke, Brett. "Utopia and the Art of the Visceral Response." In *Foods of the Gods: Eating and the Eaten in Fantasy and Science Fiction*, edited by Gary Westfahl, George Edgar Slusser and Eric S. Rabkin, 188–199. Athens, GA: University of Georgia Press, 1996.

Corbett, Elizabeth Burgoyne. *New Amazonia: A Foretaste of the Future*. London: Tower Publishing, 1889.

Delany, Samuel R. *Triton*. London: Corgi, 1977.

Dombrowski, Daniel. *The Philosophy of Vegetarianism*. Amherst: University of Massachusetts Press, 1984.

Du Maurier, George. *The Martian*. New York: Harper & Brothers, 1897.

Flammarion, Camille. *Omega: The Last Days of the World*. New York: Cosmopolitan, [c1894].

Fowler, Horace N. and Samuel T. Fowler. *The Industrial Public: A Plan of Social Reconstruction in Line with Evolution*. Los Angeles: H. N. Fowler, 1921.
Gilman, Charlotte Perkins. *Herland*. New York: Pantheon Books, 1979.
———. *Moving the Mountain*. New York: Charlton Company, 1911.
Harpur, Carpul. "Martians and Sportsmen," letter to the editor. *The Vegetarian*, October 2, 1897: 540.
The Herald of the Golden Age. "Man under Martian." August 15, 1898.
Holmes, Richard. *Shelley: The Pursuit*. London: Weidenfeld and Nicholson, 1974.
Hossain, Rokeya Sakhawat. *Sultana's Dream and Padmarag*. Translated by Barnita Bagchi. London, Penguin: 2005.
Intergovernmental Panel on Climate Change. "IPCC Special Report on Global Warming of 1.5°C: Frequently Asked Questions." 2018. http://report.ipcc.ch/sr15/pdf/sr15_faq.pdf.
Jameson, Fredric. *Archaeologies of the Future: The Desire Called Utopia and Other Science Fictions*. London: Verso, 2005.
Jones, Alice Ilgenfritz and Ella Merchant. *Unveiling a Parallel: A Romance*. Boston: Arena, 1893.
Kemp, Peter. "The Edible Predator: Wells and Food." In *H. G. Wells and the Culminating Ape*, 7–73. London: Macmillan Press, 1982.
Lake, David J. "The Truth About Weena." In *Dreaming Down Under*, edited by Jack Dann and Janeen Webb. Sydney: HarperCollins, 1998
Lane, Mary E. Bradley. *Mizora: A World of Women*. Lincoln: University of Nebraska Press, 1999.
Le Guin, Ursula K. *Always Coming Home*. London: Grafton, 1988.
———. *The Dispossessed*. Frogmore, Hertfordshire: Granada, 1981.
Macnie, John (as Ismar Thiusen). *The Diothas, or A Far Look Ahead*. New York: Putnam, 1883.
Marshall, Alan. *Ecotopia 2121*. New York; Arcade, 2016.
McCoy, John F. *A Prophetic Romance; Mars to Earth, by the Lord Commissioner*. Boston: Arena, 1896.
Milner, Andrew. *Literature, Culture and Society*. 2nd ed. New York: Routledge, 2005.
———. *Locating Science Fiction*. Liverpool: Liverpool University Press, 2012.
More, Thomas. *Utopia*. Edited by Edward Arber. Translated by Ralph Robinson. 2nd ed. London: A. Constable, 1906.
———. *Utopia*. Translated by Gilbert Burnet. Dublin: R. Reilly, 1743.
Morris, William. *News from Nowhere, or An Epoch of Rest*. New York: Longmans, Green and Co., 1908.
Moylan, Tom. *Demand the Impossible: Science Fiction and the Utopian Imagination*. Edited by Raffaella Baccolini. Revised edition. Oxford: Peter Lang, 2014.
Neville, Henry. *The Isle of Pines*. In *Three Early Modern Utopias*, edited by Susan Bruce, 187–212. Oxford: Oxford University Press, 2010.
New York Times. "Books and Authors." July 3, 1897. http://nyti.ms/2H4RVEZ.

Olerich, Henry. *A Cityless and Countryless World; An Outline of Practical Cooperative Individualism*. Holstein, IA: Gilmore & Olerich, 1893.
Parry, Jovian. "*Oryx and Crake* and the New Nostalgia for Meat." *Society and Animals* 17, no. 3 (2009): 241–256.
Piercy, Marge. *Woman on the Edge of Time*. Aylesbury: The Women's Press, 1979.
Plato. *The Republic*. Translated by Desmond Lee. London: Penguin, 2007.
Robinson, Kim Stanley. *2312*. London: Orbit, 2013.
———. *New York: 2140*. London: Orbit, 2017.
Russ, Joanna. *The Female Man*. London: Gollancz, 2010.
Sargent, Lyman Tower. "Everyday Life in Utopia: Food." In *Food Utopias: Reimagining Citizenship, Ethics and Community*, edited by Paul V. Stock, Michael Carolan and Christopher Rosin, 14–32. Florence: Taylor and Francis, 2015.
Shaw, George Bernard. *Collected Letters: 1926–1950*. Edited by Dan H. Lawrence. London: Max Reinhardt, 1988.
Shelley, Mary. *Frankenstein (1818 text)*. Oxford: Oxford University Press, 2008.
———. *The Last Man*. Hertfordshire: Wordsworth, 2004.
Shelley, Percy. *The Complete Works of Percy Bysshe Shelley*. Edited by Roger Ingpen and Walter Edwin. 10 vols. London: Ernest Benn, 1965.
Smith, David C. *H. G. Wells, Desperately Mortal: A Biography*. New Haven: Yale University Press, 1986.
Spencer, Colin. *Vegetarianism: A History*. New York: Four Walls Eight Windows, 2002.
Stapledon, Olaf. *Darkness in the Light*. London: Methuen, 1942.
———. *Last and First Men*. London: Gollancz: 2009.
United Nations Environmental Group. *Assessing the Environmental Impacts of Consumption and Production: Priority Products and Materials* (2010). http://hdl.handle.net/20.500.11822/8572.
Wells, H. G. *An Experiment in Autobiography: Discoveries and Conclusions of a Very Ordinary Brain (Since 1886)*. London: Victor Gollancz, 1934.
———. *First Men in the Moon*. London: Penguin, 2005.
———. *Guide to the New World: A Handbook of Constructive World Revolution*. London: Victor Gollancz, 1941.
———. *The Island of Doctor Moreau*. London: Penguin, 2005.
———. *Men Like Gods*. London: Sphere, 1976.
———. *A Modern* Utopia. London: Penguin, 2005.
———. *The Shape of Things to Come*. London: Penguin, 2005.
———. *The Time Machine*. London: penguin, 2005.
———. *The War in the Air*. London: Penguin, 1976.
———. *The War of The Worlds*. London: Penguin, 2005.
———. *The World Set Free*. London: Corgi, 1976.
———. *A Year of Prophesying*. New York: Macmillan, 1925.
Williams, Howard. *The Ethics of Diet: A Catena of Authorities Deprecatory of the Practice of Flesh-Eating*. Urbana: University of Illinois Press, 2003.

PART II

Environmental Ethics

CHAPTER 4

Eutopia, Dystopia and Climate Change

Andrew Milner

Introduction

In 2007, Kevin Rudd, the then leader of the Australian Opposition Labor Party, soon to be elected Prime Minister, declared that "Climate change is the great moral challenge of our generation."[1] Neither Rudd nor his Party nor Australia itself seem to have risen to this challenge.[2] But he was almost certainly right: climate change is indeed the great ethical challenge for our generation, even though there are clearly other ethical challenges still outstanding from previous generations. Science fiction (henceforth SF) is a primary mechanism—perhaps *the* primary mechanism—by which our culture imagines its possible futures, both positive and negative. And climate fiction is, as J. R. Burgmann and I have argued elsewhere, better analysed as a sub-genre of SF than as a genre in its own right.[3] In 2012, I pointed to the need for "SF that takes environmental problems as seriously as Cold War SF took the threat of nuclear war."[4] What follows is a provisional assessment of how effectively the genre has responded to the challenge of anthropogenic climate change.

A. Milner (✉)
Monash University, Clayton, VIC, Australia

Z. Kendal et al. (eds.), *Ethical Futures and Global Science Fiction*, Studies in Global Science Fiction,
https://doi.org/10.1007/978-3-030-27893-9_4

Utopia, Eutopia and Dystopia

The range of imaginative responses to global warming runs from dystopia to eutopia by way of many kinds of intervening ambiguity. I use eutopia here as the antonym of dystopia, since, as Thomas More's 1516 Greek pun in Latin made clear, utopia is neither a better nor worse place, but rather a no place.[5] Dystopia, which means bad place, is a more recent coinage, variously ascribed to Henry Lewis Younge in 1747, Noel Turner in 1782 and John Stuart Mill in 1868.[6] In Lyman Tower Sargent's definition, a "utopia (eutopia, dystopia, or utopian satire)" is "a species of prose fiction that describes in some detail a non-existent society located in time and space."[7] Academic utopian studies has thus formulated the increasingly accepted set of distinctions between utopia, referring to the general form and its general conventions; eutopia, meaning its positive variant; and dystopia, meaning its negative variant.[8]

Thus defined, eutopias are normally understood as simply good or better places, dystopias as simply bad or worse places. But, as Tom Moylan's analyses of the "critical utopia" and "critical dystopia" remind us, eutopias can be significantly dystopian in content, dystopias significantly eutopian. Moylan argued that the new American eutopias of the 1970s—Ernest Callenbach's *Ecotopia*, Sally Gearhart's *The Wanderground*, Suzy McKee Charnas's *Motherlines*, Dorothy Bryant's *The Kin of Ata Are Waiting for You*, Joanna Russ's *The Female Man*, Ursula K. Le Guin's *The Dispossessed*, Marge Piercy's *Woman on the Edge of Time* and Samuel R. Delany's *Triton*—were critical in the double sense of Enlightenment critique and the "critical mass" required to produce an explosion.[9] They were distinctive, he argued, insofar as they rejected eutopia "as a blueprint," whilst nonetheless preserving it "as a dream." They therefore focus both on the conflict between eutopia and their "originary world" and on "the continuing presence of difference and imperfection" within eutopia itself. The result is a more plausible, because recognisable and dynamic, set of alternative possibilities.[10] "In resisting the flattening out of utopian writing in modern society," he concluded, "the critical utopia has destroyed, preserved, and transformed that writing and marks the first important output of utopian discourse since the 1890s."[11]

In a subsequent account of "the dystopian turn" in late twentieth-century American SF, Moylan argued that these new critical dystopias "burrow within the dystopian tradition," but only "in order to bring utopian and dystopian tendencies to bear on their exposé of the present moment." They are thus "stubbornly" eutopian, in the sense that they do not move easily towards their own better worlds: "Rather, they linger in the terrors of the present even as they exemplify what is needed to transform it."[12] He also insisted that this was an essentially "recent development," a "distinctive new intervention," specific to the late 1980s and early 1990s.[13] And he carefully distinguished the "classical dystopia" and "critical dystopia," on the one hand, both of which are socially critical, from the "anti-utopia," "pseudo-dystopia" and "anti-critical dystopia," on the other, none of which are.[14] This determination to contextualise critical eutopia and dystopia in relation to very specific historical moments, those respectively of the rise of the American New Left in the 1960s and 1970s and the triumph of Anglo-American neo-liberalism in the 1980s and 1990s, leads Moylan into what seems to me an unnecessarily elaborate theoretical taxonomy. But this is not to suggest that historical context is irrelevant, only that it might be unwise to posit too close a connection between formal and historical levels of analysis.

In the fully revised second edition of *Demand the Impossible*, Moylan observes that "critical" can be used as "either a periodizing or an interpretive protocol, and dialectically as both." He is wary of the second of these three options, he explains, because it can easily "aestheticize" the concept into a purely formal category and thus suppress its "deep political motivation and intention."[15] Indeed, it can do so, but surely need not. The historical preconditions for such criticality can, of course, be identified and explained in terms that will necessarily be socio-political, but they are nonetheless also multiple and various and cannot therefore be tied definitively to any one time and place. Historically different but nonetheless analogous political conjunctures can produce similarly critical texts, which might thus be able to speak to each other over the decades or even over the centuries. And such similarities can be addressed in terms as strongly political as those Moylan applied to the specifics of the United States in the 1970s and 1990s.

If Moylan's position is historically over-specific, then that advanced by Margaret Atwood in her "Dire Cartographies" essay is equally over-generalising. She argues that "Dystopias are usually described as the opposite of utopias. … But scratch the surface a little, and … you see … within each utopia, a concealed dystopia; within each dystopia, a hidden utopia."[16] I doubt this is strictly true and, even if it were, eutopias can nonetheless be more or less critical, just as Moylan observes. But Atwood is certainly right to resist, at least by implication, Moylan's sense of criticality as a peculiar prerogative of American New Wave SF. Rather, these options are formally available and actually deployed, albeit discontinuously, throughout the history of the genre. As so often in utopian studies, the last word might well be left to Sargent:

> the critical utopia has a history in the genre, but, and the buts are crucial, if the term is used narrowly … there are fairly few of them, and the period that Tom identified and the authors he discussed … are very unusual in that what emerged was in fact something that reflected the specific times in which they were written.[17]

We should add, however, that some of the recent climate change eutopias and dystopias are in fact critical in precisely Moylan's and Sargent's sense of the term.

This question of eutopia and dystopia, classical and critical, can provide us with the first level of the framework for an ideal typology of contemporary climate fiction, or "cli-fi," to borrow Dan Bloom's neologism.[18] I use the term ideal type here in Max Weber's sense of:

> a one-sided *accentuation* of one or more points of view and by the synthesis of a great many diffuse, discrete, more or less present and occasionally absent *concrete individual* phenomena which are engaged according to those one-sidedly emphasized viewpoints into a unified *analytical* construct (*Gedankenbild*).[19]

In climate fiction, the range of textual variants is not, however, simply a matter of eutopia or dystopia, but also of different responses to climate change itself. Here, the range runs roughly parallel to the options available in real-world discourse. Climate policy distinguishes between mitigation and adaptation strategies and between positive and negative variants of adaptation, the former seeking possible advantages to be seized upon, the latter disadvantages to be minimised. Mitigation strategies in the strict sense are simply strategies to reduce emissions, but these are very rarely the stuff of SF narrative. Insofar as climate fiction does contemplate mitigating the effects of global warming, it often does so by way of the kind of technological fix implicated in climate engineering. For our purposes, then, mitigation and climate engineering can be considered more or less the same trope. To these three responses we can add as a fourth option various forms of climate change denial; as a fifth, the kind of deep ecological anti-humanism sometimes associated with Lovelock's "Gaia hypothesis"[20]; and, as a sixth, the kind of pessimistic fatalism that seems very common in the real world, but less so in SF.

An Ideal Typology of Contemporary Climate Fiction

Instances of all six kinds of response—denial, mitigation as engineering, positive adaptation, negative adaptation, deep ecology, fatalism—can be observed in climate fiction. Good examples of denial include Michael Crichton's *State of Fear* (2004), Liu Cixin's 三体 trilogy (2008–2010) and Nele Neuhaus's *Wer Wind sät* (2011); of the special kind of denial that calls into question the scientists rather than the science, Ian McEwan's *Solar* (2010) and Sven Böttcher's *Prophezeiung* (2011); of mitigation, Arthur Herzog's *Heat* (1977), Dirk C. Fleck's *MAEVA!* trilogy (2008–2015) and Kim Stanley Robinson's *2312* (2012) and *Aurora* (2015); of negative adaptation, George Turner's *The Sea and Summer* (1987), Michel Houellebecq's *La Possibilité d'une île* (2005), Jean-Marc Ligny's *Aqua™* (2006), Paolo Bacigalupi's *The Windup Girl* (2009), Barbara Kingsolver's *Flight Behavior* (2012) and Robinson's *New York 2140* (2017); of positive adaptation, Robinson's *Science in the Capital* trilogy (2004–2007), its 2015 omnibus edition as *Green Earth*, Margaret Atwood's *Oryx and Crake* (2004) and *MaddAddam* (2013), Bernard Besson's *Groenland* (2011) and Bacigalupi's *The Water Knife* (2015); of deep ecology, Brian Aldiss's *Helliconia* trilogy (1982–1985), Frank Schätzing's *Der Schwarm* (2004), Ligny's *Exodes* (2012) and *Semences* (2015); of fatalism, Wolfgang Jeschke's *Das Cusanus Spiel* (2005), Jeanette Winterson's *The Stone Gods* (2007), Umoya Lister's *Planetquake* (2010), Alexis Wright's *The Swan Book* (2013) and James Bradley's *Clade* (2015). These lists are by no means exhaustive, but are rather intended as examples. We should also bear in mind Weber's own cautionary reminder that in "its conceptual purity" an ideal type "cannot be found empirically anywhere in reality."[21]

Diagrammatically, we can represent our ideal typology as arranged around five measures of formal utopianism on the one hand and six measures of substantive response to climate change on the other. The six variants of climate response are those outlined immediately above. The five formal variants of utopian fiction are the classical, or simple, eutopia; the critical eutopia; the classical, or simple, dystopia; the critical dystopia; and the fiction set in a reality that is neither significantly better nor significantly worse than our own, the non-utopia we might term the base reality text. This leaves us with the grid of 30 logically possible types of climate fiction included in Table 4.1 below.

This ideal typology is confined to SF, loosely defined, and thus deliberately excludes texts conventionally regarded as instances of fantasy. There is

Table 4.1 An ideal typology of contemporary climate fiction

Content (response to climate change)	*Form (type of Utopia)*				
	Classical dystopia	*Critical dystopia*	*Base reality text*	*Critical Eutopia*	*Classical Eutopia*
Denial	三体 / *Three-Body*	*Solar*	*State of Fear*		
Mitigation	*Heat*	*Aurora*		*MAEVA!* trilogy	
Negative adaptation	*La Possibilité d'une île*	*Aqua™*	*Flight Behavior*	*New York 2140*	
Positive adaptation	*The Water Knife*	*Oryx and Crake*	*Green Earth*	*MaddAddam*	
Deep ecology	*Exodes* and *Semences*	*Der Schwarm*			*Helliconia Winter*
Fatalism	*The Stone Gods*	*The Swan Book*	*Planetquake*		

a long tradition insisting on a radical distinction between these two genres, reaching back to Hugo Gernsback, H. G. Wells, Jules Verne and Mary Shelley. This argument is very forcefully restated in Darko Suvin's insistence that the necessary and sufficient conditions for SF are "*the presence and interaction of estrangement and cognition*."[22] The prescriptive intent here was clearly to exclude myth, folktale and fantasy from the genre.[23] This emphasis on the cognitive functions of SF is accompanied in Suvin by a profound aversion to fantasy as a "proto-Fascist revulsion against modern civilization … organized around an ideology unchecked by any cognition … its narrative logic … simply overt ideology plus Freudian erotic patterns."[24] Fredric Jameson echoes Suvin when he warns that SF will add to, whilst fantasy only subtracts from, utopia's epistemological gravity and judges fantasy to be "technically reactionary."[25] The "invocation of magic by modern fantasy," Jameson concludes, "is condemned by its form to retrace the history of magic's decay and fall, its disappearance from … the disenchanted world of prose, of capitalism and modern times."[26]

And yet, much of the canon of contemporary fantasy might plausibly be claimed for climate fiction: J. R. R. Tolkien's *The Hobbit* (1937) and *The Lord of the Rings* trilogy (1954–1955), C. S. Lewis's *The Chronicles of Narnia* (1950–1956), Philip Pullman's *His Dark Materials* trilogy (1995–2000) and George R. R. Martin's still unfinished *A Song of Ice and Fire* (1996–). Indeed, both Tolkien and Lewis have already been so claimed.[27] The theological subtext in Tolkien and Lewis is, of course, Christian, in the latter case quite explicitly so; that in Pullman equally explicitly anti-Christian, but nonetheless obsessed with religion and religiosity; whilst Martin has described himself as a "lapsed Catholic" fascinated by "religion and spirituality."[28] This connection between religiosity, spirituality and fantasy should be unsurprising, insofar as both religion and fantasy are essentially forms of magical thinking. China Miéville, whose *Railsea* (2012) could also arguably be claimed for climate fiction, is the author of a kind of "weird fiction" that deliberately blends SF and fantasy. And he has vigorously rejected the necessary connectedness of SF and science. Rather than counterpose SF to fantasy, he argues that SF is best considered "a subset of a broader fantastic mode."[29] Acknowledging the difference between "not-yet-possible" estrangement effects in SF and "never-possible" effects in fantasy, he nonetheless insists that, if the predicates for a fantasy "*are treated systematically and coherently within the fantastic work*," then its cognition effects will be precisely those "normally associated with sf."[30] Hence, his eventual conclusion that "we need fantasy to think the world, and to change it."[31] There is much to be said for the inclusivity of Miéville's position, not so much an argument against SF as an argument for fantasy in addition to and alongside it. And yet there is a specific problem in relation to climate fiction. For if fantasy and religion are instances of magical thought, and if in reality magic simply doesn't work, as it clearly doesn't, then fantasy and religion will be of no use at all in responding to real-world climate changes. This isn't an argument against enjoying the pleasures of fantasy, but only against taking fantasy seriously as climate fiction. In fantasy magic can always save the day; in real life it never does. For that, we have only science and politics. And SF.

Some Preliminary Generalisations

Let me proceed to some preliminary generalisations about recent cli-fi. The first is that we need be wary of the widespread assumption that classical dystopias are necessarily either any less complex or any less rhetorically effective than critical dystopias. Classical dystopias can in fact be richly complex: the obvious examples are Liu's 三体 (*The Three-Body Problem*) and Winterson's *The Stone Gods*. Much secondary commentary in utopian studies also seems to work on the assumption that the most persuasive dystopian texts will be "critical" in Baccolini and Moylan's sense that they maintain an intratextual utopian impulse.[32] But a text might actually be all the more persuasive as extratextual warning the more completely it eliminates resistance from within itself. The distinction between classical and critical dystopias is thus certainly worth making, but not invidiously. Relatedly, we also need be wary of the equally common assumption that "genre" texts are necessarily less complex or less rhetorically effective than "literary" texts. Amitav Ghosh is right to criticise the "strange conceit that science fiction deals with material that is somehow contaminated" and to argue that "many who once bestrode the literary world like colossi are entirely forgotten while writers like Arthur C. Clarke, Raymond Bradbury, and Philip K. Dick are near the top of the list" of twentieth-century novelists.[33] But Ghosh's position remains complicit with the binary opposition between "literary" and "genre" fiction it promises to undermine. So, when he tries to come up with the names of "writers whose imaginative work has communicated a "sense of the accelerating changes in our environment," he concludes that "of literary novelists writing in English only a handful of names come to mind: J. G. Ballard, Margaret Atwood, Kurt Vonnegut Jr., Barbara Kingsolver, Doris Lessing, Cormac McCarthy, Ian McEwan and T. Coraghessan Boyle."[34] The problem is obvious: Ballard, Vonnegut and Lessing were, by any reasonable standard, "genre" SF writers; and the particular texts by the other writers Ghosh has in mind—Atwood's *MaddAddam* trilogy, Kingsolver's *Flight Behavior*, McCarthy's *The Road*, McEwan's *Solar*, Boyle's *A Friend of the Earth*—are, again by any reasonable standards, very clearly science-fictional in character. This distinction simply isn't worth the making. To put the case bluntly, Robinson's "genre" fictions like *Aurora* and *New York 2140* are at least as formally complex and rhetorically effective as Atwood's supposedly more "literary" *MaddAddam* novels.

Moving from form to substance, we can begin by noting that mitigation, including the technofix of climate engineering, seems to figure prominently mainly in the early stages of the sub-genre, as, for example, in Herzog's *Heat*. As the scientific consensus abandoned mitigation in favour of adaptation so too did SF. It is worth adding, however, that climate engineering as "terraforming" figures as an important sub-motif in Robinson's recent SF, especially *2312*, *Aurora* and *New York 2140*. If mitigation is an increasingly rare primary motif in cli-fi, negative adaptation is correspondingly increasingly common and is now almost certainly the dominant variant amongst both classical and critical dystopias. Moving from dystopia to eutopia, we can observe that there seem to be far fewer climate eutopias than climate dystopias. There is a case to be made, nonetheless, that these are likely to become more culturally significant as the climate crisis develops: when the climate heats up, so too will the writing. So far as I can tell there are no classical eutopias in contemporary climate fiction: the overall effects of anthropogenic global warming seem to be so threatening as to have ruled out this option. The one possible exception is Brian Aldiss's *Helliconia Winter*, the third volume in his critical-dystopian Gaian trilogy, which has a classically eutopian outcome on Earth if not on Helliconia, although this isn't at all central to the novel. Nonetheless, the logical possibility of classical eutopias certainly exists and it's not difficult to imagine a warming world as offering pleasant rather than unpleasant outcomes, at least for those who now live in colder climates. But by and large this option doesn't seem to have been taken up. Critical eutopias, by contrast, certainly do exist and typically focus on the socio-political corollaries of successful attempts at mitigation and adaptation. Finally, we should note that there seem to be more base reality texts than there are eutopias. These could logically include all six of our responses to anthropogenic warming, but empirically I've been able to identify instances only of denial, mitigation, negative and positive adaptation, and fatalism, but not of Gaia.

Moving from text to context, we should say something about the cultural geography of climate fiction. Drawing on Franco Moretti's application of "world systems theory" to comparative literature, I previously distinguished four "core" SF cultures (Britain, France, the US and Japan) and five "semiperipheral" cultures (Germany, Russia, China, Czechoslovakia and Poland) from the remaining "peripheral" cultures.[35] The geographical distribution of climate fiction appears to exhibit significant, albeit minor, variations from the more general structure of what I termed "the global SF field." At the core, there appears to be a comparative underproduction of cli-fi in Japan; in the semi-periphery, a comparative overproduction in Germany; in the periphery, a comparative overproduction in Canada, Finland, Australia and South Africa. How is this pattern to be explained? Here, we need to distinguish between structural and conjunctural determinants of the evolution of the sub-genre. The main structural determinants will be the world literary system and the world SF system. But their effects may be either countered or reinforced by one or more of three main conjunctural factors: the degree of perceived vulnerability to extreme climate change of any particular national political economy; the salience of Green politics within any particular national polity; and the salience of climate change within broader environmentalist discussions in any particular national culture.

All three of these pertain in part to the cultural sphere, even the first, for if the actual degree of climate vulnerability can be measured with some degree of objective accuracy—it is clear, for example, that poorer countries are generally more vulnerable than wealthier—the extent to which this is collectively perceived and understood remains culturally constructed nonetheless. Media representations of the threat of climate change, especially commentary by climate scientists, economists and ecologists, are likely to be central here. At the strictly political level, the local visibility of Green politics will depend on such factors as the electoral system and the availability or non-availability of public funding for minor parties. But it will also depend on the local balance between old and new media, the extent to which advertising revenues are dependent on carbon polluters, and so on. Our three conjunctural determinants can thus be understood as different aspects of the "greening" of the public sphere. They are not "merely" cultural, however, not even the third, since the contours of local environmentalist debates will themselves be shaped in response to wider political and economic developments. The local incidence of cli-fi, we can provisionally conclude, will be determined by the interaction between the world SF system and the (loosely defined) local Green public sphere.

Robinson himself has observed, in an article recently published in *Utopian Studies*, that:

> Climate change is inevitable—we're already in it—and because we're caught in technological and cultural path dependency, we can't easily get back out of it. … It has become a case of utopia or catastrophe, and utopia has gone from being a somewhat minor literary problem to a necessary survival strategy.[36]

In short, climate eutopias are important and we're going to need more of them. Robinson has not only identified the challenge of eutopia, however, but also made serious attempts to address it. I want to conclude, then, by examining some eutopian cli-fi texts, commencing with Robinson's own.

Climate Eutopias: Robinson, Fleck, Atwood

Green Earth is the 2015 omnibus edition of Robinson's earlier *Science in the Capital* trilogy and he himself regards it as a "realist novel," that is, a base reality text. Insofar as *Green Earth* toys with the notion of positive adaptation this is centred around the figure of Phil Chase, a fictional character loosely based on the real Al Gore. Chase's election to the US Presidency offers the promise of a form of adaptation so positive as to border on the eutopian. So, in the "Cut to the Chase" blog, written shortly after he survives an assassination attempt, Chase promises a world "*in which no one is without a job, or shelter, or health care, or education, or the rights to their own life*." This passage is from *Sixty Days and Counting* but is omitted from *Green Earth*.[37] It's difficult to know why exactly Robinson chose to delete these lines, whether he considered them "telling readers things they already knew," "extraneous details" or "excess verbiage."[38] But it strikes me as likely to be none of these. Rather, I suspect the decision arose from a growing awareness, based in the empirical experience of contemporary American realities, of just how implausible it would appear to many readers, perhaps most, that any Democrat President would ever say such things. So *Green Earth* turns out to be a realist novel after all.

New York 2140 is the first of Robinson's novels to depict a specifically eutopian outcome from global climate crisis. Initially, climate change appears to function in the novel only as its dystopian setting: by the early mid-twenty-second century, sea levels have risen by 50 feet so that the whole of Lower Manhattan has long since been flooded. Mise en scène established, the main plot seems at first to be a detective mystery about the disappearance of two "coders," Mutt and Jeff, from their temporary home on the "farm floor" of the Met Life tower on Madison Square. But this turns out to be the trigger for a more important political narrative which moves the novel towards its eventual eutopian climax. And that too is a result of climate change: Hurricane Fyodor batters the city so badly as to prompt what amounts to a popular constitutional revolution. If *Green Earth*'s Phil Chase had been an Al Gore figure, then the various inhabitants of the Met building turn out to be a composite Bernie Sanders—especially Charlotte Armstrong, the radical lawyer who calls in the NYPD to investigate the coders' disappearance. After the hurricane, she persuades her ex-husband, Larry Jackman, now head of the Federal Reserve, that bank nationalisation should be the price for a financial bailout, and runs for Congress as a Democrat. She is elected, the banks are nationalised, Congress passes a "Piketty tax" on income and capital assets, and "a leftward flurry of legislation" is "LBJed through Congress."[39] This is as eutopian an ending as any in recent climate fiction. And it is complemented by a whole series of individual happy endings: Charlotte herself strikes up a successful sexual relationship with Franklin Garr, a market trader for the aptly named WaterPrice who is 16 years her junior. The key weakness, however, is that all this happiness is far too easily bought, most especially at the political level.

Robinson's first eutopian novel, *Pacific Edge* (1990), was, as Moylan acknowledges, a late addition to the "critical utopian" canon,[40] its ecotopian El Modena of 2065 exhibiting both difference and imperfection. As Robinson himself explains, however, his friend Terry Bisson alerted him to the book's key flaw: "Stan … there are guns under the table."[41] This remark provided *Red Mars* (1993) with a chapter title; and the Mars trilogy as a whole (1993–1996) developed a detailed account of three revolutions—because, as Robinson adds, "I felt that in *Pacific Edge* I had dodged the necessity of revolution."[42] In *Green Earth*, politics is paramount, but politics of a kind that is deliberately non-revolutionary and non-eutopian, in which a charismatic Democratic President saves the world. *New York 2140* resumes that political vision, its hopes now vested in radical Democratic Congresswomen and the good fortune that the Federal Reserve might be headed by one of their ex-husbands. But these hopes are "utopian" in the pejorative sense of being hopelessly impractical. The improbability level is expanded by the fact that no significant changes to either the American constitution or the banking system have been achieved between now and 2140. The banks have been bailed out by US taxpayers three times between 2008 and 2140. How realistically likely is it that all this entrenched power could be effectively challenged as a result of one hurricane, no matter how devastating? Ultimately, the novel's eutopia is betrayed by its utopianism.[43]

Dirk C. Fleck has been perhaps the most determinedly eutopian of all cli-fi writers in contemporary Germany. An environmental activist, professional journalist and SF writer, he won the Deutscher Science Fiction Preis for best novel in 1994 for *GO! Die Ökodiktatur* (1993) and again in 2009 for *Das Tahiti-Projekt*, the first in his *MAEVA!* trilogy. The trilogy comprises *Das Tahiti-Projekt* (2008), *MAEVA!* (2011), reissued as *Das Südsee-Virus* in 2013, and *Feuer am Fuss* (2015). *Das Tahiti-Projekt* is set in 2022, *MAEVA!* in 2028 and *Feuer am Fuss* in 2035, and in combination they recount the immediate future history of a world threatened by climate collapse, but ultimately saved by the "Equilibrist" notions propounded by Maeva, originally as President of Tahiti, later as head of the "United Regions of the Pacific," later still head of the "United Regions of the Planet."[44] The URP develops initially as a loose ecotopian alternative to the UN, but soon opens itself up to sub-national regions like Alaska,

South Tyrol, Dithmarschen and Alsace.[45] Maeva is the trilogy's key political actor, whether as a concrete individual or as the "Maeva-Mythos … im Cyberspace möglich geworden" ("the Maeva-Mythos … made possible by Cyberspace").[46] But its protagonist and sometime narrator is her much older lover, Cording, a German journalist working for the British news magazine *EMERGENCY*. Their relationship can at times be read as an instance of eroticised primitivism, given that Maeva is young, beautiful and Polynesian, Cording middle-aged, cynical and European.

Maeva herself summarises the URP's aims thus:

> Zum ersten Mal in unserer Geschichte sind wir mit der selbst verursachten Zerstörung aller biologischen Grundlagen konfrontiert. Keine Generation vor uns hatte eine solche Bedrohung auszuhalten. Die eigentliche Frage, die wir uns also zu stellen haben, lautet: kollektiver Selbstmord oder geistige Erneuerung?
>
> (For the first time in history we are confronted with self-caused destruction of all biological resources. No previous generation had to deal with such danger. The question we are faced with is: collective suicide or spiritual renewal?)[47]

Spiritual renewal means essentially a return to more traditional ways of life and to the wisdom of the shamans. So the URP sends out shamans as inverse missionaries to newly recruited occidental regions, so as to advise them on how to recreate a sustainable society. As Ehawee, a Lakota-Sioux shaman, explains:

> Dies ist im Prinzip die ganze Weisheit meines Stammes. Wenn die Verschmutzung der Erde rückgängig gemacht werden soll, müssen wir als erstes die Verschmutzung in unseren Herzen und Köpfen beseitigen.
>
> (This is in principle the entire wisdom of my tribe. If the pollution of the Earth is to be made to recede, we have first to rid ourselves of the pollution in our hearts.)[48]

It is difficult not to read this as an instance of cli-fi primitivism, projecting a Western counter-ideal on to an idealised non-Western other.

By the third volume, "Die URP-Regionen nehmen inzwischen ein Viertel der gesamten Oberfläche ein" ("The URP Regions now take up one quarter of the surface of the Earth").[49] This ecotopian restructuring of the Regions radically transforms their economies. In Alsace, "Viele Bauern sind dazu übergegangen, wieder mit Pferden zu arbeiten anstatt mit Traktoren. Die Massentierhaltung ist abgeschafft, die überwiegende Zahl der Elsässer ist Vegetarier geworden." ("Many farmers had already begun working with horses instead of with tractors. Mass animal husbandry has been abolished, the overwhelming majority of Alsatians have become vegetarian.") Moreover: "Feldfrüchte und Bäume existieren … in freundlichster Partnerschaft. Auf den einst Tristen Ackerflächen wachsen nun Pappeln, Eichen und andere Bäume in langen Reihen." ("Crops and trees exist … in the friendliest of partnership. On the once sad paddocks, poplars, oaks and other trees grow in long rows.")[50] And all of this is massively encouraged and underwritten by the billionaire Malcolm Double U. In short, we are on our way to Eutopia.

It is a critical eutopia, of course, characterised by both difference and imperfection, as in the struggles between Maeva's cosmopolitanism and the shaman Rauura's localism. Indeed, Fleck's handling of alternative ecotopian visions is at times astute. One of the trilogy's more interesting sub-plots is in the encounter between the URP and the Californian ecodictatorship of ECOCA. The similarity to Callenbach's original *Ecotopia* is no doubt intentional here, as is the commentary that likens it to Kampuchea under the Khmer Rouge.[51] But, as with Robinson's *New York 2140*, the trilogy's key weakness is that its happy endings come far too cheaply, Cording's eventual death notwithstanding. In reality, not only is there no Maeva and no URP, but Tahiti hasn't yet even gained independence from France. Nor do there seem to be many billionaires willing to fund experiments like the URP. In short, this trilogy's eutopia is also "utopian" in the pejorative sense of being hopelessly impractical. That its utopianism is projected on to an exoticised and eroticised non-Western other merely compounds the problem. As Gabriele Dürbeck has observed, Fleck's "depiction of Tahiti and the Tahitians is filled with a kind of problematic exoticism that jeopardizes the ecological worth and broad applicability of the book."[52]

Atwood's *MaddAddam* (2013) provides the eutopian conclusion to her primarily dystopian cli-fi trilogy. Set in the immediate aftermath of the first two novels, *Oryx and Crake* (2003) and *The Year of the Flood* (2009), the Earth's climate is as damaged as in the earlier volumes, but the very success of Crake's "JUVE virus," which wiped out most of the planet's human population, has opened up the prospect for a new world. The novel begins with Ren and Toby, the protagonists of *The Year of the Flood*, rescuing Amanda Payne, like them a onetime "God's Gardener," but also a onetime girlfriend of Jimmy, the Snowman, the surviving protagonist in

Oryx and Crake, from two violent "Painballer" criminals. Ren, Toby and Amanda meet up with Jimmy and the genetically engineered "Crakers" and begin to build some kind of community. The narrative is initially focussed on Toby, but her love affair with Zeb, the Adam Seven of *The Year of the Flood*, leads into his distinctly "biopunk" backstory, which in turn leads into the story of how his half-brother Adam became the Adam One of the Gardeners. So Zeb tells his story to Toby and she re-tells this story to the Crakers. *MaddAddam* is a story about story-telling, then, which ends appropriately enough with two short chapters titled "Book" and "The Story of Toby," both concerned above all with the logic of story-telling.

All this story-telling has two important functions. Extratextually, it fills in for the reader the gaps left by the first two volumes, explaining the connections between Adam and Zeb, the God's Gardeners, the MaddAddamite scientists and Crake. This allows Atwood to pursue her satire on late-capitalist and—in the novel—late-human society. So, for example, in Santa Monica, "rising sea had swept away the beaches, and the once-upmarket hotels and condos were semi-flooded. Some of the streets had become canals, and nearby Venice was living up to its name."[53] Intratextually, however, it serves to demonstrate how a community between the few surviving humans, the Crakers and eventually the Pigoons—intelligent, telepathic genetically modified pigs—can be built around Toby's story-telling and the communal mythology thereby produced. As Atwood reminds us at the novel's outset, Crake had tried to rid the Crakers of symbolic thinking and music, "but they have an eerie singing style … and have developed a religion, with Crake as their creator, Oryx as mistress of the animals, and Snowman as their reluctant prophet."[54] All their religion now lacks is a written culture and a sacred book, which Toby eventually provides them. Her amanuensis will be Blackbeard, the young Craker child whimsically so named by Crake, who regards her as a font of all wisdom.[55] She teaches him to write and he will conclude the novel by reading from her Book and also "from the Story of Toby that I have written down at the end of this Book,"[56] which tells of Zeb and Toby's deaths.

Story-telling aside, in the post-apocalyptic present very little happens in *MaddAddam*. But a eutopia has been created nonetheless. As Blackbeard reads from the Book: Crake and Oryx "together they made us, and made also this safe and beautiful World for us to live in."[57] It is, of course, a critical eutopia, marked by difference—between humans, pigoons and Crakers—and imperfection—as in the sexual jealousies between Toby and the much younger, and aptly named, Swift Fox. More importantly, it is strictly speaking a posthuman, rather than human, eutopia. For, as Toby has already explained to the Crakers, Crake had understood that the "people in the chaos cannot learn … Either most of them must be cleared away while there is still an earth, with trees and flowers and birds and fish and so on, or all must die when there are none of those things left."[58] So he made the "Great Emptiness" within which their community could finally flourish.

I have criticised Robinson's *New York 2140* and Fleck's *MAEVA!* trilogy for their utopianism, which might seem an odd charge to level against deliberately eutopian texts. But in each case I mean utopian only in the pejorative sense of being hopelessly impractical. And this impracticality is a purely textual matter: it is not a question of whether or not such ideas would or would not work in the real world; but rather that there is no real intratextual plausibility to the mechanisms by which the eutopia is achieved within the novels. A comparison with William Morris's *News from Nowhere* will serve to make the point. Whatever we make of Morris's eutopia—and even if we positively prefer Bellamy's *Looking Backward*—the processes by which it is achieved through the revolution of 1952 are eminently plausible, unsurprisingly so perhaps, insofar as they both rehearse the real history of the Paris Commune and in some respects anticipate that of the Russian Revolution.[59] Neither Robinson nor Fleck manage to carry this off. But Atwood surely does. In terms of the internal logics of her trilogy, the eventual posthuman eutopian outcome makes perfect sense. As a real-world prospect it offers precious little consolation for the millions to be wiped out by Crake's BlyssPluss pill. But the fictional logic is nonetheless impeccable. As the novel itself concludes:

> Thank you.
> Now we will sing.[60]

Notes

1. Rudd, "National Climate Summit."
2. The World Resources Institute calculates total greenhouse gas emissions for Australia in 2014 as 22.36 metric tons per capita per annum, as compared to 19.84 for the USA, 14.12 for Russia, 13.38 for New Zealand, 10.08 for Germany, 10.39 for Japan, 7.64 for the UK, 5.04 for France and

4.84 for Sweden (*CAIT Climate Data Explorer*). Australia is also the world's largest exporter of coal and thus a major source of exported carbon emissions.

3. Milner and Burgmann, "Short Pre-History of Climate Fiction."
4. Milner, *Locating Science Fiction*, 194.
5. More, *Utopia*, 18–19.
6. Sargent, "In Defense of Utopia," 15; Budakov, "*Dystopia*," 86–88; Köster, "*Dystopia*"; Mill, *Hansard's Parliamentary Debates.*
7. Sargent, "Themes in Utopian Fiction," 275.
8. Sargent, "Three Faces of Utopianism Revisited," 7–10.
9. Moylan, *Demand* (1986), 10.
10. Moylan, *Demand* (1986), 10–11.
11. Moylan, *Demand* (1986), 43.
12. Moylan, *Scraps*, 198–199.
13. Moylan, *Scraps*, 188.
14. Moylan, *Scraps*, 195.
15. Moylan, *Demand* (2014), xxivn.
16. Atwood, "Dire Cartographies," 85.
17. Sargent, "Miscellaneous Reflections," 243.
18. Merchant, "Behold."
19. Weber, *Methodology*, 90.
20. Lovelock, *Gaia*.
21. Weber, *Methodology*, 90.
22. Suvin, *Metamorphoses*, 7–8.
23. Suvin, *Metamorphoses*, 7–9, 20.
24. Suvin, *Metamorphoses*, 69.
25. Jameson, *Archaeologies*, 57, 60.
26. Jameson, *Archaeologies*, 71.
27. Dickerson and Evans, *Ents*; Dickerson and O'Hara, *Narnia*.
28. Martin, "Dance."
29. Miéville, Introduction, 43.
30. Miéville, Introduction, 45.
31. Miéville, Introduction, 48.
32. Baccolini and Moylan, *Dark Horizons*, 7.
33. Ghosh, *Great Derangement*, 72.
34. Ghosh, *Great Derangement*, 124–125.
35. Moretti, *Distant Reading*; Milner, "World Systems."
36. Robinson, "Remarks," 9.
37. Robinson, *Sixty Days*, 478–479.
38. Robinson, *Green Earth*, xii.
39. Robinson, *New York 2140*, 574, 601, 602, 604.
40. Moylan, "'Utopia,'" 4.

41. Robinson, "Remarks," 3.
42. Robinson, "Remarks," 4.
43. This utopianism is genuinely Robinson's own, not merely that of the novel. As he has recently explained: "We could use the Democratic Party … to elect a majority in Congress to enact a New Deal flurry of changes. Corporations could squeal but they couldn't make the army go onto the streets against the people. In this country the corporations can't do that." Robinson and Feder, "Realism," 97. This vision of both the United States and the Democratic Party seems to me as improbable in reality as in fiction.
44. Fleck, *MAEVA!*, 209.
45. Fleck, *MAEVA!*, 209.
46. Fleck, *MAEVA!*, 170. Translation of Fleck here and below by me and David Blencowe.
47. Fleck, *MAEVA!*, 65.
48. Fleck, *Feuer*, 305.
49. Fleck, *Feuer*, 208.
50. Fleck, *Feuer*, 315.
51. Fleck, *MAEVA!*, 246.
52. Dürbeck, "Anthropocene," 326.
53. Atwood, *MaddAddam*, 168–69.
54. Atwood, *MaddAddam*, xiii.
55. Atwood, *MaddAddam*, 92.
56. Atwood, *MaddAddam*, 388.
57. Atwood, *MaddAddam*, 385.
58. Atwood, *MaddAddam*, 291.
59. Morris, *News from Nowhere*, 286–320.
60. Atwood, *MaddAddam*, 390.

Works Cited

Aldiss, Brian W. *Helliconia Spring*. New York: Atheneum, 1982.
———. *Helliconia Summer*. New York: Atheneum, 1983.
———. *Helliconia Winter*. New York: Atheneum, 1985.
Atwood, Margaret. *Oryx and Crake*. London: Bloomsbury, 2003.
———. *The Year of the Flood*. London: Virago, 2009.
———. "Dire Cartographies: The Roads to Ustopia." In *In Other Worlds: Science Fiction and the Human Imagination*. London: Virago, 2011.
———. *MaddAddam*. New York: Doubleday, 2013.
Baccolini, Raffaella, and Tom Moylan (eds). *Dark Horizons: Science Fiction and the Dystopian Imagination*. London: Routledge, 2003.
Bacigalupi, Paolo. *The Water Knife*. London: Orbit, 2015.
———. *The Windup Girl*. San Francisco: Night Shade Books, 2009.

Besson, Bernard. *Groenland*. Paris: Odile Jacob, 2011.
Böttcher, Sven. *Prophezeiung*. Köln: Kiepenheuer & Witsch, 2011.
Bradley, James. *Clade*. Melbourne: Penguin, 2015.
Budakov, V. M. "*Dystopia*: An Early Eighteenth-Century Use." *Notes and Queries* 57, no. 1 (2010): 86–88.
CAIT Climate Data Explorer. World Resources Institute. Accessed January 16, 2018. http://cait2.wri.org/historical.
Crichton, Michael. *State of Fear*. London: HarperCollins, 2004.
Dickerson, Matthew T., and Jonathan D. Evans. *Ents, Elves, and Eriador: The Environmental Vision of J.R.R. Tolkien*. Lexington, KY: University Press of Kentucky, 2006.
Dickerson, Matthew T., and David O'Hara. *Narnia and the Fields of Arbol: The Environmental Vision of C.S. Lewis*. Lexington, KY: University Press of Kentucky, 2009.
Dürbeck, Gabriele. "The Anthropocene in Contemporary German Ecothrillers." In *German Ecocriticism in the Anthropocene*, edited by Caroline Schaumann and Heather I. Sullivan. New York: Palgrave Macmillan, 2017.
Fleck, Dirk C. *Das Südsee-Virus: Öko-Thriller*. München: Piper, 2013.
———. *Feuer am Fuss*. Murnau am Staffelsee: *p*Machinery, 2015.
———. *GO! Die Ökodiktatur. Erst die Erde, dann der Mensch*. Hamburg: Rasch & Röhrig, 1993.
———. *MAEVA!* Rudolstadt: Greifenverlag, 2011.
Ghosh, Amitav. *The Great Derangement: Climate Change and the Unthinkable*. Chicago: University of Chicago Press, 2016.
Herzog, Arthur. *Heat*. New York: Simon and Schuster, 1977.
Houellebecq, Michel. *La Possibilité d'une île*. Paris: Fayard, 2005.
Jameson, Fredric. *Archaeologies of the Future: The Desire Called Utopia and Other Science Fictions*. London: Verso, 2005.
Jeschke, Wolfgang. *Das Cusanus-Spiel*. München: Droemer Knaur, 2005.
Kingsolver, Barbara. *Flight Behavior*. New York: HarperCollins, 2012.
Köster, Patricia. "*Dystopia*: An Eighteenth-Century Appearance." *Notes & Queries* 30, no. 1 (1983): 65–66.
Ligny, Jean-Marc. *Aqua*™. Nantes: L'Atalante, 2006.
———. *Exodes*. Nantes: L'Atalante, 2012.
———. *Semences*. Nantes: L'Atalante, 2015.
Lister, Umoya. *Planetquake*. Godalming: Highland Books, 2010.
Liu Cixin. 三体. Chonging: Chonging Publishing Group, 2008a.
———. 黑暗森林. Chonging: Chonging Publishing Group, 2008b.
———. 死神永生. Chonging: Chonging Publishing Group, 2010.
Lovelock, James. *Gaia: A New Look at Life on Earth*. Oxford: Oxford University Press, 2000.

Martin, George R. R. "A Dance with Dragons Interview." Interview by James Hibberd. *Entertainment Weekly*. July 12, 2011. http://ew.com/article/2011/07/12/george-martin-talks-a-dance-with-dragons/.
McEwan, Ian. *Solar*. London: Jonathan Cape, 2010.
Merchant, Brian. "Behold the Rise of Dystopian 'Cli-Fi'." *Vice: Motherboard*. June 1, 2013. https://motherboard.vice.com/en_us/article/ypp7nj/behold-the-rise-of-cli-fi.
Miéville, China. Introduction to "Symposium: Marxism and Fantasy," edited by China Miéville and Mark Bould, *Historical Materialism: Research in Critical Marxist Theory* 10, no. 4 (2002): 39–49.
———. *Railsea*. London: Del Rey, 2012.
Mill, John Stuart. Hansard's Parliamentary Debates, third series, 190, no. 1517. London: Cornelius Buck, Paternostor Row, 1868.
Milner, Andrew. *Locating Science Fiction*. Liverpool: Liverpool University Press, 2012.
———. "World Systems and World Science Fiction." *Paradoxa* 26 (2014): 1–16.
Milner, Andrew, and J. R. Burgmann. "A Short Pre-History of Climate Fiction." *Extrapolation* 59, no. 1 (2018): 1–23.
More, Thomas. *Utopia: Latin Text and English Translation*, edited by George M. Logan, Robert M. Adams and Clarence H. Miller. Cambridge: Cambridge University Press, 1995 [1516].
Moretti, Franco. *Distant Reading*. London: Verso, 2013.
Morris, William. *News from Nowhere*. In *Three Works by William Morris*. London: Lawrence and Wishart, 1977 [1890].
Moylan, Tom. *Demand the Impossible: Science Fiction and the Utopian Imagination*. New York: Methuen, 1986.
———. *Demand the Impossible: Science Fiction and the Utopian Imagination*. Edited by Rafaella Baccolini. Ralahine Classics Edition. Oxford: Peter Lang, 2014.
———. *Scraps of the Untainted Sky: Science Fiction, Utopia, Dystopia*. Boulder, CO: Westview Press, 2000.
———. "'Utopia is When Our Lives Matter': Reading Kim Stanley Robinson's *Pacific Edge*." *Utopian Studies*, 6, no. 2 (1995): 1–24.
Neuhaus, Nele. *Wer Wind sät*. Berlin: Ullstein, 2011.
Robinson, Kim Stanley. *Aurora*. London: Orbit, 2015.
———. *Blue Mars*. New York: Bantam, 1996.
———. *Fifty Degrees Below*. London: HarperCollins, 2005.
———. *Forty Signs of Rain*. London: HarperCollins, 2004.
———. *Green Earth*. London: HarperCollins, 2015.
———. *Green Mars*. New York: Bantam, 1994.
———. *New York 2140*. New York: Orbit, 2017.
———. *Pacific Edge*. New York: Bantam. 1990.
———. *Red Mars*. New York: Bantam, 1993.

———. "Remarks on Utopia in the Age of Climate Change." *Utopian Studies* 27, no. 1 (2016): 1–15.
———. *Sixty Days and Counting*. London: HarperCollins, 2007.
———. *2312*. London: Orbit, 2012.
Robinson, Kim Stanley, and Helena Feder. "The Realism of Our Time: Interview with Kim Stanley Robinson." *Radical Philosophy*, no. 2.01 (2018): 67–98.
Rudd, Kevin. "National Climate Summit." Address to Parliament House, Canberra, ACT, August 6, 2007.
Sargent, Lyman Tower. "In Defense of Utopia." *Diogenes* 53, no. 1 (2006): 11–17.
———. "Miscellaneous Reflections on the 'Critical Utopia'." In *Demand the Impossible: Science Fiction and the Utopian Imagination*, by Tom Moylan, edited by Rafaella Baccolini, 242–247. Oxford: Peter Lang, 2014.
———. "Themes in Utopian Fiction in English Before Wells." *Science Fiction Studies* no. 10 (1976): 275–282.
———. "The Three Faces of Utopianism Revisited." *Utopian Studies* 5, no. 1 (1994): 1–37.
Schätzing, Frank. *Der Schwarm*. Frankfurt am Main: S. Fischer Verlag, 2005 [2004].
Suvin, Darko. *Metamorphoses of Science Fiction: on the Poetics and History of a Literary Genre*. New Haven: Yale University Press, 1979.
Turner, George. *The Sea and Summer*. London: Faber and Faber, 1987.
Weber, Max. *The Methodology of the Social Sciences*. Translated by E.A. Shils and H.A. Finch. New York: Free Press, 1949.
Winterson, Jeanette. *The Stone Gods*. London: Hamish Hamilton, 2007.

CHAPTER 5

Evolving a New, Ecological Posthumanism: An Ecocritical Comparison of Michel Houellebecq's *Les Particules élémentaires* and Margaret Atwood's *MaddAddam* Trilogy

Rachel Fetherston

Introduction

In these early days of the twenty-first century, academics of various disciplines are still debating the significance of posthumanist theory in our understanding of what it means to be human. The pervasiveness of mind- and body-altering technologies in many spheres of life has not necessarily provided a definitive answer as to whether our species is now, or will ever be, posthuman. Possibly fostered by the increasing amount of science fiction that emphasises environmental destruction, the idea of the posthuman has become intertwined with the issue of anthropogenic exploitation and degradation of the natural world. Gian-Reto Walther et al. claimed over a decade ago that across a time span as short as 30 years, increased global temperatures have already begun to affect species

R. Fetherston (✉)
Deakin University, Burwood, VIC, Australia

Z. Kendal et al. (eds.), *Ethical Futures and Global Science Fiction*, Studies in Global Science Fiction,
https://doi.org/10.1007/978-3-030-27893-9_5

distribution, life cycles of organisms, and ecosystem dynamics.[1] Globally, the introduction of invasive species and the destruction of native habitats are causing significant declines in biodiversity, with some scientists proposing that the modern extinction crisis should be classified as the planet's sixth mass extinction event.[2] The growing importance of ecocritical research in the discussion of these issues opens a space for the arts and sciences to merge in an attempt to examine and respond to the serious impacts humankind is having on the natural world. Literary works are one means of communicating ideas about this to an audience that may not yet be fully aware of what is causing environmental devastation, and how we might begin to resolve it.

The following discussion focuses on the relationship between posthumanist and ecocritical theory in a literary studies context, and will explore why such a relationship may be a beneficial means of analysing the ecological implications of certain works of science fiction. I will bring an ecocritical mode of analysis to posthumanist theory with a view to developing an ecological posthumanism.[3] I focus on what I perceive as a work of transhumanist science fiction—Michel Houellebecq's *Les Particules élémentaires* (*The Elementary Particles*, or *Atomised*) (1998). I argue that the future transhuman species imagined in this novel is not an ideal means of understanding and resolving ecological issues, representing instead a refusal to accept our species' inherent animality. Conversely, I discuss Margaret Atwood's *MaddAddam* trilogy (2003–2013), a work of dystopian science fiction that proposes a more ecocentric perspective of the natural world. Defining the series as an ecological posthumanist text, I find that Atwood's work presents a possible solution to the modern environmental crisis in her depiction of a post-catastrophe society in which both humans and posthumans integrate with the natural environment.

Defining an Ecological Posthumanism

What does the future hold for humankind? Or, perhaps more accurately, posthumankind? While there is currently a strong academic focus on the implications for the natural world if humans continue to wreak havoc on the planet's environments, there is a need for more discussions surrounding the implications for the human species itself if we continue down this path. What do we become when the natural places and non-human beings, so intimately linked to our own evolutionary and cultural history, disappear or are irrevocably altered? I believe posthumanist theory is one means through which we can attempt to understand this.

Techno-posthumanism and multi-species posthumanism are two of the primary strands of posthumanist thought which bear relevance to my development of an ecological posthumanism.[4] While the exploration of the human-animal binary within multi-species posthumanism strongly informs ecological posthumanism, techno-posthumanism is also relevant in the context of this study due to its similarity to, and sometimes confusion with, transhumanism. What, then, are the differences between techno-posthumanism and multi-species posthumanism, and what does transhumanism bring to this discussion?

Techno-posthumanist theorists express both excitement and caution about the use of technology in altering the human species. Techno-posthumanism could most simply be described, in the words of N. Katherine Hayles, as embracing "the possibilities of technology," yet warning of a world in which human beings "regard their bodies as fashion accessories," rather than as a defining characteristic of who they are.[5] In a similar way, transhumanism supports the advancement of science and technology in their capacity to better the human mind and body. However, Elaine L. Graham insists that transhumanism contests our species' materiality, challenging our connection with "nature, finitude and ... death."[6] Techno-posthumanists exhibit a resistance to this mode of thinking, instead exploring ways in which science and technology can better engage the human species with its natural form, rather than altering it beyond recognition. The transhumanist distrust of the human body therefore brings transhumanism into conflict with both techno- and multi-species posthumanism because it reflects traditional Western humanism's mechanistic understanding of the human body, being grounded in the Cartesian view of the mind as transcendentally superior to the body.[7] It is this tradition that encourages the view that humans (or non-human animals) and machines are dualistically opposed categories; the embodied human and non-human animal remain corporeal, lacking the ability to transcend their own materiality, whereas the machine represents the potential to move past this mortal barrier. For transhumanists, the combination of the human mind and advanced machinery means that humans can forego corporeality and progress further than was possible in their embodied form.[8] In this sense, transhumanism is unlike techno-posthumanism in that it does not affirm any kind of posthuman identity that might enhance our understanding of the corporeal human, but rather seeks to alter the human form completely.[9]

Transhumanism and techno-posthumanism are subsequently defined separately by many scholars, such as Hayles, a techno-posthumanist who contends that the posthuman does not represent a physical alteration of the body and the subsequent removal of uniquely human traits. Hayles believes in celebrating human progress, but is wary of completely sacrificing natural human embodiment.[10] A significant contributor to posthumanist theory, Donna J. Haraway and her work on the "cyborg"—what she describes as "a hybrid of machine and organism, a creature of social reality as well as … fiction"—bears relevance here.[11] Haraway argues that her approach is not a transhumanist outlook on the benefits of techno-science. She is neither celebrating nor denouncing the idea of the cyborg, but is interested in the ways in which its metaphoric nature can help us to understand our humanity by challenging unitary gender, race and class stereotypes in an attempt to overcome interlinked forms of domination; this could include those forms of domination directly impacting the natural world.[12] Conversely, to techno-posthumanism and the idea of the cyborg, Graham describes transhumanism as an attempt to enhance the human body and mind, which inevitably involves removing various traits of a human that define it as such.[13] In this regard, transhumanism is less involved in engaging with the natural human form than it is seeking to alter it completely. In the context of this study, transhumanism is perhaps best identified as what ecological posthumanism is *not*, as I will later elucidate.

On the opposite side of the posthumanism spectrum lies multi-species posthumanism, described by Louise Westling as useful in its ecological exploration of the human-animal binary.[14] Jacques Derrida's later writings relate strongly to the multi-species school, with his lecture "L'Animal que donc je suis (à suivre)" ("The Animal That Therefore I Am [More to Follow]") challenging many assumptions regarding the meaning of the term "animal." Derrida asks, "L'animal en general, qu'est-ce que c'est ? … Qui est-ce ? « Ça » correspond à quoi ? … Qui répond à qui ?" ("The animal in general, what is it? … Who is it? To what does that 'it' correspond? … Who responds to whom?")[15] As Michael Naas notes, Derrida's essay discusses how Western philosophy has created a binary opposition between the human subject and the animal other, resulting in the restriction of traits (such as language and rationality) solely to human beings.[16] Derrida's philosophical critique has been brought into the posthumanist discussion by Cary Wolfe, whose work challenges aggressively human-centric ideas inherent in Western philosophy or humanism (such as human superiority over other living creatures). He highlights the way in which the construction of the supposedly normal human subject in Western humanist philosophy "grounds discrimination against nonhuman animals and the disabled" because it suggests that one must possess specific traits to be considered truly human.[17]

It could be said that an ecological posthumanist approach is an expansion of multi-species posthumanism in regard to the problem of anthropocentrism and the analysis of the human-animal binary. The ecological posthumanism that I put forward here is a theory that expounds humankind's historical and biological connection to the natural environment via an exploration of how the posthuman contains elements of what makes us human, as well as elements of the otherness that would normally lead many of us to perceive the non-human world as inferior. It is through Maurice Merleau-Ponty's suggestion that humans can connect with "le monde du silence" ("the world of silence") that a difference between multi-species and ecological posthumanism comes to light.[18] Although proponents of the multi-species school question the boundaries between humans and other animals, some maintain that an "abyss" still exists between the two.[19] However, even in a world devoid of human language—a "world of silence"—humans can still work towards understanding those beings which communicate differently to ourselves. Ecological posthumanism suggests that within the human mind lies the ability to begin to (although not completely) understand the non-human other of the natural world.

Two important philosophical movements that could inform the development of ecological posthumanism include deep ecology and ecofeminism. According to Carolyn Merchant, deep ecology promotes the idea of "a new ecological paradigm": a complete reworking of humankind's view of the environment to replace the dominant human-centric paradigm.[20] This human-centric worldview is entangled with mechanistic understandings of material embodiment; as transhumanist theory upholds, the human mind is superior to the body, and thus also superior to the non-human organisms that exist in material form. I believe that the ecological paradigm put forward by deep ecologists is incredibly pertinent to ecological posthumanism, as it suggests a future which challenges the dominant values of these traditional mechanistic views of material embodiment and human superiority, bringing to the fore the needs of the non-human as well as those of the human.

Val Plumwood's theories of ecofeminism are also of relevance here. Plumwood's ecofeminist theory focuses on the widespread acceptance of the male-female hierarchy, and how this hierarchy has in turn normalised the oppressive treatment of other human groups, especially the many

Indigenous peoples around the world whose cultures are often strongly linked to the natural world. Plumwood explains that Western society in particular has utilised a strongly gendered view of nature as feminine and passive in promoting an intensely hierarchical society. She condemns the traditional Western dualisms of "human/nature, mind/body, male/female" for helping to create "a logic of interwoven oppression" that has affected the rights of women and other social groups, resulting in the oppression of nature as well.[21] In a similar vein to Haraway's cyborg theory, Plumwood puts forward ecofeminism as a means by which to address and challenge issues of interlinked domination.

However, it is Erika Cudworth's "complexity of difference" approach that provides the most encompassing ecophilosophical lens through which to develop a theory of ecological posthumanism.[22] Cudworth argues that all systems of domination "interrelate in complex and contradictory ways."[23] She opts for an intersectional understanding of different types of domination, including an exploration of the role of non-human beings within what Kate Rigby describes as "intermeshing systems."[24] It is this more inclusive line of thinking that I will bring to the analysis of the distinct modes of transhumanism and posthumanism exemplified in the novels of Houellebecq and Atwood, as both works are concerned with a range of issues pertaining to both human and non-human difference.

A Transhumanist Perspective of Michel Houellebecq's *Les Particules élémentaires*

Described by Douglas Morrey as a French writer equal in fame to the likes of Albert Camus, Houellebecq is often viewed as a highly controversial author due to the extreme sexualisation of his fictional female characters and his novels' explorations of contentious racial and religious issues.[25] His novel *Les Particules élémentaires* was a global bestseller, described by one reviewer as "the great end-of-millennium novel," with English translations by Frank Wynne being published in 2001 (as *The Elementary Particles* in the US and *Atomised* in the UK).[26]

Les Particules élémentaires depicts the slow and irrevocable disintegration of a society challenged by the widespread, self-destructive neo-individualism that confronts much of Western society today. Houellebecq depicts the struggles of half-brothers Bruno Clément and Michel Djerzinski, both of whom are attempting to find meaning in their love lives and careers, each vastly different to the other in his aspirations and life choices. The novel is set in France in the late 1990s and features various analepses which describe events from Bruno and Michel's pasts, and those of their parents. Bruno is obsessed with sex and longs to be accepted as an adequate sexual partner by the women he interacts with, while Michel pursues wider meaning through his perspective on intimate love, his knowledge of science, and his eventual creation of what I argue is a transhuman species. The narrative perspective of *Les Particules élémentaires* is also quite unclear, the reader contemplating whether the narrative voice is that of Bruno, Michel, the implied or actual author or, after reading the epilogue narrated by a transhuman being from the year 2079, a future transhuman. Angela C. Holzer summarises Houellebecq's story as an imagining of "an ambivalent future world of eternal, asexual, and genetically identical ... 'gods'" that critiques the 1960s sexual emancipation as a self-indulgent movement that has resulted in a highly sexualised, consumer-driven existence.[27] "Existential angst" also features prominently in the novel, where it is linked to the competitiveness and everyday malaise of the age of information.[28] For Bruno and Michel, the conventionally advantageous individualism of modern Western society represents a relentless tirade of sexual pressures. Like a marketable product, the sexual appeal of a person often depends on characteristics such as weight and other signs of physical health, as well as age, emotional maturity and the number of sexual partners one has engaged with.[29] According to Carole Sweeney, intercourse in Houellebecq's world is simply another means of expressing one's individuality and narcissism.[30] Sex is akin to a business transaction devoid of intimacy, and this is one reason why Michel and his cohorts create a species that no longer requires sex to reproduce. For Michel, ejaculation is like "un petit suicide" ("a little suicide"), implying waste, despair and a lack of necessity; he is not interested in producing life when it only results in another human who hates living.[31]

The disincarnate character of the new species that Michel creates points in a transhumanist direction, informed by an apparent hostility towards human corporeality. The transhumanist theme of Houellebecq's text does not allow for what Westling deems a more ecological understanding of humans' participation in the natural world.[32] It instead contemplates the inadequacies of the animal nature of the human species, and denounces what the narrator perceives as a primitive, sexual competitiveness. The issue of whose views are presented throughout the novel raises the question of whether it is Houellebecq, the implied author, or the transhuman narrator who is condemning the idea of the human animal. This cannot be answered definitively, as it is difficult if not impossible to determine if or when the focalisation of the story shifts from the future, transhuman narrator reflecting on the past to the implied authorial intent of the work as a whole.

Michel's transhumanist utopia (a utopia at least for him) materialises the traditionally humanist belief that humans must rise above the physical, sexuate body. This contrasts with the more ecologically posthumanist approach suggested by Plumwood, wherein "a post-Cartesian reconstruction of mind" enables humans to view the mind as inherently enmeshed with our corporeality.[33] For the transhuman, asexually reproducing being created by Michel, exclusion from sex and a lack of desire are no longer problematic. As Sweeney explains, this future species is now "able to fully autoeroticise," meaning that sexual relations with another are neither desired nor necessary.[34] Additionally, Michel manages to erase "un des éléments fondamentaux de la personnalité humaine" ("one of the fundamental elements of human individuality") by creating all transhuman beings with the same genetic code (*LPE*, 312; *A*, 375). Sweeney argues that this more effectively achieves "the emancipatory goal of sexual egalitarianism" through a fairer means than the actual sexual freedom ideologies of the 1960s.[35]

Houellebecq has also been strongly critiqued for his depiction of the human body, particularly the female body, in its most grotesque and disturbing form. Martin Crowley contends that many of Houellebecq's depictions of bodily disgust are aimed at women, in particular those who are ageing or supposedly unattractive and overweight.[36] The narrator refers specifically to the supposedly disgusting nature of women's shapeless breasts and "pendants" ("sagging") human tissue (*LPE*, 142; *A*, 169). A feminist critique of the novel might argue that Houellebecq's writing is simply misogynistic, which would not be incorrect. Nevertheless, Houellebecq's emphasis on the decaying and unattractive human form also plays a role in his overall portrayal of how appealing a transhumanist future might be to those wishing to escape their body. Morrey, thus, describes this focus on the human body as going beyond a "superficial misogyny," and in fact representing something much deeper and more disturbing: an immense fear of organic embodiment that extends to the personal anxiety of the author or narrator themselves.[37] Additionally, the male protagonists are also involved in many disturbing bodily activities; it is not only the female body that is portrayed as revolting. Bruno's killing of a cat—"Le crâne du chat a éclaté, un peu de cervelle a giclé autour" ("The cat's skull shattered and some of its brains spurted out") (*LPE*, 70–71; *A*, 82)—after he masturbates over his mother's naked body is one of the most shockingly graphic events in the novel. The symbolism of this scene is not clear until much later, when Bruno attempts to kill a snake at his mother's funeral, shouting: "La nature je lui pisse à la raie..." ("Nature? I wouldn't piss on it if it was on fire...") (*LPE*, 262; *A*, 314). Significantly, Bruno's reactions to both his mother's naked body and dead body involve a condemnation of the non-human other. Bruno develops a hatred of the natural because, in his mind, the natural is truly grotesque. These descriptions portray the humans of Houllebecq's novel as biologically inadequate, further explaining Michel's creation of a new species that allows for an escape from this bodily entrapment.

One notable instance that highlights the transhumanist nature of Michel's future species is when the epilogue's narrator describes the role of the scientist, Frédéric Hubczejak—"le premier, et pendant des années le seul, à défendre cette proposition radicale issue des travaux de [Michel]" ("the first, and for many years the only, defender of the most radical of [Michel's] proposals") (*LPE*, 308; *A*, 370–371)—in trying to garner support for the development of "une nouvelle espèce, asexuée et immortelle..." ("a new species which was asexual and immortal...") (*LPE*, 308; *A*, 371). The genderless narrator states that "il est par contre plus surprenant de noter que les partisans traditionnels de l'humanisme réagirent par un rejet radical" ("it is perhaps more surprising to note that traditional humanists also rejected the idea out of hand") (*LPE*, 309; *A*, 371). This can be understood in relation to the characteristically humanist values of "*liberté individuelle* [et] ... *dignité humaine*" ("'personal freedom' [and] 'human dignity'") that are lacking in the society of this futuristic race (*LPE*, 309; *A*, 371).[38] This wariness shown by humanist groups suggests that Michel's research project rejects the profoundly unique and admirable traits that humanists hold to define our species. However, the narrator's description of the humanists' wariness as surprising speaks to the idea that the traditional humanist movement in fact supports the Cartesian separation of mind and body—something which Michel achieves in the development of this transhuman species. Transhumanist theory is said to build on this traditional humanist rejection of the natural human form, so it could be said that the humanist groups rejecting this transhuman species are in fact ironically responsible for it.[39] It is therefore surprising that they would not be in full support of such an endeavour. Michel's transhumanist utopia radicalises Cartesian dualism and materialises the traditionally humanist belief that humans must rise above the physical, sexuate body and instead focus on the mind. By contrast, Plumwood's post-Cartesian model of the mind enables humans to remain corporeal, appreciating the idea of the mind as something that extends beyond what we know to be human.[40] Following Graham's definition, then, the novel's conclusion is indeed transhumanist, standing in stark contrast to Plumwood's post-Cartesian thinking.

However, it is important to keep in mind that one of the more intriguing aspects of Houellebecq's work is the question of what should or should not be taken at face value. Does the narrator or implied author truly contend that a transhumanist future is the best and only answer to the age-old problem of human suffering? Is it a dystopia or a utopia? Life prior to the transhumanist ending is indeed far from ideal for both Michel and Bruno, while life after could be perceived as a utopia for the two half-brothers (although they may not be around to enjoy it). In an ecologically posthumanist context, though, this future is far from utopian, representing a message of hatred for the natural and a desire for a life free from the constraints of our bodily nature. If we reject the reality of human birth, sex, decay and death, then can we ever begin to understand the lives of the earthly non-humans that also have these experiences?

One possibility is that Houellebecq is encouraging the reader to embrace the uniqueness of humanity, and that even in the light of our species' greatest flaws, a transhumanist society is not the solution.[41] The novel as a whole is instead an extremely negative portrayal of the current trajectory of the human species that should be feared and, if possible, remedied. Although it appears utopian in many ways, *Les Particules élémentaires*, like many great dystopian works, has the potential to frighten readers into action or, at the very least, self-reflection when it comes to Western society's egocentric individualism—the same individualism that many argue is the cause of so much environmental devastation.

The epilogue evokes this sentiment exactly, with the transhuman narrator stating that "l'ambition ultime de cet ouvrage est de saluer cette espèce infortunée et courageuse qui nous créés" ("the ultimate ambition of this book is to salute the brave and unfortunate species which created us"); this same species that was at once "douloureuse et vile" ("vile [and] unhappy"), yet in possession of "tant d'aspirations nobles" ("such noble aspirations") (*LPE*, 316; *A*, 379). The final sentence of the novel provokes a sense of disappointment and shame as we are told: "Ce livre est dédié à l'homme" ("This book is dedicated to mankind") (*LPE*, 317; *A*, 379). In this context, then, transhumanism represents a terrifying solution to a problem that could instead be alleviated through an appreciation of humans' corporeal nature as a trait that grants a better understanding of the natural and the non-human.

Ecological Posthumanism in Margaret Atwood's *MaddAddam* Trilogy

A prolific author of both fiction and non-fiction, Atwood boasts a literary career spanning more than five decades.[42] She has been involved in a large array of causes through her writing career, with her main focuses often being environmentalism and human rights. Her concern for these issues is on display throughout the *MaddAddam* trilogy, which emphasises things that "we've done ... , we're doing ... , or we could start doing ... tomorrow."[43] The series follows the lives of several characters embroiled in what the God's Gardeners—the novels' environmental religious group—deem "the waterless flood": an apocalyptic event that signifies the disintegration of capitalist society and the subsequent rejuvenation of the natural environment which has suffered under capitalism.[44] The novels are primarily concerned with environmental issues and the idea of a non-superior human living in and respecting nature—ideas that Shannon Hengen has found surfacing throughout Atwood's oeuvre.[45]

The issue of environmental injustice during times of ecocatastrophe is highlighted through Atwood's use of narrative focalisation, which shifts significantly over the course of the series and emphasises the many ways in which environmental destruction affects those of varying wealth and privilege. *Oryx and Crake* centres on Snowman (or Jimmy, as he was formerly known), who is unknowingly involved in bringing about the flood through the development of the "BlyssPluss Pill": a contraceptive pill that scientist Crake uses as a means to spread a "plague" throughout the world to eradicate humankind.[46] Once this is accomplished, Snowman takes on the responsibility of guiding Crake's genetically engineered Crakers to the outside world. The Crakers are an ecologically adept posthuman species, engineered by Crake so that the planet will be re-populated with a new form of human capable of integrating into and respecting the natural world. *The Year of the Flood* alternates between a third-person narration focalised on Toby, a senior member of the God's

Gardeners group, and the first-person narration of Ren, a sex worker who spends time with God's Gardeners as a child. Both are attempting to survive the epidemic. *MaddAddam* is told primarily in the third person, centring on the interactions between Toby and Zeb and the post-catastrophe utopia that has been left to them. Arguably the first novel, *Oryx and Crake,* presents a dystopia from the perspective of Snowman, who reflects on the consumer-driven world ravaged by climate change that existed pre-catastrophe.

The pre-apocalyptic society of the trilogy is one in which the impact of environmental exploitation has reached devastating heights; Jayne Glover describes this world as one that is "ecologically unrecognisable."[47] There has been an increase in unpredictable weather, such as thunderstorms and UV radiation, as demonstrated when Toby dons a "top-to-toe" outfit and a "broad hat and ... sunglasses" when venturing outside (*YofF*, 433, 434). The references to "droughts in Texas" and flooding due to hurricanes demonstrate the wide-reaching effects of climate change, as well as the environmental injustice that occurs as a result of severe social divisions (*YofF*, 110, 111). Like Plumwood and Cudworth, Atwood emphasises the inextricable connections between environmental devastation and social injustice. As privileged compound citizens, Jimmy and Crake are sheltered from the detrimental effects of environmental disaster and have access to "nose cones ... to filter out microbes" when traveling outside of the compounds—as Crake explains, "the air was worse in the pleeblands" (*O&C*, 338). Those inhabiting the pleeblands do not have the same privileges, while citizens of developing societies suffer the impacts of climate change even more severely. Oryx is one character who experiences these harsher effects of environmental catastrophe. Although later living in the gated compounds after becoming the teacher and carer of the Crakers, Oryx's childhood in an unnamed developing country demonstrates how the poor and underprivileged suffer most from the consequences of climate change. When describing her previous life, she explains how "the weather had become so strange [that] crops were suffering," forcing villagers to sell their children to maintain a livelihood (*O&C*, 136). This is indeed a world in great need of change.

As an ecological posthuman species, the Crakers' purpose and characteristics are in stark contrast to the transhumans of *Les Particules élémentaires*. Glen, or Crake as he is known for most of the series, is portrayed as a rising star in the science industry of the privileged gated compounds. He is assigned the task of "working on immortality," yet uses this project as a guise for his true objective: to create a new species of human that possesses no "destructive features," and is "perfectly adjusted to their habitat" (*O&C*, 356, 358, 359). The Crakers are not racist because they do "not register skin colour"; they do not hunt, farm or encourage hierarchies because they lack "the king-of-the-castle hard-wiring" that drives territory and resource disputes; they do not engage in sexual abuse because men and women come "into heat at regular intervals" (*O&C*, 358, 359). Group mating also occurs at the behest of the women, so if men are not chosen to mate, their "sexual ardour ... dissipates immediately" and feelings of exclusion or jealousy are not experienced (*O&C*, 194). Most importantly, they are what Crake considers immortal because they do not perceive the notion of death and are "programmed to drop dead at age thirty," eliminating the problem of ageing and overpopulation (*O&C*, 356).

The Craker species is thus, in Gerry Canavan's words, "more ecological, rational and sustainable" than humans.[48] In contrast to Houellebecq's transhumans, the Crakers symbolise an embracement of the human animal and a return to humankind's natural heritage, rather than a denial of the natural body in favour of a supposedly better, less animalistic life. Atwood's appraisal of the human animal has been recognised by many critics, with Ronald B. Hatch commending "her rejection of the anthropomorphic viewpoint" and suggesting a similarity between her writing and that of ecocritics who seek to "re-position humanity" as one with nature rather than as superior to it.[49] The Crakers are a celebration of our own animality, and demonstrate its usefulness in better connecting humans to natural environments, in turn enabling us to conserve them.

However, Atwood is in no way proposing the Crakers as an actual solution to potential environmental catastrophe; as Crake himself states, "they represent the art of the possible" (*O&C*, 359). The "possible," in this context, is what Canavan perceives as the changes that need to be made within *our* species that could help to avert such an apocalypse.[50] Reversing the way humans value money, property and power, and de-value the natural environment, could be the beginning of a more ecologically aware culture that recognises our dependence on and interconnectedness with nature. This species does not need to be the posthuman Crakers—it can indeed be the very *human* us.

There are also two major issues associated with what Glover describes as Crake's ambition to achieve "an ecotopian world": first, how to eliminate the current human race, and second, what would be irrevocably lost if this were to happen.[51] If given the choice, should one save the planet's natural world and non-human life in exchange for the total annihilation of our own species? Perhaps instead Atwood is suggesting that by engaging with and understanding the human animal, there could be a shift towards wanting to protect the non-human animal just as much as we want to protect the human. Character names are largely representative of this: "Oryx" and "Crake," for example, allude to the modern extinction crisis, as these animals no longer exist in the time of the novels. Diana Brydon suggests that the characters' adoption of extinct species names symbolises a growing recognition that humankind is inextricably connected to these creatures, and that with their extinction comes our own.[52] The Crakers also demonstrate this in their special affinity with the pigoons: highly intelligent pigs, genetically engineered in the compounds to be exploited for organ transplants, and which Snowman believes could have taken over the world "if they'd had fingers" instead of trotters (*O&C*, 314). In both *Oryx and Crake* and *The Year of the Flood*, pigoons are primarily viewed as a threat to the remaining humans, yet in *MaddAddam* their role is more central and much less menacing. In this final novel, one of the young Crakers finds himself capable of "talk[ing] with the Pig Ones" through telepathy.[53] It is through this connection that the remaining humans establish a relationship with the pigoons and consequently defeat "the bad men," surviving gang members whose violent behaviour threatens to affect the new, ecocentric society (*MA*, 292).

In *The Year of the Flood* and *MaddAddam*, Crake's ideal world fails and is replaced with a small society of both ecologically minded and rational humans who interact with the Crakers. By the conclusion of *Oryx and Crake,* the reader is aware that Snowman is not the only human left on Earth; we discover in the later novels that Ren, Toby, and other members of God's Gardeners and MaddAddam (the radical group of intellectuals who protest and undermine the authorities and are later recruited by Crake to develop the Crakers) are still alive. Although their survival ultimately ruins the fatalist utopian prospects of Crake's human-free world, *MaddAddam* recounts the work that these humans do in order to complete Crake's goal "of clearing away the chaos" (*MA*, 290). That is, to eradicate the humans that pose a threat to this new, ecologically balanced society. Unlike Houellebecq's transhumanist future, Atwood's ecological posthumanist one is integrated with existing humans—humans like us—suggesting that we need aspects of both the posthuman and the human to survive an ecocatastrophe.

Like Houellebecq, though, Atwood emphasises the negative aspects of consumer culture throughout the series. The privileged citizen's choice to be a consumer represents the everyday human's awareness of environmental exploitation, but their unwillingness to do anything about it. Jimmy's time working at the "AnooYoo compound" (*O&C*, 291), before becoming Snowman, reveals one example of the way consumer culture pervades society in *Oryx and Crake*. Those who hire Jimmy at AnooYoo tell him that their work, much like Crake's, represents "the art of the possible" (*O&C*, 289); their job is to reveal their clients' inadequacies because "what people want is perfection. ... In themselves" (*O&C*, 288). Companies like AnooYoo make profit by inventing reasons for people to view themselves as imperfect, encouraging them to purchase products that have "no guarantee" of solving a problem that did not really exist in the first place (*O&C*, 289). Disconcertingly reminiscent of modern commercialism, the pre-apocalyptic world of the compounds revolves around material value. Jimmy's time working in this industry is also a largely egotistical experience; he can afford to "buy new toys" such as "sweat-eating" exercise outfits and "a talking toaster," while his social life consists of sleeping with married women, often dating "several of them at once" (*O&C*, 294). Such egotism and the shallow, materialist evaluation of fellow humans are problems that humankind must overcome if we are to more deeply connect with the non-human other. This consumer culture is a result of traditional humanist values that still pervade Western society, emphasising the Cartesian separation of mind and body, and the supposed imperfections involved in natural ageing and other bodily functions. Merle Jacob argues that the discordant understanding of our species as superior to the natural environment originates in this perception of mind and body as dualistic opposites.[54] If humans negatively perceive their own natural embodiment, how can they change their view of other species that are supposedly inferior because of their inherent animalism? It is only after Crake's epidemic has eradicated most of humankind that Snowman begins to perceive his own reliance on the natural world. In the absence of consumer culture and corporate bureaucracy, Snowman better understands the human connection with nature. As birds begin to nest in abandoned buildings, Snowman recognises that his now useless watch symbolises the irrelevance of material items in a world that is now at "zero hour" (*O&C*, 3).

The Role of Liminality in Our Ecological Future

In contrast to the ecological posthumanism portrayed in Atwood's writing, Houellebecq's *Les Particules élémentaires* presents a reductive understanding of the human animal and is not a helpful, nor in many ways ethical, means of coming to terms with the human species' impact on the natural world. Michel's transhumanist solution leaves little room to consider the non-human other, as it does not attempt a broader exploration of the animal that lies within the human. As much as Atwood portrays a dismal and terrifying future, the *MaddAddam* trilogy reveals the potentially constructive, ecological relationship that humankind could develop with non-human others and the natural world.

Lee Rozelle intimates the significance of Jimmy's post-apocalyptic pseudonym in relation to this argument, suggesting that by adopting the persona of "the Abominable Snowman," Jimmy characterises himself as "liminal" (*O&C*, 8). That is, he represents the ambiguous state of liminal landscapes: places that were once doomed, but that have now been given a chance to flourish due to the "convergence of humans and ecosystems."[55] However, Snowman also symbolises the human potential to decide the fate of the planet. Like the mythical Abominable Snowman, he is "existing and not existing, … [an] apelike man or [a] manlike ape," with the ability to cross the boundaries between the human and the animal (*O&C*, 8). Snowman can begin life anew, either by attempting to return society to what it was, or by enacting a new, ecocentric mindset. Atwood is implying that we, too, have the same choice in reality: do we continue to live as we do now, forsaking the natural world for selfishly human endeavours, or do we begin again by adopting a more ecologically minded lifestyle that ensures the survival of our own species and of others?

Again, the idea of the ecological posthuman is vital. Although not a perfect solution, the Crakers embody the potential direction that societies must take if there is to be any significant change. They bring together the apelike human and the humanlike ape, demonstrating the benefits to the natural world *and* to our species if the human animal is truly embraced. Crake is wrong when he assumes that his elimination of the human race means that "it's game over forever" (*O&C*, 262). Indeed, the opposite occurs, and for the better. The survival of several humans post-epidemic means that the lesson does not go unlearned, and that alongside the Crakers, the remaining survivors hopefully prevent a return to an anthropocentric, environmentally degraded society.

Notes

1. Walther et al., "Ecological Responses," 394.
2. Sax and Gaines, "Species Diversity," 561; Barnosky et al., "Has the Earth's," 51.
3. For more on this ecocritical approach, see: Rigby, "Ecocriticism," 154.
4. Westling, "Literature," 29.
5. Hayles, *How We Became Posthuman*, 5.
6. Graham, *Representations*, 9.
7. Westling, "Literature," 29.
8. Graham, *Representations*, 9.
9. Hayles, *How We Became Posthuman*, 4.
10. Hayles, *How We Became Posthuman.*
11. Haraway, "Cyborg Manifesto," 149.
12. Haraway, *Companion Species Manifesto*, 4; Gandy, "Persistence of Complexity," 43.
13. Graham, *Representations*, 9.
14. Westling, "Literature," 30.
15. Derrida, *L'Animal*, 75; Derrida, *The Animal*, 51.
16. Naas, "Derrida's Flair," 220.
17. Wolfe, *What is Posthumanism?*, xvii.
18. Merleau-Ponty, *Le Visible et L'Invisible*, 190; Merleau-Ponty, *The Visible and the Invisible*, 145.
19. Westling, "Literature," 38.
20. Merchant, *Radical Ecology*, 85.
21. Plumwood, "Ecosocial Feminism," 211.
22. Cudworth, *Developing Ecofeminist Theory.*
23. Cudworth, *Developing Ecofeminist Theory*, 126.
24. Rigby, "Book Review," 129.
25. Morrey, *Michel Houellebecq*, 1.
26. Karwowski, "Michel Houellebecq," 41.
27. Holzer, "Science," 2.
28. Karwowski, "Michel Houellebecq," 41.
29. Morrey, *Michel Houellebecq*, 50.
30. Sweeney, *Michel Houellebecq*, 127.
31. Houellebecq, *Les particules élémentaires*, 276. Henceforth, cited in text as *LPE*; Houellebecq, *Atomised*, 330. Henceforth, cited in text as *A*.
32. Westling, "Literature," 32.
33. Plumwood, "Intentional Recognition," 398.
34. Sweeney, *Michel Houellebecq*, 125.
35. Sweeney, *Michel Houellebecq*, 101.
36. Crowley, "Houellebecq," 20.

37. Notably, Morrey argues that Houellebecq's attacks on the ideals of the 1968 sexual revolution are in fact quite similar to those made by feminists themselves. He further posits that Houellebecq's critique of the sexual revolution in *Les Particules élémentaires* and his other writings is often confused with an aggressive opposition to feminism. However, Morrey also suggests that Houellebecq's understanding of feminism, as demonstrated in interview, is limited, which suggests that critics cannot completely exclude the issues of misogyny and anti-feminist thought from their criticism of Houellebecq's work. Morrey, *Michel Houellebecq*, 17, 20–21.
38. Emphasis in original text.
39. Westling, "Literature," 29.
40. Plumwood, "Intentional Recognition," 398.
41. See: Rosendahl Thomsen, *New Human*, 203.
42. "Margaret Atwood Chronology," xiii–xvi.
43. Atwood, *Writing with Intent*, 92.
44. Atwood, *The Year of the Flood*, 72. Henceforth, cited in text as *YofF*.
45. Hengen, "Margaret Atwood and Environmentalism," 74.
46. Atwood, *Oryx and Crake*, 345, 381. Henceforth, cited in text as *O&C*.
47. Glover, "Human/Nature," 54.
48. Canavan, "Hope," 146.
49. Hatch, "Margaret Atwood," 181.
50. Canavan, "Hope," 152.
51. Glover, "Human/Nature," 55.
52. Brydon, "Atwood's Global Ethic," 449.
53. Atwood, *MaddAddam*, 292. Henceforth, cited in text as *MA*.
54. Jacob, "Sustainable Development," 478.
55. Rozelle, "Liminal Ecologies," 61.

Works Cited

Atwood, Margaret. *MaddAddam*. New York: Doubleday, 2013.

———. *Oryx and Crake*. London: Virago, 2004.

———. *The Year of the Flood*. New York: First Anchor Books, 2010.

———. *Writing with Intent—Essays, Reviews, Personal Prose: 1983–2005*. New York: Carroll & Graf Publishers, 2005.

Barnosky, Anthony D., Nicholas Matzke, Susumu Tomiya, Guinevere O. U. Wogan, Brian Swartz, Tiago B. Quental, Charles Marshall et al. "Has the Earth's Sixth Mass Extinction Already Arrived?" *Nature* 471, no. 7336 (2011): 51–57.

Brydon, Diana. "Atwood's Global Ethic: The Open Eye, The Blinded Eye." In *Margaret Atwood: The Open Eye*, edited by John Moss and Tobi Kozakewich, 447–458. Ottawa: University of Ottawa Press, 2006.

Canavan, Gerry. "Hope, but Not for Us: Ecological Science Fiction and the End of the World in Margaret Atwood's *Oryx and Crake* and *The Year of the Flood*." *Literature Interpretation Theory* 23, no. 2 (2012): 138–159.

Crowley, Martin. "Houellebecq—The Wreckage of Liberation." *Romance Studies* 20, no.1 (2002): 17–27.

Cudworth, Erika. *Developing Ecofeminist Theory: The Complexity of Difference*. New York: Palgrave Macmillan, 2005.

Derrida, Jacques. *L'Animal que donc je suis*. Edited by Marie-Louise Mallet. Paris: Éditions Galilée, 2006.

———. *The Animal That Therefore I Am*. Edited by Marie-Louise Mallet. Translated by David Wills. New York: Fordham University Press, 2008.

Gandy, Matthew. "The Persistence of Complexity: Re-reading Donna Haraway's 'Cyborg Manifesto.'" *AA Files* 60 (2010): 42–44.

Glover, Jayne. "Human/Nature: Ecological Philosophy in Margaret Atwood's *Oryx and Crake*." *English Studies in Africa* 52, no. 2 (2009): 50–62.

Graham, Elaine L. *Representations of the Post/Human*. New Brunswick: Rutgers University Press, 2002.

Haraway, Donna J. "A Cyborg Manifesto: Science, Technology, and Socialist-Feminism in the Late Twentieth Century." In *Simians, Cyborgs, and Women: The Reinvention of Nature*, 149–181. New York: Routledge, 1991.

———. *The Companion Species Manifesto: Dogs, People and Significant Otherness*. Chicago: Prickly Paradigm Press, 2003.

Hatch, Ronald B. "Margaret Atwood, the Land, and Ecology." In *Margaret Atwood: Works and Impact*, edited by Reingard M. Nischik, 180–201. Rochester, NY: Camden, 2000.

Hayles, N. Katherine. *How We Became Posthuman: Virtual Bodies in Cybernetics, Literature and Informatics*. Chicago: University of Chicago Press, 1999.

Hengen, Shannon. "Margaret Atwood and Environmentalism." In *The Cambridge Companion to Margaret Atwood*, edited by Coral Ann Howells, 72–85. Cambridge: Cambridge University Press, 2006.

Holzer, Angela C. "Science, Sexuality, and the novels of Huxley and Houellebecq." *CLCWeb: Comparative Literature and Culture* 5, no. 2 (2003). https://doi.org/10.7771/1481-4374.1189.

Houellebecq, Michel. *Atomised*. Translated by Frank Wynne. London: Vintage, 2001.

———. *Les particules élémentaires*. Paris: Éditions J'ai Lu, 2010.

Jacob, Merle. "Sustainable Development and Deep Ecology: An Analysis of Competing Traditions." *Environmental Management* 18, no. 4 (1994): 477–488.

Karwowski, Michael. "Michel Houellebecq: French Novelist for Our Times." *Contemporary Review* 282, no. 1650 (2003): 40–46.

"Margaret Atwood Chronology." In *The Cambridge Companion to Margaret Atwood*, edited by Coral Ann Howells, xiii–xvi (Cambridge: Cambridge University Press, 2006).

Merchant, Carolyn. *Radical Ecology: The Search for a Livable World*. New York: Routledge, 1992.

Merleau-Ponty, Maurice. *The Visible and the Invisible*. Edited by Claude Lefort. Translated by Alphonso Lingis. Evanston: Northwestern University, 1968.

———. *Le Visible et L'Invisible*. Paris: Éditions Gallimard, 1964.

Morrey, Douglas. *Michel Houellebecq: Humanity and its Aftermath*. Liverpool: Liverpool University Press, 2013.

Naas, Michael. "Derrida's Flair (for the Animals to Follow…)." *Research in Phenomenology* 40, no. 2 (2010): 219–242.

Plumwood, Val. "Ecosocial Feminism as a General Theory of Oppression." In *Ecology*, edited by Carolyn Merchant, 207–219. New Jersey: Humanities Press International, Inc., 1994.

———. "Intentional Recognition and Reductive Rationality: A Response to John Andrews." *Environmental Values* 7, no. 4 (1998): 397–421.

Rigby, Kate. "Book Review: *Developing Ecofeminist Theory: The Complexity of Difference*." *Feminist Theory* 8, no. 1 (2007): 128–130.

———. "Ecocriticism." In *Introducing Criticism at the Twenty-First Century*, edited by Julian Wolfreys, 151–178. Edinburgh: Edinburgh University Press, 2002.

Rosendahl Thomsen, Mads. *The New Human in Literature: Posthuman Visions of Changes in Body, Mind and Society after 1900*. London: Bloomsbury Academic, 2013.

Rozelle, Lee. "Liminal Ecologies in Margaret Atwood's *Oryx and Crake*." *Canadian Literature*, no. 206 (2010): 61–68.

Sax, Dov F., and Steven D. Gaines. "Species Diversity: From Global Decreases to Local Increases." *Trends in Ecology and Evolution* 18, no. 11 (2003): 561–566.

Sweeney, Carole. *Michel Houellebecq and the Literature of Despair*. London: Bloomsbury Academic, 2013.

Walther, Gian-Reto, Eric Post, Peter Convey, Annette Menzel, Camille Parmesan, Trevor J.C. Beebee, Jean-Marc Fromentin, Ove Hoegh-Guldberg and Franz Bairlain. "Ecological Responses to Recent Climate Change." *Nature* 416, no. 6879 (2002): 389–395.

Westling, Louise. "Literature, the Environment, and the Question of the Posthuman." In *Nature in Literary and Cultural Studies*, edited by Catrin Gersdorf and Sylvia Mayer, 25–47. New York: Rodopi, 2006.

Wolfe, Cary. *What is Posthumanism?* Minneapolis: University of Minnesota, 2010.

CHAPTER 6

The Perverse Utopianism of Willed Human Extinction: Writing Extinction in Liu Cixin's *The Three-Body Problem* (三体)

Thomas Moran

Introduction

Liu Cixin's highly acclaimed novel *The Three-Body Problem* (三体) (2006) is undoubtedly one of the most compelling twenty-first-century speculations on the notion of utopia. In this chapter, I explore the text's perverse utopianism, in which the willed extinction of the human species is understood not as a dystopian drive, but rather as a means of overcoming outmoded or unworkable forms of utopian thought. I begin by providing a brief background on *Three-Body* and an outline of the text's narrative, focusing explicitly on the manner in which human extinction is framed as a desirable outcome. This focus is then used to explicate the book's critique of existing utopian models, both in terms of the socialist projects of the twentieth century and the technocapitalism of the present. I argue that this focus represents a broader critical engagement in the text with utopianism as it has been presented in Chinese science fiction and Chinese

T. Moran (✉)
Monash University, Melbourne, VIC, Australia

Z. Kendal et al. (eds.), *Ethical Futures and Global Science Fiction*,
Studies in Global Science Fiction,
https://doi.org/10.1007/978-3-030-27893-9_6

socialist realism, both of which have understood technological progress as key to the emancipation of the species. I then outline a tension within the text between an attempt to maintain the project of technological utopianism and the attempt to write extinction. I argue that the drive to write extinction is a utopian attempt to overcome the limits of contemporary utopian thought and to find, within this universe without humanity, a true "no place," devoid of anthropic concerns and open to the charms of the void. In *Three-Body*, the thought of extinction pushes the allegorical potential of the science fiction form to its limits, but it is only at this limit that utopia can be written.

The Utopian-Dystopian Dialectic in *The Three-Body Problem*

Three-Body is the first novel in the *Remembrance of Earth's Past* trilogy (*地球往事*), which was serialised in China in 2006 and published in English in 2014. It was translated by Chinese-American science fiction author Ken Liu and was the first translated novel to win the Hugo Award. Given the popularity and significant impact of this translation, which brought Liu Cixin's novel to a new global audience, I will be referring to Ken Liu's English translation throughout this chapter.[1] The text has drawn praise from an unlikely array of sources, from science fiction authors such as Kim Stanley Robinson to former American President Barack Obama and Facebook CEO Mark Zuckerberg.[2] *Three-Body* has been understood to be part of a "new wave" of Chinese science fiction. While I will discuss the literary context of this "new wave" in greater detail in section two, Liu Cixin has been described as part of a Chinese science fiction renaissance which also includes Wan Jinkang and Han Song.

Liu Cixin was born in 1963 in Yangquan in the Shanxi province, notorious for its dangerous coal mining industry in which both his parents worked.[3] During the Cultural Revolution (1966–1976), he was sent to live in Luoshan County in Henan where, in 1970, he witnessed the launch of the first Chinese satellite, *东方红一号* (*The East Is Red 1*). Liu writes that he was filled with an "indescribable curiosity, a yearning," but also a literal hunger due to the food shortages that year. In 1975, he began to read physics books written for children, describing the "druglike euphoria" that accompanied his discovery of the concept of a light year—a comingling of terror and awe he would later describe as his first taste of the sublime.[4] The same year, Liu witnessed the "Great Flood of August," where multiple dams collapsed and 141,000 people died in the deluge.[5] Seeing the landscape filled with refugees, he later said, "I thought I was looking at *the end of the world*."[6] In a postscript to *Three-Body*, Liu notes that it is these moments, or "primal scenes" from childhood, which shaped his fiction's intricate blend of the historical and the speculative.[7]

The dialectic between the utopian historical project of Chinese communism and the dystopic culmination of this project in the Cultural Revolution is central to *Three-Body*. The book begins in 1967 at the height of the turmoil, in what Liu titles "The Madness Years."[8] Liu describes in grisly detail the scientist Ye Zhetai (叶哲泰) being beaten to death by teenage Red Guards for refusing to renounce the "reactionary" nature of the theory of relativity. The scientist's daughter, Ye Wenjie (叶文洁), watches on, horrified, as her father dies in front of her. Her family affiliation leaves her politically suspect and she is one of the "sent down youth" who are assigned to the country for political re-education.[9] Here, as a member of the Inner Mongolia Production Corp, she witnesses the large-scale deforestation of the deep Northern provinces. These events leave Ye with an intense misanthropy. Liu writes, "it was impossible to expect amoral awakening from humankind itself, [Ye thought] just like it was impossible to expect humans to lift off the earth by pulling up on their own hair. To achieve moral awakening required a force outside the human race."[10]

Ye is eventually recruited to the Red Coast Base in 1971, an enormous satellite used to send and receive signals from outer space. The project is part of the Chinese state's search for extraterrestrial intelligence. As the text notes, this is part of a Cold War strategy, citing a series of fictional government reports which outline the plan to weaponise first contact. Liu writes, "If such contact were monopolised by one country or political force, the significance would be comparable to an overwhelming advantage in economic and military power."[11] Working at the base, Ye develops a theory that the energy of the sun could be used as a super antenna to amplify a signal and propel it into the deepest reaches of space. The impact of the Cultural Revolution, however, means that aiming a message at the sun could be interpreted as politically suspect. Mao is symbolically "the Red Sun in the hearts of the people of the world."[12] This image of Mao as the sun is a recurrent image in socialist realist aesthetics, from poster art to poetry and literature[13] He is not a tool to contact extraterrestrials. Nonetheless, Ye ignores her superiors and attempts the experiment in secret. It seems to have had no effect, becoming just another unreceived signal in the "vast desolation" of the night.[14] But eight years later, in 1979, Ye receives a peculiar signal while on duty. It is a reply. Deciphering the waves, it reveals to her surprise a warning: "Do not answer! I am a pacifist in this world. It is the luck of your civilisation that I am the first to receive your message. I am warning you do not answer! Your planet will be invaded. Your world will be conquered! Do not answer!"[15] Ye, however, ignores the warning, and in fact greets it with joy. She sends back an invitation, "Come here! I will help you conquer this world. Our civilisation is no longer capable of solving its own problems. We need your force to intervene."[16]

So far, it seems that the text presents us with a variation on a familiar theme: beware the utopian impulse. The attempt to transform society (evidenced in the utopianism of the Chinese socialist project) or the attempt to transform nature (evidenced in the mass deforestation) is doomed to failure. Not only will it end in chaos and bloodshed, but it may even cause a person to betray the human race. This is a familiar historicisation of the failed socialist projects of the twentieth century.[17] More specifically, it addresses the failed utopianism of the Maoist project and the violent excesses of the Cultural Revolution. In this interpretation, the text appears to embody Fredric Jameson's definition of the "anti-utopian" as a reaction against the potential to imagine a social and economic alternative to capitalist economic organisation. Jameson writes that such works support the "universal belief that the historic alternatives to capitalism have been proven unviable and impossible, and that no other socioeconomic system is conceivable, let alone practically available."[18] This definition of the anti-utopian is echoed by Tom Moylan, who writes that "Anti-Utopia … found its most powerful vocation in shaping the hegemonic reaction against communism and socialism."[19] Ye's desire to annihilate the human race in this interpretation is therefore not only a reaction to the violence of Maoist socialism but also its dialectical counterpoint, part of the same violent impulse which created chaos and catastrophe in the first place.

What complicates this conceptual definition of *Three Body* as an anti-communist anti-utopia is the book's critique of the post-socialist Chinese present. *Three-Body* portrays this present as a similarly disastrous period, in which ecological catastrophe and political violence continue to devastate the planet on a global scale. The novel's unique internal criterion for the measurement of dissent thus becomes evident. The ubiquitous desire for human extinction indexes a dissatisfaction with the world as it is currently organised. Since Ye's invitation to the extraterrestrials, the number of

Earth's inhabitants welcoming the coming invasion has grown. Continued communication with the extraterrestrials has revealed that they come from a planet called Trisolaris, which is on the verge of being swallowed by its three suns and whose inhabitants are seeking a new home world. The Earth-Trisolaris Organisation (ETO) is a friendship society for the coming invasion: part political conspiracy, part doomsday cult. Liu writes, "the most surprising aspect of the Earth-Trisolaris Movement was that so many people had abandoned all hope in human civilisation, hated and were willing to betray their own species, and even cherished as their highest ideal the elimination of the entire human race, including themselves and their children."[20] It is telling that the book's portrayal of interplanetary colonisation and warfare was interpreted by members of China's enormous high technology industry as an allegory for the vicious competition between corporate interests.[21] Liu presents Chinese post-socialist society, with its combination of state ownership and ruthless free market competition, as invoking a similar anti-human disgust as the world of socialist China.[22] The book extends this disgust to the entire planet. The ETO is an international organisation, a shadow United Nations, working to undermine people's faith in the future of humanity. Liu includes various direct internal theorisations to explain the growing popularity of the ETO, which I discuss in more depth below, but one key example is an overwhelming "force of alienation."[23] Echoing the work of the young Marx, alienation from their labour and their fellow members of the species generates a desire amongst the scientific elite for the annihilation of the species.

The politics of utopia and the will to extinction are central forces within *Three-Body*. The novel is incredibly ambitious in scope and much remains to be written on its historicisation of scientific developments in China and the West, as well as its exploration of virtual environments through the "Three-Body" online game, which the ETO uses to recruit new members. This chapter limits its focus to the political allegories at work within the book, seeking to draw out a series of critical questions or conceptual impasses that Liu's work elaborates. Overwhelmingly, this chapter discusses the manner in which pessimism becomes involved in a complex dialogue with the concept of utopianism, which, following the work of Ernst Bloch, has been understood primarily as the capacity and practise of hope.[24] I will argue that instead of hope, Liu's perverse utopianism places despair—precisely a despair that seeks to annihilate human life—at the heart of its disturbing but entrancing project.

The Utopian Legacy: Science Fiction and Socialist Realism

In order to understand the politics of utopia within *Three-Body*, we must briefly contextualise it within the history of Chinese science fiction. This historicisation will affirm a broader claim, made by Mingwei Song, that the text is part of an ongoing attempt to reformulate and speculate upon the question of utopia within Chinese science fiction.[25] As noted above, *Three-Body* has most frequently been discussed in English-language criticism as part of a "new wave" of Chinese science fiction.[26] This is, in part, the result of the relative scarcity of translations of older Chinese science fiction texts but also because of a very real historical limitation on the production of Chinese science fiction during the height of the Maoist period.[27] Nonetheless, as David Wang der-Wei argues, science fiction or "scientific fantasies" were central to what he describes as the "repressed literary modernity" of the late Qing dynasty.[28] Wang contests the notion that Chinese modernist literature emerges with the May Fourth Movement of 1919 by outlining a number of genres, including science fiction, that were produced during the late nineteenth and early twentieth century. He argues that these genres were ignored or actively dismissed as reactionary or regressive by the leftist literary modernists of the May Fourth Movement, who instead sought to establish a realist literature along the lines of the Western "social novel" of the nineteenth century.[29] In contrast, Wang notes in particular the work of Liang Qichao who published *The Future of New China* (新中国未来记) in 1902. Wang's work in many respects anticipates the reformist desires of the May Fourth Movement while propounding an alternate formal mode in which to articulate them. The text imagines a future China in which technological innovation and social transformation have created a prosperous and exciting new nation.[30] This concern with national development has, as Mingwei Song argues, been central to Chinese science fiction and continues to be a topic of inspiration and critique in contemporary texts of the genre.[31] Chinese utopianism since Liang has been explicitly linked to technological modernity and its potential for unlocking new capacities of both the human species and the nation.

A utopian embrace of the potentials of technological modernity has been central to Chinese science fiction from its inception. As Rudolf B. Wagner notes in his survey of Chinese socialist science fiction, this concern reflects the technological utopianism of the "hard" science fiction genre more generally. "Hard" science fiction refers to science fiction works which draw upon the "hard" natural sciences, as opposed to the "soft" social sciences, and are characterised by a highly technical focus. Like the literature of both the United States and the Soviet Union, hard science fiction in China has been linked to the popularisation of scientific thought and the development of a mass scientific pedagogy.[32] Along with this instrumental function, hard science fiction in China echoed the utopianism present within the ideology of Maoist socialism.[33] During the 1950s, writers such as Zheng Wenguang adopted elements of soviet socialist realist science fiction to describe the potential for space exploration and futuristic technologies in a newly liberated China. Therefore, it is somewhat misleading to suggest, as scholars like Wang or Mingwei Song have done, that science fiction has always existed in China as a repressed form.[34] Echoing science fiction scholarship more broadly, studies of Chinese science fiction often position the genre in opposition to the official discursive form of literary realism. But even during the Cultural Revolution, when science fiction or indeed any non-revolutionary literature was banned, we can interpret socialist realist literature as a science fiction form. Combining romantic revolutionary nationalism with in-depth analysis of Marxist social theory and obsessive descriptions of the technical structures of the production process, socialist realist literature contained an important utopian quality. While the novels were purportedly set in the present, they actually imagined a future China in which full communism had already been achieved.[35] Socialist realist literature, therefore, has a strange temporal quality; it projects a utopian future onto the present while simultaneously hoping to generate this future through the production and consumption of the literary work. This complex time structure imagines literature as a form of time travel, one that has the capacity to generate futures through its evocation of an alternate world existing within the present. Thus, while science fiction as a form may have been repressed, the desire for social transformation that it engendered continued to bloom in the literature of socialist realism.

Three-Body is engaged in a direct dialogue with the utopianism of Chinese science fiction and socialist realism. After the Cultural Revolution, science fiction re-emerged as a distinct literary form where it contributed, as Wagner argues, to the rehabilitation of the scientific community in the post-Maoist period. It also once more served a pragmatic function of popularising and promoting scientific practise which was required for the growth of the post-socialist economy. *Three-Body* describes this period as the "dawn of a new life" after the Cultural Revolution. Liu writes "science and technology were the only keys to opening the door to the future … it was the season of rebirth and renewal for China's battered science establishment."[36] The text can be understood to be historicising these periods through their relationship to science and technology. *Three-Body* is heir to the utopianism of both science fiction and socialist realist traditions, all the while seeking to problematise this legacy. It could therefore be argued that for all its obsession with military technology, hypothetical astrophysics and interplanetary colonisation, *Three-Body* participates in a deconstruction of the technological utopianism present in the work of these earlier Chinese writers and encoded in the genre of hard science fiction in general. Rather than bringing about a socialist utopia on earth or its extension, a communism in the stars, scientific progress in *Three-Body* has devastated the planet and contributed to what the author describes as the alienation of the technical and scientific classes. I discuss both these elements in more depth in the following sections, but I want to suggest here that the text presents neither a wholehearted rejection of scientific rationality nor a wholehearted affirmation of technological progress. It operates instead by maintaining this contradiction at the heart of the text, creating a space in which both possibilities are continuously offered and then foreclosed. In *Three-Body* utopianism and dystopianism remain in constant conflict and are perhaps indissolubly joined.

Utopian Reason and Its Nihilistic Fate

The utopian-dystopian configuration of *Three-Body* coalesces around the question of technological and scientific development. We can note this contradiction in the author's own statements on the relationship between scientific thought and literature. In an interview appropriately titled "Why Are Humans Worth Rescuing," Liu suggests that he is "a technologist and committed to scientism," while in another interview he says that the science fiction author must "create and depict a world like God."[37] Indeed this relationship has underpinned Liu's own life; before he became a successful author, he worked as a computer engineer and wrote science fiction by night. If, indeed, the role of science fiction is to be the handmaiden to the sciences, then we can understand the role of the author to be part of the process of disseminating reason. Science fiction would operate in Schiller's terms as a means of "spiritualising reason."[38] Clearly science fiction no longer continues to serve that instrumental purpose, but nonetheless the relationship between reason and utopianism is central to the utopianism of *Three-Body*.

The dialectic between the rational and the utopian emerges acutely within the narrative through the description of the Earth Trisolaran Organisation. The group aims to sabotage Earth's potential to resist alien colonisation by undermining people's faith in scientific development. The book controversially attributes global warming and ecological catastrophe to the ETO's work to discourage techno-scientific progress. This quasi-conspiratorial dismissal of climate catastrophe within the text is at odds with the narrative's constant return to the spectre of environmental collapse, which I will discuss in greater depth below. The eventual culmination of the plan to undermine scientific development is initiated by the Trisolarans themselves, who send proton-sized nanocomputers called "sophons" to Earth. Earth's relatively stable climate is contrasted with Trisolaris's incredibly unstable one caused by its three suns. Due to this stability, scientific development on Earth can accelerate, whereas Trisolaran technical development is relatively constant. The sophons prevent Earth accelerating beyond Trisolaris during the time it takes for the invading fleet to arrive.[39] Indeed, here it seems as if the text, for all its complexity, poses a rather traditional hard science fiction faith in the utopian potential of science to overcome all obstacles. Except in *Three-Body* it is precisely the absence of future technical development that spells humanity's doom.

The relationship between reason and utopianism, however, is shown to be more complex. Liu writes that most of the ETO are members of the scientific class and welcomed the invasion because they had "already begun to consider issues from a perspective outside the human race."[40] This non-anthropocentric position is contrasted with that held by the masses, or "common people," who "still felt an overwhelming, instinctual identification with their own species."[41] Here, the text begins to complicate any simplistic distinction between rationality and religious belief. The ETO is described as an "organisation of spiritual nobles" and is divided into two factions, the Adventists and the Redemptionists.[42] The Adventists betrayal "was based only on their despair and hatred of the human race. … This despair began with the mass extinctions of the Earth's species caused by modern civilisation."[43] Conversely, the Redemptionists "developed spiritual feelings toward the Trisolaran civilisation."[44] As Liu writes, "the human race was a naïve species, and the attraction posed by a more advanced alien civilisation was almost irresistible."[45] Within the ETO, a combination of hope and despair motivate the will to annihilate humanity, but the Adventists' pessimistic hatred of humanity is the mirror image of the Redemptionists' hope in a new race, while both are produced by an alienation which the text attributes to scientific reason.

The theme of alienation is pursued obsessively throughout the text. The term itself is a shifting placeholder which comes to stand for a number of related concepts. At one level, it corresponds to Marx's theory of alienated labour, which attracts the hyper-specialised technical class to the ETO. But it also corresponds to an alienation produced by reason itself, which generates a pessimistic view of the irrational impulses that continue to destabilise human civilisation. Liu writes: "The ETO concluded that the common people did not seem to have the comprehensive and deep understanding of the highly educated about the dark side of humanity."[46] The intractability of this "dark side" is what has led to ecological catastrophe and produced the violence of the Cultural Revolution. In this Hobbesian sense, a deep irrationality lies at the heart of human nature. Jameson rightly notes that this trope is central to utopian fiction, in which "evil" is reified as part of human nature and a future society is constructed around repressing or neutralising this threat to human flourishing.[47] In Hobbes' *Leviathan* (1651), it is the state or sovereign that prevents humanity from falling back into this natural condition.[48] He characterises this state as *bellum omnium contra omnes* or the war of all against all.[49] The ancient Chinese philosopher Mozi (468–391 BCE) similarly theorised the necessity for the state to suppress the individualistic and destructive nature of its citizens.[50] For the ETO, the solution is a little more drastic: human annihilation. This Hobbesian view is extended to the depiction of extraterrestrial life, which is ultimately constructed as hostile and violent. Yet, there is another sense in which we can understand the alienation of the ETO: it is the alienation caused by the absence of fixed values, produced by the rational overcoming of traditional beliefs and structures. In an aphorism titled, "Decline of Cosmological Values," Friedrich Nietzsche describes the nihilism attendant upon the advent of scientific reason. He writes: "The feeling of valuelessness was reached with the realisation that the overall character of existence may not be interpreted by means of the concept of 'aim.' … Existence has no goal or end."[51] This nihilism leads members of the scientific community to join the ETO, finding spiritual fulfilment in the prospect of a political or religious project, even if this project is the annihilation of humanity.

The problem of nihilism initiated by reason must be historicised in relation to the specific experience of Chinese modernity.[52] As Liu himself notes, "the kinds of technological and social changes that took societies in the west centuries to move through have sometimes been experienced by a mere two generations in China. … The anxiety of careening out of balance, of being torn by parts moving too fast and too slow, is felt everywhere."[53] The nihilism diagnosed by *Three-Body* has a dual character. It is not only a response to the "disenchantment of nature" produced by reason, but is intensified by a comparative political disenchantment of socialist values. During the socialist period, scientific development was conjoined with the historical project of building socialism. In this sense, the onset of nihilism was tempered by the values of Marxism-Leninism that developed a utopian ideological mission as such for science. Mao writes that "the knowledge which grasps the laws of the world, must be redirected to the practice of changing the world, must be applied anew in the practice of production, in the practice of revolutionary class struggle and revolutionary national struggle and in the practice of scientific experiment."[54] Indeed, Marxism-Leninism itself was construed as a science of history that interpreted historical and scientific development in a teleological manner.

Literature, specifically science fiction literature, becomes a form in which the relationship between reason and nihilism can be explored or, as in *Three-Body*, further intensified. Within the text, the ETO project is explicitly linked to aesthetic production. A character notes: "Recent big budget films all have rustic themes. …To use the words of the directors, they represent 'the beautiful life before science spoiled nature.' " The neo-pastoral mode is shown to be an elaborate aesthetic trap, presenting an illusory communion with nature which flees into the pre-Industrial past to avoid the terrifying spectre of environmental degradation in the present. Imagining dystopian futures is also part of the ETO's plan to generate global pessimism, with the same character noting a science fiction contest "with a top reward of five million for the person who imagined the most disgusting possible future."[55] This internal commentary on the glut of dystopian science fiction operates as a *mise-en-abyme* for the text itself. *Three-Body* itself feeds on and encourages a catastrophic pessimism while simultaneously offering a solution to it in the form of scientific reason. But given that it is reason itself that appears to have produced this nihilistic condition, such a solution is also foreclosed. Within the text itself, scientific reason cannot prevent the coming extinction of humanity, as the sophons have successfully neutered the Earth's capacity for resistance through technological progress. If we are to take seriously Liu's conception of the author-as-God, we must ask why this deity was driven to create such an uncomfortable universe, governed by such harsh and contradictory laws.

The text constructs a further encounter with the nihilism of reason in which even reason's capacity for prediction is undone. The text describes the online game "Three-Body" created by the ETO to recruit new members. The game is set in a world of radical contingency where Earth's scientific laws hold no traction. The game models the environment of Trisolaris, with its three suns and unstable climate, with players attempting to generate a theory to predict when the suns' orbits will ravage the planet. Players play as characters from the history of Western and Chinese science, such as the aforementioned Mozi or Galileo, while attempting to solve the "Three-Body Problem" of the movement of the planet's three suns. This is a variation on the three-body problem of classical mechanics, replacing the common referents of the Earth, the Moon and the Sun, with the three suns of Trisolaris.[56] The game is ultimately unwinnable, however, and ends by stating that the problem has no solution.

The game world dramatises the traumatic possibility that anthropic science has only local validity—in other words, that the physical laws that govern life on Earth, embodied here by the three-body problem, are rendered null and void when faced with a different planet's structure. The inability to solve the three-body problem also presents us once again with the spectre of reason uncoupled from a utopian project. A player encounters an enormous monument which is explained as "a monument to Trisolaris as well as a tombstone ... for an aspiration, a striving that lasted through almost two hundred civilisations ... as of now it's completely over."[57] The text of *Three-Body* is itself a kind of tombstone, chronicling the death of a particular idea, that of the utopian project of the rational overcoming of tyrannical irrationality—a project which today seems to have been proven unsolvable and thus abandoned.

Allegorising Extinction and the End of Science Fiction

In order to comprehend the internal structure of *Three-Body*, we must return to the historical conditions which give rise to the book's focus on human extinction. Through an investigation into the allegories at work in the novel, we can propose a limit in the capacity for science fiction to allegorise extinction. At an immediate level, alien invasion serves to allegorise existing threats to the human species. Within the text these threats are understood to reflect an ethical failing on the part of humanity and most significantly as a failure to maintain an ethics of care in relation to the environment. These threats are what Nick Bostrom describes as "global catastrophic risks."[58] One of the most pressing risks which both Bostrom and *Three-Body* confront is the ecological catastrophe posed by runaway climate change. Ecological concerns have been at the forefront of Chinese politics in the last ten years. From the much-publicised smog in the Beijing biosphere to Xi Jingping's recent commitment to clean energy, there has been a rise in ecological politics in China both in terms of state policy and mass activism.[59] The coming invasion also allegorises the ongoing extinction of animal species. Ye, for example, reads a smuggled copy of Rachel Carson's classic work *Silent Spring* (1962) during the Cultural Revolution. This book, which details the disastrous effects of DDT and other pesticides, had been banned by the Communist authorities and Ye is accordingly punished. The other extinction threat directly confronted within the text is the potential catastrophe caused by a nuclear strike. Ye's message to the Trisolarans occurs during China's acceleration of their nuclear capacities in the 1970s. Mao famously stated in 1955 that the Chinese people could not be cowed by the atom bomb: "even if the U.S. atom bombs were so powerful that, when dropped on China, they would make a hole right through the earth, or even blow it up, that would hardly mean anything to the universe as a whole, though it might be a major event for the solar system."[60] This is precisely the transformation of reason into unreason that Theodor Adorno and Max Horkheimer diagnosed.[61] The apocalyptic atmosphere of Cold War politics is another reason for Ye to send her message for the invasion of Earth. The extinction threats serve as a feedback loop; Ye invites the extinction of humanity because we have already placed ourselves under the threat of extinction and thus ought to be annihilated.

Both these extinction threats are linked in the text to the history of Chinese socialism. As previously noted, this is not merely a straightforward denunciation of Maoism. Both the deforestation projects and the nuclear programme had a utopian quality. They were part of the project of building communism and were the ecstatic subject of posters, books and films. Socialist realism had a stable allegorical structure in which representations of the production process became a means of representing the abstract historical process of building communism. Extinction in *Three-Body* is therefore a mode of addressing the end of the utopianism promised by the socialist project. Not only has the project of building communism been jettisoned but the prior projects or achievements have been denounced.[62]

We can therefore understand *Three-Body* to be part of a larger canon of post-socialist artworks that seek to reflect on or comprehend the historical past. Works such as Mo Yan's 生死疲劳 (*Life and Death Are Wearing Me Out*) (2006) use the allegorical mode to address that which is impossible to face head on, both in the sense that open references to the past could provoke censorship, but more crucially because this past appears to exceed the capacity for realist representation. Paradoxically, such texts draw on the allegorical imagery of socialist realism to formulate an oblique criticism of the past. *Three-Body* departs from this approach. History is confronted directly in the brutal accounts of the violence of the Cultural Revolution, but the decline of the utopian idea is addressed allegorically.

The after-life of socialist utopianism is glimpsed when Ye meets the former Red Guards who murdered her father. They have no remorse and complain that they have been forgotten by the nation to which they sacrificed their youth. One of the former cadres complains: "It's a new age now. Who will remember us? Who will think of us, including you? Everyone will forget all this completely."[63] Liu locates a double forgetting in the cultural amnesia concerning the Cultural Revolution. In order to repress the horrors, the utopianism which motivated them must be ignored as well. Extinction becomes a mode of releasing the horrors of the past, and in doing so releasing the utopianism which motivated them, if only to put it to death once more. After this meeting, Ye is firm in her resolve: "The small sliver of hope for society that had emerged in her soul had evaporated like a drop of dew in the sun."[64] The sun, which as noted earlier was an allegory for the figure of Mao, has annihilated all hope in the future of humanity. The extinction allegorised in *Three-Body* is the extinction of the utopian idea, or at least a utopianism tied to a socialist imaginary which can only be dimly conceived as a terrifying nightmare.

The final extinction threat allegorised in *Three-Body* is the heat death of the universe. The colonisation project of the Trisolaran civilisation is motivated by the destruction of their planet by its three suns. In a sense, the arrival of the Trisolarans is an arrival from Earth's distant future, when our planet will be swallowed by an expanding sun.[65] As Ray Brassier argues, the inevitable extinction of terrestrial life poses a problem for philosophical thought, or in this case for literature. Brassier writes: "How does thought think a world without thought? Or more urgently: How does thought think the death of thinking?"[66] *Three-Body* forces a confrontation with a future universe without humanity. The text dramatises this traumatic encounter with extinction through its depiction of a suicide epidemic amongst Earth's scientists. For Brassier, the nihilism invoked by this conclusion is intimately linked with scientific reason as it is through the process of reason that thought reaches this terrifying conclusion. Yet, for Brassier, the nihilism of reason is not something to disavow but to affirm, in order to encourage philosophical and aesthetic speculation on a non-anthropocentric universe.

Three-Body then becomes a mode of attempting to accept this terrifying possibility, a mode with which to apprehend extinction not as a threat to be avoided but as the inevitable destiny of terrestrial life. The extinction of humanity as a result of human failing, through hubris or violence, is capable of being allegorised. Extinction in this sense has an eschatological quality; it is the result of a moral failure on the behalf of humanity, as evidenced above in the discussion of the religious nature of the ETO. But the extinction of the Earth is of a fundamentally different character and cannot be anthropomorphised or construed as an act of divine judgement. This poses a serious problem for allegorical writing and hence allegorical interpretation. Walter Benjamin, in his discussion of the allegory, notes the recurrent necessity to express historical phenomena in the frozen features of the human face. He writes: "in allegory the *facies hippocratica* ['Hippocratic face,' the sunken and pinched features exhibited by the dying] of history lies before the eyes of the observer as a petrified, primordial landscape. Everything about history that, from the very beginning, has been untimely, sorrowful, unsuccessful, is expressed in a face—or rather in a death's head."[67]

While the horrors of history may emerge as a written sign, the contemplation of extinction resists narrative inscription. This becomes particularly acute in *Three-Body* in a description of the night sky, when the character Yang learns of the coming invasion. Liu writes, "the universe was a cramped heart, and the red light that suffused everything was the translucent blood that filled the organ."[68] There is a need to render the coming extinction in recognisable anthropomorphic terms, in this case Benjamin's "primordial landscape" is glimpsed as a human heart. However, the text does not resolve this contradiction between a historical and cosmological understanding of extinction. It is precisely in the cosmological notion of extinction as solar death that we glimpse the historical failures of previous utopian ideals. This failure to allegorise species death may be construed alongside larger discussions of the difficulty of discovering a written form adequate to the discussion of climate catastrophe. Against calls for a return to an epic or mythic form, as famously argued by Amitav Ghosh, *Three-Body* pushes the rationalist hard science fiction form to its allegorical limit.[69] It is at this limit that the text begins to become interesting, calling not for the ethical salvation of humanity but for the radical annihilation of existing forms of political and ethical commitment. *Three-Body* offers no comforting returns to prior ethical systems or modes of being but instead presents only an unbearable coldness. This coldness may be the only way to begin conceptualising environmental catastrophe, without resorting to a thinly veiled language of divine judgement or agonising self-recrimination. The text's exploration of the limits of reason is not a flight into irrationalism. Rather, it suggests the futility of the fantasy of a purely rational and technical solution to planetary ecocide. The solution must be technical and social, as neither is capable of being implemented successfully without the other.

The sun of Mao's head is precisely the same sun which will annihilate terrestrial life, which is thus used to call for the invasion and extinction of the human race. In *Three-Body* there is very little sense of the melancholy or mourning that Benjamin associates with allegorical mode of reflection on the historical past.[70] Instead, writing extinction is a mode of negating existing utopianisms. It is a way of suggesting that the only utopia worth imagining is one without the human race in it. Or at least the human race as it has hitherto been conceived. Bostrom suggests that "anthropic bias" makes it impossible to adequately assess global catastrophic risks because we cannot imagine a world without us in it. But *Three-Body* forces a confrontation with the spectre of a world without humanity. In the final pages of the text, the human race is compared with a plague of locusts. The Trisolarans beam a message, "You are bugs!" (虫子) and from the perspective of a superintelligent colonising species we are indeed bugs.[71] This may seem an unlikely place to look for a utopian model. But harnessing the swarming, destructive potential of the species is closer to Liu's pessimistic conception of the human species than a fantasy of a sudden ethical awakening. At once inhuman and bug-like, the plague of the Earth may still yet be the most reasonable way of working against the destruction of the world, as pack of locusts which will work to blot out the sun.

Conclusion: Utopia Must Die

In *Three-Body*, the unbearable memory of the historical past is impossible to disentangle from the monstrous forgetting which will engulf human civilisation once it has been destroyed forever by the expanding sun. But what appears to be the most terrifying thought is the only mode of utopian thinking worth pursuing. Reason is no longer tied to a utopian project of building socialism. At its worst, it remains bound to the blind irrationalism of capital and the value form. At its best, it reveals the nihilistic truth of human extinction. The nihilism engendered by reason is not to be sutured by a belief in human dignity or progress. Liu writes, "The ultimate goal and ideal of the ETO is to lose everything. Everything that now belongs to the human race, including us."[72] Yet the text reveals the ETO to be still constrained by religious thought, reifying evil as human nature and conceiving of extinction in terms of divine (or extraterrestrial) judgement. Environmental politics, with their attendant aesthetics, which remain tied to these modes, have little purchase on the magnitude of the cataclysm to come. Instead, in *Three-Body,* the goal of eliminating the human race is a way of conceiving a new subjectivity and the destruction of the earth is a way of imagining a new world. Utopia must be reconceived not as the eschatological destiny of humanity but as an abyssal nothingness which cannot be apprehended by thought. Thomas More's conception of utopia as a "no place" re-emerges in *Three-Body* as a radically unrepresentable future. It cannot be allegorised, but it can be glimpsed through its negation. The book is a tombstone, a death's head, and a cramped heart for utopia. Nietzsche writes: "We possess art lest we perish of the truth."[73] *Three-Body* is a text which revels in our perishing. Literature, and science fiction more specifically, forces a confrontation with the devastating truth of extinction, putting this truth at the forefront of the artwork. It is the grim truth that utopia's past cannot be revived and the traumatic attempt to write a future is so alien it defies inscription. It is, in short, the grim resolution that utopia must die.

Notes

1. For a discussion of the translation of the text, see Ken Liu, "Translator's Postscript."
2. Obama, "Transcript"; Feloni, "Why Mark Zuckerberg"; Mingwei Song, "Three-Body Trilogy."
3. For a study of the Chinese coal mining industry, see Wright, *Political Economy.*
4. Liu Cixin, "Author's Postscript," 427.
5. Long et al., "Typhoon Nina," 451–472.
6. Liu Cixin, *Three-Body*, 428.
7. Liu Cixin, "Author's Postscript," 428.
8. Liu Cixin, *Three-Body*, 3.
9. Rosen, *Role.*
10. Liu Cixin, *Three-Body*, 24.

11. Liu Cixin, *Three-Body*, 187.
12. Mittler, *Continuous Revolution*, 271.
13. Mittler, *Continuous Revolution*, 271.
14. Liu Cixin, *Three-Body*, 295.
15. Liu Cixin, *Three-Body*, 296
16. Liu Cixin, *Three-Body*, 300.
17. Berlin, "Decline."
18. Jameson, *Archaeologies*, xii.
19. Moylan, *Scraps*, 131.
20. Liu Cixin, *Three-Body*, 344.
21. Liu Cixin, "Worst," 362.
22. For a theorisation of the notion of post-socialism see Zhang, "Postmodernism."
23. Liu Cixin, *Three-Body*, 344.
24. Bloch, *Principle of Hope*.
25. Mingwei Song, "Representations"; Mingwei Song, "Variations."
26. Manthorpe, "Cixin Liu"; Barnett, "People."
27. For a discussion of the translation of Chinese science fiction, see Jang, "Translation." For a discussion of the status of science fiction writing during the Cultural Revolution and its aftermath, see Wagner, "Lobby Literature."
28. Wang, *Fin-de-siècle Splendor*, 253.
29. Wang, *Fin-de-siècle Splendor*, 253.
30. Wang, *Fin-de-siècle Splendor*, 303.
31. Mingwei Song, "Variations," 87.
32. Wagner, "Lobby Literature," 19.
33. For a study of the utopianism of Mao's thought, see Meisner, *Marxism, Maoism and Utopianism*.
34. Wang, *Fin-de-siècle Splendor*; Mingwei Song, "Representations."
35. For an analysis of such works, see King, *Milestones*. For a collection of works written during the Maoist period, see McDougall, *Popular Chinese Literature*.
36. Liu Cixin, *Three-Body*, 322.
37. Mingwei Song, "Representations," 12.
38. Schiller, *On the Aesthetic*.
39. Liu Cixin, *Three-Body*, 390.
40. Liu Cixin, *Three-Body*, 344.
41. Liu Cixin, *Three-Body*, 344.
42. Liu Cixin, *Three-Body*, 344.
43. Liu Cixin, *Three-Body*, 346.
44. Liu Cixin, *Three-Body*, 346.
45. Liu Cixin, *Three-Body*, 346.
46. Liu Cixin, *Three-Body*, 344.
47. Jameson, "Politics of Utopia," 36.

48. Hobbes, *Leviathan*.
49. Hobbes, *Man and Citizen*, 118.
50. Mozi, *Mozi*.
51. Nietzsche, *Will to Power*, 13. For a study of the popularity of Nietzsche's philosophy in China in the twentieth century, see Kelly, "Highest Chinadom."
52. For a study of Western theories of technology and the relevance of their application to Chinese modernity, see Hui, *Question*.
53. Barnett, "People."
54. Mao Tse-tung, "On Practice," 304.
55. Liu Cixin, *Three-Body*, 145.
56. Valtonen, *Three-Body*.
57. Liu Cixin, *Three-Body*, 255.
58. Bostrom, "Existential," 17.
59. Deng, Farah and Wang, "China's Role." See also Ligang Song and Wing Thye Woo, *China's Dilemma*.
60. Mao Tse-tung, "Chinese People," 152.
61. Horkheimer and Adorno, *Dialectic of Enlightenment*.
62. For Deng Xiaoping's 1981 denunciation of the Cultural Revolution, see Vogel, *Deng Xiaoping*. For the current State's official opinion on the Cultural Revolution, see "Society Firmly Rejects."
63. Liu Cixin, *Three-Body*, 329.
64. Liu Cixin, *Three-Body*, 329.
65. Schröder and Smith, "Distant Future."
66. Brassier, *Nihil Unbound*, 223.
67. Benjamin, *Origin*, 166.
68. Liu Cixin, *Three-Body*, 137.
69. Ghosh, *Great Derangement*.
70. Benjamin, *Origin*, 185.
71. Liu Cixin, *Three-Body*, 418.
72. Liu Cixin, *Three-Body*, 272–273.
73. Nietzsche, *Will to Power*, 435.

Works Cited

Barnett, David. "People Hope My Book Will Be China's *Star Wars*: Liu Cixin on China's Exploding Sci-fi Scene." *The Guardian*. December 14, 2016. https://www.theguardian.com/books/2016/dec/14/liu-cixin-chinese-sci-fi-universal-the-three-body-problem/.

Benjamin, Walter. *The Origin of German Tragic Drama*. Translated by John Osborne. London: Verso, 2003.

Berlin, Isiah. "The Decline of Utopian Ideas in the West." In *The Crooked Timber of Humanity: Chapters in the History of Ideas*, edited by Henry Hardy, 20–48. London: Pimlico, 2003.

Bloch, Ernst. *The Principle of Hope*. Translated by Neville Plaice. Cambridge: MIT Press, 1995.

Bostrom, Nick. “Existential Risk Prevention as Global Priority.” *Global Policy* 4, no.1 (2013): 15–31.

Brassier, Ray. *Nihil Unbound: Enlightenment and Extinction*. Basingstoke: Palgrave Macmillan, 2007.

Deng Haifeng, Paolo Davide Farah, and Anna Wang. “China’s Role and Contribution in the Global Governance of Climate Change.” *The Journal of World Energy, Law and Business* 8, no. 6 (2015): 581–599.

Feloni, Richard. “Why Mark Zuckerberg Wants Everyone to Read this Chinese Science Fiction Novel.” *Business Insider*. October 22, 2015. https://www.businessinsider.com.au/mark-zuckerberg-recommends-the-three-body-problem-2015-10.

Ghosh, Amitav. *The Great Derangement: Climate Change and the Unthinkable*. Chicago: University of Chicago Press, 2016.

Hobbes, Thomas. *Leviathan*. London: Penguin Books, 1985.

———. *Man and Citizen*. Indianapolis: Hackett, 1991

Horkheimer, Max, and Theodor Adorno. *Dialectic of Enlightenment*. Translated by Edmund Jephcott. Stanford: Stanford University Press, 2002.

Jang, Qian. “Translation and the Development of Science Fiction in Twentieth Century China,” *Science Fiction Studies* 40, no.1 (2013): 116–132.

Jameson, Fredric. *Archaeologies of the Future: Utopia and other Science Fictions*. London: Verso, 2005.

———. “The Politics of Utopia,” *New Left Review* 25 (2004): 35–54.

Kelly, David. “The Highest Chinadom: Nietzsche and the Chinese Mind, 1907–1989,” In *Nietzsche and Asian Thought*, edited by Graham Parkes, 151–174. Chicago: University of Chicago Press, 1991.

King, Richard. *Milestones on a Golden Road: Writing for Chinese Socialism, 1945–80*. Vancouver: University of British Columbia Press, 2013.

Liu Cixin. “Author’s Postscript for the American Edition.” In *The Three-Body Problem*, by Liu Cixin, translated by Ken Liu. London: Head of Zeus, 2016.

———. *The Three-Body Problem*. Translated by Ken Liu. London: Head of Zeus, 2016.

———. “The Worst of All Possible Universes and the Best of All Possible Earths: *Three-Body* and Chinese Science Fiction.” In *Invisible Planets: Thirteen Visions of the Future from China*, edited by Ken Liu, 361–368. London: Head of Zeus, 2016.

———. 三体. Chengdu: 科幻世界 Kehuan Shijie, 2006).

Liu, Ken. “Translator’s Postscript.” In *The Three-Body Problem*, by Liu Cixin, translated by Ken Liu. London: Head of Zeus, 2016.

Long, Yang, Maofeng Liu, James Smith, and Fuqiang Tian. “Typhoon Nina and the 1975 Flood over Central China.” *Journal of Hydrometeorology* 18, no. 2 (2017): 451–472.

Manthorpe, Rowland. "Cixin Liu Is the Author of Your next Favourite Sci-fi Novel." *Wired*. October 29, 2016. http://www.wired.co.uk/article/chinese-sci-fi-writer-cixin-liu/.

Mao Tse-tung. "The Chinese People Cannot be Cowed by the Atom Bomb." In *Selected Works of Mao Tse-tung: Volume V*, 152–153. Peking: Foreign Language Press, 1977.

———. "On Practice," In *Selected Works of Mao Tse-tung: Volume I*. 295–310. Peking: Foreign Language Press, 1965.

McDougall, Bonnie. *Popular Chinese Literature and Art in the People's Republic of China, 1949–1979*. Berkeley: University of California Press, 1984.

Meisner, Maurice. *Marxism, Maoism and Utopianism*. Madison: University of Wisconsin Press, 1982.

Mittler, Barbara. *A Continuous Revolution: Making Sense of Cultural Revolution Culture*. Cambridge: Harvard University Press, 2012.

Moylan, Tom. *Scraps of the Untainted Sky: Science Fiction, Utopia, Dystopia*. Oxford: Westview, 2000.

Mozi. *The Mozi: A Complete Translation*. Translated by Ian Johnston. Hong Kong: The Chinese University Press, 2010.

Nietzsche, Friedrich. *The Will to Power*. Translated by Walter Kaufmann and R. J. Hollingdale. New York: Vintage Books, 1968.

Obama, Barack. "Transcript: President Obama on What Books Mean to Him," by Michiko Kakutani. *The New York Times*. January 16, 2017. https://www.nytimes.com/2017/01/16/books/transcript-president-obama-on-what-books-mean-to-him.html.

Rosen, Stanley. *The Role of Sent-Down Youth in the Chinese Cultural Revolution*. Berkeley: University of California Press, 1981.

Schiller, Friedrich. *On the Aesthetic Education of Man*. Translated by Keith Tribe. London: Penguin Books, 2016.

Schröder, K. P., and Robert Connon Smith. "The Distant Future of the Sun and Earth Revisited." *Monthly Notices of the Royal Astronomical Society* 386 (2008): 155–163.

"Society Firmly Rejects the Cultural Revolution." *People's Daily Online*. May 17, 2016. http://en.people.cn/n3/2016/0517/c90780-9058793.html/.

Song, Ligang, and Wing Thye Woo. *China's Dilemma: Economic Growth, The Environment, Climate Change*. Canberra: Australian National University, 2008.

Song, Mingwei. "Representations of the Invisible." In *The Oxford Handbook of Modern Chinese Literatures*, edited by Carlos Rojas Andrea Buchner and Mingwei Song, 546–565. Oxford: Oxford University Press: 2016.

———. "Variations on Utopia in Contemporary Chinese Science Fiction." *Science Fiction Studies* 40, no. 1 (2013): 86–102.

———. "The Three-Body Trilogy: The Three-Body Problem, The Dark Forest, Death's End," review of *The Three Body Problem*, by Cixin Liu. *Ohio State University Modern Chinese Literature and Culture*. December, 2015. http://u.osu.edu/mclc/book-reviews/mingweisong.

Valtonen, Mauri. *The Three-Body Problem.* Cambridge: Cambridge University Press, 2006.
Vogel, Ezra. *Deng Xiaoping and the Transformation of China.* Cambridge: Harvard University Press, 2011.
Wagner, Rudolf. "'Lobby Literature: The Archaeology and Present Function of Science Fiction in China." In *After Mao: Chinese Literature and Society*, edited by Jeffrey Kinkley, 17–62 (Cambridge: Harvard University Press, 1985).
Wang, David Der-wei. *Fin-de-siècle Splendor: Repressed Modernities of Late Qing Fiction, 1849–1911.* Stanford: Stanford University Press, 1997.
Wright, Tim. *The Political Economy of the Chinese Coal Industry.* Abingdon: Routledge, 2012.
Xudong, Zhang. "Postmodernism and Post-Socialist Society: Cultural Politics in China After the 'New Era.'" *New Left Review* 237 (1999): 77–105.
Yuk, Hui. *The Question Concerning Technology in China: An Essay in Cosmotechnics.* Falmouth: Urbanomic, 2016.

CHAPTER 7

Ecopocalyptic Visions in Haitian and Mexican Landscapes of Exploitation

Giulia Champion

Introduction

The omnipresence of apocalyptic imagery, especially in popular culture, trivialises its rhetorical power, as people imagine the zombie apocalypse with more enthusiasm than dread. As Eddie Yuen writes:

> The spectre of the apocalypse haunts the world today. Every political, cultural, and aesthetic field we look at is replete with talk of catastrophe. This poses a particular challenge for environmentalists and scientists who are tasked with raising awareness about what is unquestionably a genuinely catastrophic moment in human and planetary history.[1]

The banalisation of apocalyptic discourse is problematic, first, because "environmentalists and scientists must compete in this marketplace of catastrophe, and find themselves struggling to be heard over the din" and, second, because fear itself cannot properly be a strong enough incentive to actively change one's lifestyle and political views in the attempt to avoid

G. Champion (✉)
University of Warwick, Coventry, UK

Z. Kendal et al. (eds.), *Ethical Futures and Global Science Fiction*, Studies in Global Science Fiction,
https://doi.org/10.1007/978-3-030-27893-9_7

climate change.[2] This is true not only because fear is a tool of authoritarianism and exclusionary rhetoric and politics, but because people who can actually make a difference in trying to prevent the ecological apocalypse, or "ecopocalypse," do not necessarily feel concerned by it, granted they actually accept the science of climate change in the first place. As Yuen remarks, "political and economic elites *do* see themselves as exempt from the crisis."[3] Indeed, the security and comfort to which these elites have grown accustomed appears as a constant that cannot be threatened, owing to the power of their wealth as well as the fact that most of them have not been confronted with the multiplying and tangible consequences of climate change, unlike many in "postcolonial" geographical zones. One need simply think of the unseasonable heavy rain which destroyed crops in North India in 2015, the many hurricanes that devastated several islands in the Caribbean (and southern regions of the United States) since 2017, the sinking islands of the Pacific, or the droughts which continue to plague the Caribbean today.[4] The list goes on, and these are not merely "recent" issues related to climate change: some countries, including Haiti and Mexico, have been living with the ecological consequences of colonialism and imperialism for years.

This chapter investigates the ethical concerns of the ecological and social deteriorations of rural and urban spaces, and the framing of these with apocalyptic language and scenarios, in Jacques Roumain's 1944 novel *Gouverneurs de la rosée* (*Masters of the Dew*), and Homero Aridjis's 1993 *La Leyenda de los soles* (The Legend of the Suns) and 1995 *¿En quién piensas cuando haces el amor?* (Of Whom Do You Think When You Make Love?).[5] Despite their inherent linguistic and cultural differences, as well as their dissimilar national and historical contexts, they converge in their depiction of ecopocalyptic crises as consequences of years of colonial and neo-imperialist exploitation of lands and communities. As Yuen notes, "the 'end times' arrived for millions in the Western Hemisphere with the arrival of Columbus and countless species and eco-systems were condemned to make way for the 'progress' enjoyed by the Global North."[6] These "end times" were both human and environmental, as David Watts observes in his study on the West Indies, thereby aligning the environmental trauma of this region to the human ones:

> Social scientists often characterise these islands as having been shaped by two of the most severe human traumas of global significance to have taken place within the last four centuries: first, the virtually total and rapid removal of a large aboriginal population following initial European contact; and,

> later, the forced transference into them of many hundreds of thousands of Africans from their homelands under conditions of slavery to support a system of plantation agriculture. The environmental degeneration inferred above, which thus far has not been detailed in print, may be deemed to form a third trauma, which is perhaps now of equal and growing importance to the inhabitants.[7]

The equalisation of human and environmental trauma in this region, as well as other ones that have been despoiled and exploited after years of imperial colonisation and neo-imperial abuse, is critical in the context of postcolonial environmental ethics in relation to the novels studied in this chapter.[8] Indeed, these works subvert the objectification and the commodification of bodies and landscapes by aligning humans with the environment and thereby contesting the "othering" of nature (and people) which is inherent to the "colonisers"' language.[9]

Postcolonial environmental ethics, I contend, cannot be understood separately from the "moral extensionism" fundamental to environmental ethics, which promotes the expansion of the circle of moral concerns beyond the human species.[10] In these novels, I argue, there are two movements that overcome the human/nature binary: first, there is an anthropomorphising and feminisation of nature, recalling early colonial discourse, and thereby aligning the colonial mistreatment of people and land; and, second, characters are described with natural features and likened to animals to emphasise the affinity and belonging of the human *with* the natural world. This is critical in the context of climate change because, as Val Plumwood notes, the fact that "western culture has treated the human/nature relation as a dualism ... explains many of the problematic features of the west's treatment of nature which underlie the environmental crisis, especially the western construction of human identity as 'outside' nature."[11] Consequently, the dynamic of these novels—reinstating humanity's membership to the natural world to subvert this dichotomy—aims to remind us how climate change is affecting (and will affect) us all, even if it might affect certain groups of people and places before others. This reinstatement is only possible through the expansion of our moral concerns to these often-ignored people and places, and ecosystems in general. Nevertheless, as I will show, this moral expansionism appears to enter into conflict with most of the novels' dystopian and futuristic societies' inhabitants, with only a few characters seeming to understand the importance of having more ethical relationships amongst each other and towards nature.

Furthermore, the use of ecopocalyptic imagery in these novels differs from its use in "Western" popular culture since it does not serve to warn and frighten, but rather to describe past, present and future states of Haiti and Mexico, each impacted by human and environmental traumas and catastrophes, the ongoing exploitation of their natural resources, and contemporary models of industrialisation and development. This move away from mainstream apocalyptic imagery constitutes a return to the etymology of the word itself, as well as its biblical nature, which Stefan Skrimshire explains thus:

> If the mediatised language of catastrophe is problematic it is because it is "apocalyptic" only in the Hollywood sense: it is devoid of ethical content. It says nothing of who we are and where we are going. This is something of a paradox: "apocalypse" is derived from the Greek word for revelation, or the unveiling of divine truth to mortal man. To many people apocalyptic literature (including biblical texts) represented the imaginative attempt to portray the corruption of the present in order to inspire radical social transformation. In contrast, populist catastrophism today represents a form of veiling, or clouding, of the ethical and political question of climate change.[12]

Hence, I contend that the ecopocalyptic visions of Roumain's and Aridjis's novels constitute a literary depiction of real and tangible issues of social and environmental degradation in rural Haiti and urban Mexico through dystopian imagery and the science fiction genre. The apocalyptic scenarios in the novels, I shall argue, return to the etymological and biblical foundations of revelation and distance themselves from the extravagant catastrophism of mainstream popular culture. This movement also permits the consideration of ethical concerns in relation to the environmental and social degradation depicted. These ecopocalypses are not merely warnings, but reflect on very real and pressing matters in Haiti and Mexico, and most of Latin America and the Caribbean, as well as other more urgently "endangered" parts of the world.

The main argument of this chapter is that these novels' underlying utopian impulse is sustained by a hope attainable only by advocating more ethical relationships with nature and amongst people—relationships that would achieve community and therefore move away from the estranged and alienated associations formed under the economy of capitalism and neo-liberal politics. Through a comparative analysis of the three novels, I will establish how they advocate these more ethical relationships and consider the legacies of colonialism and imperialism while depicting degrading landscapes and social bonds. It is important to note that all three novels begin as dystopias with a sense of hopelessness and scenes of overwhelming social and environmental decline, and all three close on a transformed landscape and community with a sense of hope, renewal, and change, as well as a promise of life.

Mythic Time, Ecopocalypse, and Visions of Hope

The uplifting ending of each novel is possible because in all three works time is not perceived in a linear manner in accordance with the Western Gregorian calendar, which would make the apocalypse the end point, but is cyclical, promising a spirit of renewal. In Roumain's *Gouverneurs de la rosée*, this is brought forth by the character of Manuel, Délira's and Bienaimée's son, who had left the rural Haitian community to work on a Cuban sugar plantation, only returning after many years. His understanding of nature and his determination to find water, as the village is afflicted by an extreme drought, restore hope to the whole community. The novel begins with Délira's hopelessness: "Nous mourrons tous... –et elle plongea sa main dans la poussière; la vieille Délira Délivrance dit: nous mourrons tous" ("We all die... –and she plunged her hand in the dust; the old Délira Délivrance said: we all die").[13] Nature and people alike are dying because of the drought and, as it will be shown, the dichotomy between the two eventually disappears, as it does in the other two novels. Délira contemplates inescapable death: on the one hand, because the drought has turned the soil to dust, and on the other hand, she believes that her son, Manuel, who left many years earlier, is most likely dead.[14]

However, Manuel returns, and he is determined to find water. Even though he dies shortly after he discovers a new source, he ensures that the only other person to whom he has shown it, his fiancée Annaïse, will guide the rest of the village to it. The last lines of the novel depict water flowing back into the rural town, and although Délira laments the death of her son, Annaïse tells her that Manuel lives on through her: "Elle prit la main de la vieille et la pressa doucement contre son ventre où remuait la vie nouvelle" ("She took the old woman's hand and gently pressed it against her stomach where new life moved").[15] Time is thus cyclical, and the renewal of life and the regeneration of hope allow the community to come together again after years of enmity arising from a family feud. Manuel is dead, but water and his child live on to secure the prosperity of the village, which explains the pluralisation of the title. Indeed, *Gouverneurs de la rosée*, literally "governors of dew," refers to the Haitian Creole expression *gouvènè rouzé*, which is the title given in rural areas to the labourer responsible for the watering and irrigation, and more broadly anything concerning water distribution.[16] In the novel, water can only be brought to the village if everyone works together again, in the spirit of community and by putting old hatreds aside. Only with a more ethical outlook on nature, like Manuel's, can the village of *Gouverneurs de la rosée* live at peace with each other and the world around them.

Aztec mythologies and calendars become central in Aridjis's novels, where the conception of time is partly cyclical. His earlier novel *La leyenda de los soles* (The Legend of the Suns) refers to Aztec chronicles and foundational myths, in which periods are framed by the lives and deaths of different Suns.[17] While the Suns themselves are not cyclical, since it is always a new and different Sun that replaces the last and they are limited to five, the conception of rebirth inherent to the legend allows a conception of the apocalypse as something other than the end point of linear time.[18] I contend that the presence of a Sixth Sun, the novel's Sun of Nature, reflects the texts' utopian impulse. It also reflects Aridjis's understanding of apocalypticism, as he ends his collection of essays on the apocalypse concluding that "quizás sólo un *Deus ex machina*, o un ángel apocalíptico con conciencia ambiental podrá salvar lo que debe ser salvado" ("perhaps only a *Deus ex machina*, or an angel of the apocalypse with an environmental awareness will be able to salvage what needs to be saved").[19]

Aztec mythology becomes more central when a character, Cristóbal Cuauhtli, travels through time to enlist the help of one of the protagonists, Juan de Góngora. He journeys to the year 2027—a date chosen by Aridjis because the Aztec calendar identifies it as the year of the Fifth Sun's death.[20] Cristóbal tasks Juan with finding the missing page from the *Códice de los Soles* (*Codex of the Suns*), the book containing all the myths the Aztecs have gathered and written. The page was stolen by General Carlos Tezcatlipoca, who appears to be the tyrannical chief of police of Ciudad Moctezuma (the renamed 2027 Mexico City), although, he is really the embodiment of the jaguar divinity, born from the darkness when the gods created the Fifth Sun.[21] Cristóbal demands Juan's help: "debes recobrar la hoja del *Códice de los Soles* antes de que aparezcan los *tzitzimime*, monstrous del crepúsculo que han de venir a la Tierra para devorar a los hombres cuando muera el Quinto Sol" ("you must recover the page of the

Codex of the Suns before the appearance of the *tzizimime*, the monsters of the twilight, fated to come to earth and devour men when the Fifth Sun dies").[22] When faced with Juan's incredulity, Cristóbal informs him of the two possible outcomes after the death of the Fifth Sun:

> Si mañana muere Tezcatlipoca, la diosa azul, que vive en el Iztac Cíhuatl, será la diosa del Sexto Sol, el Sol de la Naturaleza. Si muere ella, el general Carlos Tezcatlipoca tomará el poder para instaurar el reino del terror nocturno, el Sol del Espejo Humeante.
>
> (If tomorrow Tezcatlipoca dies, the blue goddess, who lives in the volcano Iztac Cíhuatl, will be the goddess of the Sixth Sun, the Sun of Nature. If she dies, the general Carlos Tezcatlipoca will seize power to establish the reign of nocturnal terror, the Sun of the Smoking Mirror.)[23]

Throughout the novel, Tezcatlipoca is presented as the exact opposite of the blue goddess because he not only abducts and abuses young women, but also is a murderer. In a passage of *La leyenda* coming directly after Juan and Cristóbal's meeting, Tezcatlipoca pays a visit to his half-sister, Natalia, an *ecoguerrillera*. He forces her to watch while he kills all the animals she has been saving from extinction in her sanctuary and then murders her.[24] This passage, strategically placed after Juan and Cristóbal's conversation, emphasises Tezcatlipoca's violence against animals and women alike, indicating his status as the antithesis of the blue goddess.

Cristóbal grants Juan the ability to pass through walls to facilitate his quest for the missing page, which allows him to observe the social degradation, both public and private, of the decaying city while he travels across it. Like Roumain's *Gouverneurs*, both of Aridjis's novels begin with hopelessness and death but end with renewal. *La leyenda* opens on a day in November when Tezcatlipoca has died, and establishes the scarcity of water as a major environmental issue:

> Entraba el mundo en el signo de Escorpíon, era día lunes, había anochecido y en la Ciudad de México no había agua. El sábado, a la hora del crepúsculo, el general Carlos Tezcatlipoca, jefe de la Policía, había muerto en un accidente de coche.
>
> (The world was entering the sign of the Scorpion and it was a Monday, it was getting dark and there was no water left in Mexico City. On Saturday, when the sun was setting, the general Carlos Tezcatlipoca, the Chief of Police, died in a car accident.)[25]

Nonetheless, because Tezcatlipoca is not human, he comes back to life. The novel ends after Juan and his girlfriend, Bernarda Ramírez, have escaped the city, which has been destroyed during an earthquake, and at its outskirt they see the blue goddess of nature, suggesting that Tezcatlipoca has truly died this time. The last line of the novel informs the reader that it is "el primer día del Sexto Sol" ("the first day of the Sixth Sun").[26]

En quien piensas is not precisely a sequel to *La leyenda* as it narrates the same period of time but follows a different set of characters. Unlike *La leyenda*, which is related by a third-person omniscient narrator, *En quien piensas* is told from the perspective of one of the protagonists, Yo Sánchez. Her name refers to the Spanish personal pronoun 'I', emphasising the immersion of the reader in a first-person point of view. She lives with three sisters and works with them in their theatre. The novel opens with the death of Rosalba, María's twin, and the only sister who did not work in the theatre. María and Rosalba are characterised as loving blue and often wearing that colour, which is reminiscent of what the sky used to look like before its decaying present state. The sky is often described as covered in smog and "coffee-coloured," as well as unnaturally devoid of light and sun, and notably Rosalba's name can be translated to "pink dawn," something the characters have not beheld in a long time.[27] Furthermore, María and Rosalba have an affinity with nature, María being good at taking care of plants and Rosalba of birds.[28] Moreover, during Rosalba's funeral, Yo remarks at the presence of a species of butterflies she believed to be extinct:

> Delante de nosotros pasaron docenas de mariposas monarcas. No las veía desde mi infancia. Desorientadas anduvieron entre las tumbas y los árboles muertos, quizás en busca de agua. Una de ellas, como sobreviviente de la extinción biológica y como fantasma de migraciones pasadas, fuera de lugar y de tiempo, se posó en el pelo de María.
>
> (In front of us flew a dozen of monarch butterflies. I had not seen any since my childhood. They fluttered disoriented between the tombs and the dead trees, perhaps looking for water. One of them, like a survivor of the biological extinction and like a ghost of past migrations, out of place and time, landed on María's hair.)[29]

All these elements outline Yo's and the sisters' association with nature, as well as their ethical relationships with it.[30]

Not unlike the two other novels, *En quien piensas* ends on a symbol of hope in the midst of the earthquake that announces the end of the Fifth

Sun. After Rosalba's death, Yo and the sisters bring her many birds to their home in the centre of Ciudad Moctezuma, and when the earthquake strikes, Yo notices that the birds begin to sing: "lo más curioso de todo es que en ese momento de destrucción masiva, de confusión general, de estremecimientos y estruendos, animados por las luces confundidas, todos los pájaros se pusieron a cantar, creyendo que era el alba" ("the strangest thing was that in that very moment of massive destruction, of general confusion, of shudder and tumult, animated by the confusing lights the birds began to sing as if it were dawn").[31] And it is indeed the dawn of a new Sun: the Sixth Sun, the Sun of Nature. This passage emphasises the importance of Rosalba's birds, which appear to be the last few real animals left in the city, since throughout both of Aridjis's novels the decaying flora and fauna are extremely rare. This final scene contrasts strongly with an earlier description of artificial and mechanical nature: "en el centro de la plaza surgió un árbol de metal. En sus ramas tubulares estaban cantando pájaros autómatas, que abrían y cerraban el pico y las alas a cada trino" ("in the middle of the square emerged a metallic tree. On its tubular branches robotic birds were singing, opening and closing their beaks and wings with each chirp").[32] The grotesque image, reminiscent of a disproportionate and incongruous cuckoo clock, is one amongst many of the ecopocalyptic visions to be found in the novel, as well as throughout *Gouverneurs* and *La leyenda*.

Crisis of Representation

The crucial difference between the novels is that Roumain's focuses on a decaying rural landscape, while Aridjis's depict a dystopian urban space. Aridjis depicts a future where technology and corruption are displacing nature and community, while the city faces the consequences of extreme pollution, including drought and rising sea levels.[33] When observing the environmental and social degradations in Ciudad Moctezuma in *La leyenda,* Juan de Góngora wonders how one can depict and represent artistically the absence of devastated nature:

> —Por esa avenida, venía un río, ¿cómo pintar ahora su ausencia, su cuerpo entubado, su carga de aguas negras?, ¿cómo pintar la desesperación de un río, el grito silencioso de la Naturaleza en agonía?—se preguntó, delante de su cuadro—. ¿Cómo pintar la soledad del último conejo teporingo que se extingue en la falda de un volcán?

(On this avenue, there used to be a river. How can one paint its current absence, its piped body, its load of black water? How can one paint the river's despair, the silent cry of agonising Nature?"—He wondered in front of his painting, "How can one paint the loneliness of the last teporingo rabbit going extinct on the hillside of a volcano?)[34]

This passage expresses the artist's anxiety and difficulty in representing the environmental degradation brought about by climate change. While here Juan attempts to depict it visually, his last name, reminiscent of the Spanish Baroque poet Luis de Góngora, broadens the scope of this challenge to other media, such as literature, and thus refers to the narratological crisis over the representation of rapid climate change and environmental deterioration. This brings to mind Amitav Ghosh's nonfictional work *The Great Derangement* (2016), where Ghosh argues that contemporary literature has failed to address climate change. However, I believe that literature emerging from former colonies, like Aridjis's and Roumain's works, has indeed been addressing such issues, since these countries have already been confronted with the consequences of climate change.[35]

This metafictional passage from *La leyenda* demonstrates how a reflection on the representation of decaying nature itself becomes an act of effectively representing it. Juan's repeated rhetorical questions become efforts to grasp and articulate the challenge in narrating climate change. Questions over how to paint "the river's despair" or "the loneliness" of the last survivor of a near-extinct species recall the blending of senses proper to synaesthesia and emphasise the aesthetic and narratological crises faced by artists, as does the oxymoron "silent cry of agonising Nature," which evokes the title of Rachel Carson's seminal work *Silent Spring* (1962), all the while anthropomorphising nature and animals in attributing them human feelings. Like Carson's study, Aridjis's novels aim to examine the dire conditions and increasingly serious issues that result from climate change. The environmental problems in the novels might appear fictional and inflated to conform to dystopian speculative fiction, yet the scenes of decaying nature are based on tangible consequences of industrialisation and climate change that we have already been witnessing.[36] In 1997, two years after the publication of *En quien piensas*, Aridjis published a collection of critical essays titled *Apocalipsis con figuras*, in which he presents an analytical and historical overview of the "apocalypse" and the many traditions surrounding this concept. The next apocalypse, he claims, will be ecological and man-made, and he presents examples of such catastrophic consequences of climate change, both antecedent and contemporary.[37] He notes that "irónicamente, la sobrepoblada México-Tenochtitlan-Distrito Federal, que según el mito se fundó en el agua, morirá por falta de agua" ("ironically, the overpopulated Mexico-Tenochtitlan-Federal District, which according to the myth was founded on water, will die due to its scarcity").[38] Indeed, the city was once dubbed the "American Venice," as a passage from *La leyenda* reminds the reader:

> Un olor nauseabundo flotaba en la ciudad, gatos, perros, gorriones y ratas aparecieron muertos en la calle, en los sótanos, en los patios, en las azoteas y en las trastiendas. Los únicos que corrieron con puntual fetidez fueron los ríos de aguas negros y los basureros líquidos, reminiscencias viles de lo que un día fue la Venecia americana.
>
> (A nauseating smell floated in the city, cats, dogs, sparrows and rats turned up dead on the streets, in basements, in courtyards, on terraced roofs and in stores' backrooms. Only the rivers of black water and the liquid garbage dumps flowed with pungent stench, vile reminders of what was once the American Venice.)[39]

The water canals are replaced by rivers of polluted water and waste.

The futuristic sceneries depicted in Aridjis's novels are grounded in the hydraulic challenges that Mexico City has been facing for over a decade, which themselves originate in the violence and destruction of Spanish colonisation. As Joel Simon notes in *Endangered Mexico* (1997):

> If Mexicans want to assign blame for their current water woes, they might as well look to Hernán Cortés. Cortés marveled at Tenochtitlan's beauty but what he inherited at the end of the two-year campaign was a pile of rubble. Tenochtitlan was sacked, burned annihilated. The Spanish siege specifically targeted the hydraulic infrastructure. The dikes were dismantled to make room for the Spanish brigantines; the aqueducts were destroyed in order to deprive the city of fresh water; the canals were filled in to allow passage for the Spanish cavalry. ... After three centuries of abuse, the valley's hydrology had been permanently and irreparably damaged.[40]

Similarly, the decaying rural landscape portrayed in Roumain's *Gouverneurs* is a result of the plantation economy in Haiti, which required mass deforestation, and has been affecting the island's water cycles ever since. Roumain's novel was written in 1944 and Haiti has only faced increasing catastrophes since then, including droughts in the 1970s and the 2010 earthquake. As Martin Munro notes "nothing could have prevented the earthquake itself, but human, historical, and social forces were to large extent responsible for the terrible scale of destruction and the great loss of life."[41] Munro acknowledges the legacy of imperialism and uneven development in Haiti, from French colonisation to the period of American occupation, which have affected the country not only environmentally, but also socio-politically and economically. At the beginning of Roumain's novel, Bienaimé, Manuel's father, observes the ecopocalyptic scenery around him, where an extreme drought after deforestation has consumed the landscape:

> Les érosions ont mis à nu de longues coulées de roches: elles ont saigné la terre jusqu'à l'os. Pour sûr qu'ils avaient eu tort de déboiser. Du vivant encore de défunt Josaphat Jean-Joseph, le père de Bienaimé, les arbres poussaient dru là-haut. Ils avaient incendié le bois pour faire des jardins de vivres.
>
> (Erosions had lay bare long slide of rock and bled the earth to the bone. They surely had been wrong to clear the forest. In the time of the deceased Josaphat Jean-Joseph, Bienaimé's father, trees once grew thick up there. They had burned the woods to plant food.)[42]

Hence, in addition to the mass deforestation caused by the plantation economy on the island, many parts were also cut down to create space for agricultural fields, cattle enclosures and settlements.

Bienaimé then remembers the villagers' *coumbite* (a collective agricultural effort organised during harvest time amongst neighbouring labourers) to plant these new fields, recalling the hard work undertaken by the

> travailleurs de la terre qui savaient qu'ils ne pourront porter un morceau à la bouche s'ils ne l'ont extrait du sol par un labeur viril. Et la terre avait répondu: c'est comme une femme qui d'abord se débat, mais la force de l'homme, c'est la justice, alors, elle dit: prends ton plaisir…
>
> (workers of the land who knew that they would not be able to eat if they do not extract the sustenance from the soil by manly labour. And the earth had answered: it was like a woman who first struggles, but then, as man's strength is justice, she says: enjoy yourself…)[43]

The use of "manly" as an adjective to describe the physical work on the fields, and the comparison of the earth with a woman being raped, reflects the feminisation of the Americas and the Caribbean in early colonial discourse. Indeed, even the name "America" is a feminised derivation of

Amerigo Vespucci and there is a long history of allegorical depictions of the new land as a naked woman. This is not an isolated example in Roumain's novel,[44] and as Martin Munro argues, "Manuel frames the deforestation that has led to the long drought around a narrative of a mistreated, feminized land, naked and without protection."[45] At the end of the novel, however, it is not male mastery over the land that will resolve the drought—only the community, brought back together by Annaïse and Délira after Manuel's death, can restore water to the land. The village had been divided by a blood feud and Gervilien, the character who kills Manuel because he cannot overcome his hatred, flees after the murder, but the dying Manuel asks his mother not to avenge his death and to ensure that the community comes together, in a *coumbite*, to prepare the land and guide the water to the village: "vous avez offert des sacrifices aux loa, vous avez offert le sang des poules et des cabris pour faire tomber la pluie, ça n'a servit à rien. Parce que ce qui compte, c'est le sacrifice de l'homme" ("you offered sacrifices to loa, you offered hens' and young goats' blood to make rain fall, it was useless. Because what matters is man's sacrifice").[46]

The feminisation of the land is also present in Ardjis's novels, which likewise present an additional layer of meaning in relation to the early colonial rhetoric. Indeed, the earth and women are similarly associated through, on the one hand, the city's streets and monuments which can metaphorically be interpreted as scars of colonialism on the land and, on the other hand, with the treatment of the abducted and raped young women in the novels. The names of the streets and establishments in *La leyenda* and *En quien piensas* are often derived from unexpected historical figures or events. Indeed, as Miguel López-Lozano explains:

> Both novels point to how the historic landmarks of the two cities function as a parody of nationalism as they pay homage to the dubious deeds and heroes of post-revolutionary governments, being named after economic crises, national disasters, and questionable heroes such as the Plaza of the Devaluated Peso, Monument to the Unknown Bureaucrat, Avenue of the Only Party of the Revolution, Monument to Hernán Cortés.[47]

There are also the "Café Colón" ("Columbus's Coffee"), the "Avenida de los Narcopolíticos" ("Avenue of the Narcopoliticians"), and, in *En quien piensas* particularly, "el Paseo de la Malinche" ("the Avenue of la Malinche") is very present, being a main street near the protagonists' home as well as a major artery of the city.[48] La Malinche, also known as Malintzin Tenepal and baptised "Marina" by Cortés upon his arrival, was his translator and later became his "wife." In Mexican folklore, she is often seen as a betrayer of her people and named *la chingada*, the violated woman. La Malinche is a symbol of the *mestizo* people representing the encounter between the Amerindian and the Spanish, and over time she has come to represent the "bad" woman in Mexican *machista* culture, as opposed to the "pure" Virgen de Guadalupe.

Ecopocalypse and Social Degradation

This is critical in the new corrupt and crime-ridden context of Aridjis's 2027 Ciudad Moctezuma, in which young women are being abducted and raped on such a major scale that a group, "Raped Anonymous," announces each evening the disappearances of the day, and sadly few of them are ever found alive.[49] Two ineffective and corrupt detectives are put on the case, and only a scapegoat is caught and attributed responsibility. However, the attentive reader understands that the person responsible for the abductions and rapes is no other than General Carlos Tezcatlipoca. Indeed, at the beginning of the novel it is said that "un rasgo característico de él era que siempre llevaba los ojos ocultos tras lentes negros y la dentadura de oro reluciente" ("a specific feature of his was that he always hid his eyes under dark sunglasses and wore shiny golden dentures").[50] When Ana Violeta, Bernarda Ramírez's daughter, goes missing, she is certain that the serial abductor and violator called the "Tláloc" is responsible. However, the young woman is able to escape and when she is found, at the same time that the scapegoat is accused and executed, she tells Bernarda and Juan that "lo único que recuerda de ese individuo es que llevaba lentes negros y dentadura de oro" ("the only thing she remembers about this individual is that he wore black sunglasses and golden dentures").[51]

The corruption in the dystopian urban space of Ciudad Moctezuma is paramount and police brutality knows no bounds. In *La leyenda*, the president of Mexico sees Tezcatlipoca as an excellent "preventive policeman":

> pues arrestaba al delincuente antes de que pensara cometer el crimen, y muchas veces lo ejecutaba antes de que hubiese tenido la oportunidad no sólo de cometerlo, sino de pensarlo. Presidente y general hablaban el mismo lenguaje de violencia indirecta.

> (since he arrested the criminal before he could think about committing the crime, and many times he executed him before the criminal had the opportunity not only to commit the crime, but also to think about it. The president and the general spoke the same language of indirect violence.)[52]

The irony in this passage is not only emphasised by the use of the singular and definite article to qualify "the criminal" and "the crime" in order to show the repetition of that type of unlawful policing, but also by the contradiction inherent to the passage: can one be considered a criminal if one has not even thought about committing a crime yet? According to dystopian and SF conventions, Ciudad Moctezuma is a police state, plagued by violence and corruption, with invasive technology. The canals of the "American Venice" are replaced by "ríos de automóviles" ("rivers of cars") and the city is overpopulated, with the crowds described as an anonymous amalgam of people: "miles de gentes andaban en el Paseo de la Malinche, más como un organismo múltiple que como cuerpos independientes, más como fantasmas del presente que como seres reales" ("thousands of people were walking along the Avenue of la Malinche, more like a multiplied organism than independent bodies, more like ghosts of the present than like real beings").[53] In the dystopian urban space, it is everyone for themselves, unlike in Roumain's novel, where the tradition of collective work, *coumbite*, creates community as all the labourers come as one to help each other survive, eat and live.

In both *La leyenda* and *En quién piensas* there appears one of the typical features of the dystopian SF city: the overwhelming presence of media. In Aridjis's work, futuristic television plays a crucial role:

> La Circe de Comunicación había convertido a los seres humanos en puercos mentales. El prójimo puto caníbal pasaba las horas y los años dormido con los ojos abiertos devorando las imágenes y los sonidos que la Circe le arrojaba a él y a su progenie sin cesar
>
> (The Circe of Communication had converted human beings into mental hogs. The fellow bloody cannibal spent hours and years asleep with eyes wide open devouring the images and sounds that the Circe incessantly hurled at them and their offspring.)[54]

Here, sensational and omnipresent media merge with Homer's epic, adding another layer to the mythical intertextuality already present with the Aztec legends that pervade and structure the narrative. The viewers, compared to cannibals, are mentally overdosing on the images of other people without actually creating any real social bonds. They are entrapped in unethical relationships of voyeurism and passivity rather than forming a community. Ironically, as it is noted in the novel, the "Circe de Comunicación había incomunicado a la gente entre sí" ("Circe of Communication had isolated people from each other").[55] This time it is not a witch transforming people into hogs and trapping them like in the *Odyssey*, rather it is their own technology that serves this purpose.

Bridge the Human/Nature Divide

In *En quien piensas*, there appears to be the possibility of a small community in the theatre in which Yo works. However, despite the fact that all these women care for each other, Yo feels isolated. For example, one day Yo notes that it is her birthday, but that the sisters do not know as they have never asked, and it is not for lack of interest in celebrations since all of their birthdays are ostensibly marked and circled on their calendar. One of the reasons for her inner loneliness is her size, indeed she has been taller than everyone else since she was 14, as is narrated in a flashback from her childhood.[56]

In one of the flashbacks, Yo tells the reader of a visit to the zoo in which she found herself face to face with a caged giraffe, the last one of her kind:

> El caso es que ante la presencia de mi doble natural sentí vértigo y apreté los párpados. Oí su voz, semejante a un ronquido, pero no comprendí lo que me dijo. Cuando abrí los ojos, noté que ella había entrecerrado los suyos, mareada, tal vez, por percibir su doble humano.
>
> (The fact is that in the presence of my natural double I felt dizzy and closed my eyes. I heard its voice, like a snore, but I could not understand what it was telling me. When I opened my eyes, I noticed that the giraffe had closed its eyes perhaps also confused to perceive its human double.)[57]

Here, Yo perceives the giraffe as her "natural double," but when the animal attempts to communicate with her, she cannot understand it. Interestingly, in another flashback Yo describes being physically unable to scream, and all she can produce when trying are snores ("ronquidos").[58] This furthers the association between Yo and the giraffe, while emphasising their inability to have a voice in the narrative. Others also appear to perceive Yo and the giraffe's similarity since she realises that a crowd has gathered around the two of them.[59] Yo appears bothered by the fact that onlookers associate her with the giraffe, thus demonstrating the difficulties and challenges of "moral extensionism" in environmental ethics. Indeed, Yo later describes herself as an urban animal, uprooted from nature, who breathes contaminated air and drinks polluted water, and although her first impulse is to identify with nature, societal conventions mean she feels insulted when associated with the animal. As Plumwood notes, the dualisms on which we base our understanding of identity and difference relegate animality and the natural to the world of otherness, reflecting the othering that occurs in exclusionary discourse such as colonialism, racism and sexism.[60] Yet the instinctive impulse to identify with nature, which is curbed by social standards, is represented by the novels, I argue, as a way to create more ethical relations with nature—relations which acknowledge that climate change is not merely destruction of the earth, but of all of us, since we are all part of one ecosystem.

This dynamic can also be seen in Roumain's novel. Gervilien's hands, for example, are described as a series of roots, while his hair is said to rest on his forehead like small bushes.[61] Likewise, when Manuel meets Nérestant, the former is so impressed by his stature that he thinks:

> Quel bûcheron il faudrait pour ébrécher et abattre un tel homme songeait Manuel le regardant venir. ... Il offrait sa main gigantesque. Manuel la prit. Une force terrible dormait dans ces doigts épais et rugueux comme l'écorce.
>
> (What a lumberjack it would take to chip and cut down such a man, thought Manuel as he saw him walk towards him. ... He offered his hand. Manuel took it. A terrible strength slept in his fingers rough and coarse like bark.)[62]

Moreover, Annaïse tells Manuel that, as with water, one needs to search deeply in his words to find their meaning.[63] The association of human with nature is further reinforced by Manuel who sees the land as a part of himself: "je suis ça: cette terre-là, et je l'ai dans le sang. Regarde ma couleur: on dirait que la terre à déteint sur moi et sur toi aussi" ("I am this: this earth, this soil, and it is in my blood. Look at the colour of my skin: one could say that the earth has run on me and on you too").[64]

Conclusion

The difficulty of understanding how completely climate change will affect us, just as it will affect other living organisms and beings on earth, is challenged in these works by Roumain and Aridjis. In these novels, environmental and social deterioration are interlinked and cannot be addressed separately.

Because of this, the concept of environmental ethics must be supplemented by a postcolonial environmental ethics, particularly in the case of countries that have been impacted by imperialism and more recent exploitative practices, which keep them in states of extreme poverty and uneven development. Advocating more ethical relationships amongst people and with nature is a way to counter, or at least slow down, the effects and consequences of climate change brought about by both colonialism and neo-liberalism. Many places today are suffering the consequences of industrialisation: centuries of environmental abuse have led to more devastating floods, hurricanes, cyclones and earthquakes, while animal and plant species are disappearing every day. However, as Doug Henwood explains: "I keep reminding myself that recovering a utopian sensibility is about the most practical thing we could do right now. Dystopia is for losers."[65] Ecopocalyptic visions must therefore be understood not only as exciting catastrophism to warn and frighten masses into action, but as descriptions of the current decaying state of nature and society. Yet a utopian impulse often underlies such narratives, which share a message of hope in new possibilities.

Notes

1. Yuen, "Politics of Failure", 15.
2. Yuen, "Politics of Failure," 20.
3. Yuen, "Politics of Failure," 36.
4. See "Heavy rain, hailstorms destroy crops in north India," *Times of India*, March 17, 2015, https://timesofindia.indiatimes.com/india/Heavy-rain-hailstorms-destroy-crops-in-north-India/articleshow/46591081.cms; Debbie Ransome "Barbuda Seeking to Define 'Normal' after a Devastating Hurricane," *Caribbean Intelligence*, n.d. https://www.caribbeanintelligence.com/content/barbuda-seeking-define-normal-after-devastating-hurricane; Mark Hand, "Puerto Rico Escapes Direct Hit, but Remains Vulnerable to Climate Change's Impacts," *Think Progress*, September 7, 2017, https://thinkprogress.org/puerto-rico-hurricane-irma-6d6a40db2591/; Julia Belluz, "It's not just Puerto Rico: 6 Other Caribbean Island Nations are in Crisis after the Hurricanes," *Vox*, October 3, 2017, https://www.vox.com/science-and-health/2017/9/26/16367410/hurricane-maria-2017-puerto-rico-caribbean-barbuda-dominica-virgin-islands-cuba-st-martin; Matt Young, "Pacific Island Nations Urge World Leaders to act as Island Expect to Sink," *News*, November 15, 2017, http://www.news.com.au/technology/environment/pacific-island-nations-urge-world-leaders-to-act-as-islands-expected-to-sink/news-story/9416ac1726d1f8d02a1ae435924e364f; "New Concerns about Drought in the Caribbean," *The New York CaribNews,* n.d., http://www.nycaribnews.com/caribbeandrought.html; Michael Mann and Andrew E

Dessler "Global heating made Hurricane Dorian bigger, wetter – and more deadly," *The Guardian*, September 4, 2019, https://www.theguardian.com/commentisfree/2019/sep/04/climate-crisis-hurricane-dorian-floods-bahamas

5. All translations are mine unless specified otherwise.
6. Yuen, "Politics of Failure," 36.
7. Watts, *West Indies*, 3.
8. I use this term as it was coined and defined by Graham Huggan and Helen Tiffin: "A re-imagining and reconfiguration of both the nature of the human and the place of the human in nature—that is, a postcolonial environmental ethic—necessitates an investigation of the category of the 'human' itself, and of the multiple ways in which this anthropocentrist construction has been, and is, complicit in racism, imperialism and colonialism, from the moment of conquest to the present day." Huggan and Tiffin, "Green Postcolonialism," 6–7.
9. Indeed, as Val Plumwood notes, the exclusionary rhetoric of otherness found in colonisation and the objectifying of nature both come from a similar discourse based on dichotomies: "Human domination of nature wears a garment cut from the same cloth as intra-human domination, but one which, like each of the others, has a specific form and shape of its own. Human relations to nature are not only ethical, but also political." Plumwood, *Feminism*, 11.
10. I use the terms "moral expansionism" and "environmental ethics" here as discussed by Christine E. Gudorf and James E. Huchingson: "To be true to its object, environmental ethics must expand our circle of moral standing to allow for the inclusion of other animals, plants, and systems of plants and animals, not to mention mountains and rivers. Commitment to the project of this 'moral extensionism' is the fundamental challenge and a distinguishing feature of environmental ethics." Gudorf and Huchingson, *Boundaries*, 8–9.
11. Plumwood, *Feminism*, 2.
12. Skrimshire, "Curb Your Catastrophism."
13. Roumain, *Gouverneurs de la rosée*, 7.
14. Roumain, *Gouverneurs de la rosée*, 20–21.
15. Roumain, *Gouverneurs de la rosée*, 198–199.
16. Hoffmann, "Complexité linguistique," 156.
17. López-Lozano explains that "the Aztecs held that the Fifth Sun was marked for destruction as well, this time by earthquake and famine, and just as in the case of the previous four, after its destruction, a new stronger, sun would emerge." López-Lozano, *Utopian Dreams*, 177.
18. Brundage, *Fifth Sun*, 27.
19. Aridjis, *Apocalipsis con figuras*, 385.
20. Aridjis, *Apocalipsis con figuras*, 269.

21. Aridjis, *La leyenda de los soles*, 38.
22. Aridjis, *La leyenda de los soles*, 39.
23. Aridjis, *La leyenda de los soles*, 39–40. As Burr Cartwright Brundage explains, "the name Tezcatlipoca has generally been translated as Smoking Mirror. … Tezcatlipoca was said to have stolen this mirror and secreted it for a time, thus withholding relief from a serious famine then in progress. But even more sinister [are the] powers of the magic mirror in fact, the eerie Tezcatlipoca carried the title Tezcatlanextia, He Who Causes Things to Be Seen in the Mirror. Thus, Tezcatlipoca's mirror was one of his most trenchant weapons; with it, if he so desired, he could undermine any truth." Brundage, *Fifth Sun*, 81. The idea that he can undermine any truth is crucial to the narrative as he is always the opposite of what he appears to be: he is a deity, not just the Chief of Police, but he is also a corrupt official, a murderer; he kills his sister and most likely he has also killed some of the young women that he regularly abducts and rapes.
24. Aridjis, *La leyenda de los soles*, 69–72.
25. Aridjis, *La leyenda de los soles*, 11.
26. Aridjis, *La leyenda de los soles*, 198.
27. Aridjis, *En quién*, 14, 109, 135, 159, and 169.
28. Aridjis, *En quién*, 59.
29. Aridjis, *En quién*, 29.
30. For a discussion of these female characters and ecofeminism, see López-Lozano, *Utopian Dreams*, 175–229.
31. Aridjis, *En quién*, 273.
32. Aridjis, *En quién*, 48.
33. And that is not only a fictional element concerning Mexico City, as Joel Simon states in his work: "Air pollution receives so much attention because it is so obvious. It is everywhere and its effects are immediate. The water threat is long term—and it takes a trained eye to see the damage. The only visible evidence that the city is running out of water is the fact that it is sinking." Simon, *Endangered Mexico*, 60.
34. Aridjis, *Le leyenda de los soles*, 164.
35. Ghosh, *Great Derangement*, 72, 84.
36. In fact, Mexico was devastated by earthquake in 1985, Aridjis's novels are both published ten years later, and thus the presence of such a major catastrophe can be read between the lines of this text. As Joel Simons explains, while the catastrophe might not have been man-made, it is certainly related to years of environmental abuse: "Certainly the earthquake was a 'natural disaster'. The impetus was a cataclysmic event that could not have been controlled or predicted. But a great deal of tragedy was also manmade, a result of centuries of environmental abuse in the Valley of Mexico." Simon, *Endangered Mexico*, 88–89.
37. Aridjis, *Apocalipsis con figuras*, 139.

38. Aridjis, *Apocalipsis con figuras*, 377
39. Aridjis, *La leyenda de los soles*, 19.
40. Simon, *Endangered Mexico*, 64–69.
41. Munro, "Introduction," 1.
42. Roumain, *Gouverneurs de la rosée*, 9
43. Roumain, *Gouverneurs de la rosée*, 9
44. Roumain, *Gouverneurs de la rosée*, 18–19.
45. Munro, "Disaster Studies," 513.
46. Roumain, *Gouverneurs de la rosée*, 164.
47. López-Lozano, *Utopian Dreams*, 206.
48. Aridjis, *La leyenda de los soles*, 141; Aridjis, *En quién*, 115.
49. Aridjis, *La leyenda de los soles*, 21.
50. Aridjis, *La leyenda de los soles*, 25.
51. Aridjis, *La leyenda de los soles*, 178.
52. Aridjis, *La leyenda de los soles*, 26.
53. Aridjis, *La leyenda de los soles*, 18, 46.
54. Aridjis, *En quién*, 176.
55. Aridjis, *En quién*, 177.
56. Aridjis, *En quién*, 101.
57. Aridjis, *En quién*, 87.
58. Aridjis, *En quién*, 98.
59. Aridjis, *En quién*, 88.
60. Plumwood, *Feminism*, 42–43.
61. Roumain, *Gouverneurs de la rosée*, 48.
62. Roumain, *Gouverneurs de la rosée*, 158.
63. Roumain, *Gouverneurs de la rosée*, 115.
64. Roumain, *Gouverneurs de la rosée*, 70.
65. Henwood, "Dystopia is for Losers," xv.

Works Cited

Aridjis, Homero. *Apocalipsis con figuras: El hombre milenario*. Mexico City: Taurus Pensamiento, 1997.

———. *En quién piensas cuando haces el amor*. Mexico City: Alfaguara, 1995.

———. *La leyenda de los soles*. Mexico City: Fondo de Cultura Económica, 1993.

Brundage, Burr Cartwright. *The Fifth Sun: Aztec Gods, Aztec World*. Austin: University of Texas Press, 1979.

Ghosh, Amitav. *The Great Derangement: Climate Change and the Unthinkable*. Chicago: The University of Chicago Press, 2016.

Gudorf, Christine E., and James E. Huchingson. *Boundaries: A Casebook in Environmental Ethics*. Washington, D.C.: Georgetown University Press, 2010.

Henwood, Doug. "Dystopia is for Losers." In *Catastrophism: The Apocalyptic Politics of Collapse and Rebirth*, edited by Sasha Lilley, David McNally, Eddie Yuen, and James Davis, ix–xv. Oakland: Specter, 2012.

Hoffmann, Léon-François. "Complexité linguistique et rhétorique dans *Gouverneurs de la Rosée* de Jacques Roumain." *Présence Africaine* 2, no. 98 (1976): 145–161.

Huggan, Graham, and Helen Tiffin. "Green Postcolonialism." *Interventions* 9, no. 1 (2007): 1–11.

López-Lozano, Miguel. *Utopian Dreams, Apocalyptic Nightmares: Globalization in Recent Mexican and Chicano Narrative*. West Lafayette: Purdue University Press, 2008.

Munro, Martin. "Disaster Studies and Cultures of Disaster in Haiti." *French Studies: A Quaterly Review* 69, no. 4 (2015): 509–518.

———. "Introduction: Fall and Rise." In *Haiti Rising: Haitian History, Culture and the Earthquake of 2010*, edited by Martin Munro, 1–6. Liverpool: Liverpool University Press, 2010.

Plumwood, Val. *Feminism and the Mastery of Nature*. London: Routledge, 1993.

Roumain, Jacques. *Gouverneurs de la rosée*. Paris: Zulma, 2013.

Simon, Joel. *Endangered Mexico: An Environment on the Edge*. San Francisco: Latin American Bureau, 1997.

Skrimshire, Stefan. "Curb your Catastrophism." Last modified July 5, 2008. https://www.redpepper.org.uk/curb-your-catastrophism/.

Watts, David. *The West Indies: Patterns of Development, Culture and Environmental Change since 1492*. Cambridge: Cambridge University Press, 1987.

PART III

Postcolonial Ethics

CHAPTER 8

Postcolonial Science Fiction and the Ethics of Empire

Bill Ashcroft

Introduction

Postcolonial science fiction has provided a platform for radical new critiques of colonialism and such critiques begin with the genre of science fiction itself.[1] In *Postcolonialism and Science Fiction*, Jessica Langer notes the scarcity of postcolonial science fiction as "the elephant in the room" or more precisely, in Nalo Hopkinson's words, the "elephant-shaped hole" in the room.[2] Postcolonial literatures and science fiction seek alternate futures for the human race, both look beyond the joint nightmare of colonial modernity, both are profoundly involved in future thinking, and both offer a clear platform for the utopian. This vision almost inevitably involves an ethical examination of imperialism itself.

John Rieder's *Colonialism and the Emergence of Science Fiction* exposes the extent to which the images of the inhuman alien and of the distant planet ripe for settlement reprise the motifs of colonial demonisation and invasion that drove the imperial enterprise. The twin myths of colonialism, the native stranger and the strange land—which offer both a utopian

B. Ashcroft (✉)
School of the Arts and Media, University of New South Wales,
Sydney, NSW, Australia

Z. Kendal et al. (eds.), *Ethical Futures and Global Science Fiction*,
Studies in Global Science Fiction,
https://doi.org/10.1007/978-3-030-27893-9_8

destination and a chance to create a utopia by subduing them—are also the twin myths of science fiction. But rather than shy away these from colonial tropes,

> these twin giants of the science fiction world, postcolonial science fiction hybridizes them, parodies them and/or mimics them against the grain in a play of Bhabhaian masquerade. The figure of the alien comes to signify all kinds of otherness, and the image of the far-away land, whether the undiscovered country or the imperial seat, comes to signify all kinds of diaspora and movement, in all directions. Their very power, their situation at the centre of the colonial imagination as simultaneous desire and nightmare, is turned back in on itself.[3]

Indeed, an observation of the range of literature that calls itself postcolonial science fiction perpetually hovers around these two objects of critique, sometimes more concerned with writing back to the colonialist orientation of the genre than speculating on the utopian possibilities of places distant in time and space. But the richest examples of the genre raise both utopian possibilities and problematic philosophical questions about the species and the nature of human life itself.

Octavia Butler's *Xenogenesis* trilogy (1987–1989), now published as *Lilith's Brood* (2000), appears to reproduce the functions of imperial world construction in the alien Oankali species, who sustain themselves by "trading" genetic material with other species throughout the universe. The human resisters' insistence on biological and reproductive independence plays out some familiar themes of anti-colonial discourse. Yet what sets Butler's work apart is the haunting ambivalence that attends the ethical dimensions of the Oankali/human encounter. Ethical issues around choice, colonial dominance, racial difference, cloning, interspecies ethics and bioethics rehearse the dynamics of colonial contact in a futuristic setting but with less easily answered questions. Because the imperial tendencies of science fiction are based on human dominance—dominance of the alien other, of the alien space, even the building of a galactic empire—Butler's radical table turning, in which the humans are more dominated than dominating, allows a more nuanced contemplation on the ethics of power. Humans have the opportunity for genetic continuation by interbreeding with another species, and consequently the issue of the posthuman raises significant philosophical questions about the nature of life and the utopian potential of biological transformation. If race is a construct of racism rather than a genetic reality, what does the prospect of species hybridisation say about the integrity of human life?

Fundamentally, the examination of ethics in postcolonial science fiction follows the trajectory of such considerations in historical imperialism and colonialism. For some, the concept of the ethics of empire is a non-sequitur and in December 2017 a controversy broke out over the establishment of the Ethics and Empire Project in the McDonald Centre of Oxford University. Under the leadership of Professor Nigel Biggar, the principal purpose of the project was to examine the history of ethical critiques of "empire," to test these against the "historical facts," and thus to "develop a nuanced and historically intelligent Christian ethic of empire" which would enable "a morally sophisticated negotiation of contemporary issues such as military intervention for humanitarian purposes in culturally foreign states."[4]

An open letter signed by 59 Oxford scholars who "work on histories of empire and colonialism and their after-effects," rejected this project's legitimacy, particularly the "balance sheet" view of ethics in which

> the British empire's abolition of the slave trade stands simply as a positive entry in a balance-book against (for example) the Amritsar massacre or the Tasmanian genocide. Abolition does not somehow erase the British empire's own practice of slavery and the benefits it continued to reap from the slave trade long after it ended—such as railway investments in the UK or cotton imports from the US South. Nor can historians accept the simple claim that imperialism "brought order" without examining what that actually meant for those subject to it.[5]

Despite the project's attempts at a nuanced view of imperial ethics, the "ethical" examination of empire revolves around a tiresomely persistent question: Is colonialism justified if it improves the lot of the colonised? That is, do the ethics of imperialism rely on results rather than principles? Of course, such a question assumes an enormous degree of license, centring on issues of justice, freedom and morality and the assumption that only through colonisation could colonial states enter modernity and international community. To postcolonial scholars this question is risible. Does the building of the Indian rail system compensate for colonial oppression and alienation, the destruction of structures of governance and the withholding of food in WWII which led to widespread and catastrophic famine?

In a trenchant article written in 1931 by Ben Azikiwe, the extent and depravity of the colonial benefit myth is laid bare:

> Jules Ferry, in a speech made in the French Chamber of Deputies in 1884, said that superior races had a right over inferior races because it is their duty to civilize them. Harris approves the claim that great nations are destined to rule the earth. On the other hand, Dr. Paul Rohrbach is more radically inclined. He maintains that so far as progress is concerned, primitive peoples have no right which the white man could respect. Thus he excuses the right to exploit the weaker races on the camouflage that they are benefited by the amenities of western civilization. But Lord Lugard is more reactionary in his imperialistic ideals. He views imperialism as an economic necessity. In *The Dual Mandate in British Tropical Africa*, quoted by Oldham, he says that since raw materials are essential to industrial society, civilized humanity has the right to develop undeveloped resources.[6]

These are sentiments with which historians of imperialism are familiar and they underlie contemporary suggestions that African countries should be recolonised. An interesting reflection of the difficulty of anti-colonial responses in the 1930s was Azikiwe's suggestion of "the dual mandate principle," by which colonial and native administrations would have equal power, a principle that fails to support his fascinating analysis with an active anti-colonialism.[7]

Arjun Appadurai, commenting on Biggar's Empire and Ethics project, questions the rise of ethics studies themselves:

> But the most serious issue here is not the absurdity of the new centre's "balance sheet" approach to historical periods and institutions, which is rightly derided by its opponents. It is the extension of the rubric of ethics to patently violent, brutal and exploitative regimes and practices. The troubling tectonic process here is not the ethics of Empire but the growing Empire of ethics.[8]

Appadurai's objection is that despite good work on ethics in social sciences, political theory and history, the turn to ethics has also opened the gates to apologists of various stripes. However, a long-standing interest in ethics in literature has revealed the capacity of literary interventions such as Octavia Butler's to provide a nuanced, complex and thought-provoking series of ethical dilemmas.

The ethical conundrum occurs in a genre that extends, unselfconsciously, the ethical problems of imperialism, a problem summed up in the concept of "unenforceable obligations." According to Hoy, "The ethical resistance of the powerless others to our capacity to exert power over them is therefore what imposes unenforceable obligations on us. The obligations are unenforceable precisely because of the other's lack of power."[9] However, does the idea of "unenforceable obligations" extend the problem by overlooking the agency of the "powerless" other? It is within this dispensational binary that the ethics of postcolonial science fiction emerge. Are ethics given or taken? Are ethics a function of power or resistance, that is, are they more constraining on the powerful than the powerless? Do ethics even apply to intelligent non-human species? Do such species have "unenforceable obligations"? May their conceptions of justice mean that their relations with humans are completely just and ethical? How might postcolonial science fiction address these questions?

The prime achievement of postcolonial literatures has been to provide a space for the voices of the colonised, marginalised and oppressed. This has led to an enormously powerful exercise of transformative agency as postcolonial writers have taken the language of the colonised and transformed it into a vehicle of self-representation. While the issue of resistance has been prominent in postcolonial writing, a more subtle and recurrently overlooked dimension of such writing has been the presence of a comprehensive utopianism that imagines a different world. Such transformative utopianism is a prominent characteristic of the genre of postcolonial science fiction, which goes beyond the idea of ethics as "unenforceable obligation" to a utopian ethics of possibility that by its very nature displays ethical rights as *taken* rather than *given*. The richest examples of the genre raise both utopian possibilities and problematic philosophical questions about the species and the nature of human life itself.

Above all, postcolonial science fiction offers, potentially, a revivified vision of Ernst Bloch's concept of *Heimat*, a concept present in all postcolonial literatures but radically emphasised in science fiction. Exactly where Butler stands in this environment is an interesting question because the literary mode enhances the ambivalence of the ethical landscape in which humans are the powerless and the "ethical" considerations of the Oankali species may look very different from human ethics. Furthermore, the literary mode may generate affective responses in which ethical identification may extend to the powerful alien other rather than to the dangerous and aggressive human society. This certainly seems to be the case as the narrative of the trilogy proceeds, where this identification becomes stronger as the struggle of the Oankali for survival is made clear. The *Xenogenesis* series demands that we navigate what amounts to a complex series of ethical dilemmas that hinge on the "benefits of colonialism" myth but turn it upside down because the colonisers are extra-terrestrial (and arguably morally superior) and not necessarily accountable to the principles of human ethics.

Octavia Butler's *Xenogenesis*

The trilogy opens on an Oankali "ship" in which the central character, Lilith, has been kept in suspended animation for over 250 years since the humans' destruction of the earth in nuclear war. "Humanity in its attempt to destroy itself had made the world unlivable. … She had considered her survival a misfortune—a promise of a more lingering death" (15).[10] She has been saved by the Oankali because they live to trade genetic material with other species, indeed they are compelled to do this for their survival. They value biological diversity above everything, and they have come to an earth on which those humans surviving nuclear devastation are either sterile or give birth to horrific offspring. The trade the Oankali offer is to make humans fertile and whole and physically far superior in return for the opportunity of mating with them. The children thus formed will live for centuries, be able to heal themselves but will no longer be fully human beings, but in fact posthuman. We can see already the ways in which colonial ethics are deeply complicated, in a situation where the Oankali have not only "improved" human life but have also enabled human beings to survive. Here the problem of colonial ethics emerges: are ethics determined by favourable consequences or fundamental principles, even if the principle consequence is the survival of the human race?

The Oankali family system involves the mating of male and female, whether human or Oankali with the intermediation of an ooloi, a sexless being capable of astonishing genetic feats. The progeny of such mating are called "constructs," whether from human or Oankali parents, and have all the benefits of Oankali survival and longevity. Because of this habit of genetic mutation the fate of the Oankali is to travel the universe looking for trade:

> We, Oankali and construct, we're space-going people, as curious about other life and as acquisitive of it as Humans were hierarchical. Eventually we

> would have to begin the long, long search for a new species to combine with to construct new life-forms. Much of Oankali existence was spent in such searches. (531)

Lilith has been especially chosen by the Oankali for what is, to them, a fascinating genetic condition—her cancer. They have saved Lilith but at what ethical and psychological cost? A common response to the situation is: "Although the Oankali are not responsible for having separated Lilith from her family, they behave like slave owners who viewed the breeding of their slaves as their prerogative."[11] Indeed, "She did not own herself any longer. Even her flesh could be cut and stitched without her consent or knowledge" (6). But the problem is that they have restored the world and are infinitely more advanced than humans: "How can you teach us to survive on our own world?" she asks, "How can you know enough about it or about us?" "How can we not?" replies the Oankali, "We've helped your world restore itself. We've studied your bodies, your thinking, your literature, your historical records, your many cultures. … We know more of what you're capable of than you do" (32). The Oankali are fascinated by genetic complexity and regard Lilith's cancer as "beautiful" for its capacity for genetic multiplication (22). They can and do cure the cancer in humans but employ the genetic material for other forms of life-saving mutation. "Your bodies are fatally flawed. The ooloi perceived this at once. At first it was very hard for them to touch you. Then you became an obsession with them. Now it's hard for them to let you alone" (38).

Butler complicates the colonial benefits thesis in a number of ways, not the least of which is the utter moral inferiority of humans to the Oankali. Human beings suffer from what the Oankali call "the human contradiction," that is, they are highly intelligent, but hierarchical:

> "You are hierarchical. … It's a terrestrial characteristic. When human intelligence served it instead of guiding it, when human intelligence did not even acknowledge it as a problem, but took pride in it or did not notice it at all. … That was like ignoring cancer. I think your people did not realize what a dangerous thing they were doing." (39)
>
> Humans had evolved from hierarchical life, dominating, often killing other life. Oankali had evolved from acquisitive life, collecting and combining with other life. (564)

This hierarchical nature makes humans aggressive and domineering. The humans released from suspended animation who choose to live on earth as humans, even though they are sterile and disease ridden, living in villages at war with one another. This aggressive hierarchical nature is underpinned by their fear of difference "Oankali crave difference. Humans persecute their different ones." (329)

> In addition to this, humans are duplicitous:
>
> Humans said one thing with their bodies and another with their mouths and everyone had to spend time and energy figuring out what they really meant. And once we did understand them, the Humans got angry and acted as though we had stolen thoughts from their minds. (548)

The Oankali habit of always telling the truth (or saying nothing) is a great strength, "as though catching them in lies would make them vulnerable," thinks Lilith, "As though it would make the thing they intended to do less real, easier to deny" (59). The innate Oankali truthfulness and utter communality of decision making are predicated on their biology:

> "We acquire new life—seek it, investigate it, manipulate it, sort it, use it. We carry the drive to do this in a minuscule cell within a cell—a tiny organelle within every cell of our bodies. Do you understand me? … That's the way we perceived your hierarchical drives at first." He paused. "One of the meanings of Oankali is gene trader. Another is that organelle—the essence of ourselves, the origin of ourselves. Because of that organelle, the ooloi can perceive DNA and manipulate it precisely." (41)

They are vegetarian, because killing sentient beings "was not simply wasteful to the Oankali. It was as unacceptable as slicing off their own healthy limbs. They fought only to save their lives and the lives of others. Even then, they fought to subdue, not to kill" (564). Despite their apparent enslavement of humans, no human is forced to interbreed or prevented from living on earth. Where humans are terra-centric the Oankali are nomadic, deterritorial, in love with genetic difference and complexity and willing to travel anywhere for it. So, in terms of their relationship to life, humans appear profoundly less ethical than the Oankali, yet the power relationship between the two species foregrounds the issue of slavery.

Slavery

In virtually every respect slavery is a non-starter in the ethics stakes. It is recognised as so obviously immoral that there can be no argument about any putative benefits. Throughout history, but particularly in the industrial level, empire-driven enslavement of Africans over at least three centuries, slavery has been characterised by mental, psychological and physical abuse, hard labour, underfeeding and a wealth of inhumane practices leading to early death. But what if there were forms of slavery beneficial to, or perceived to be beneficial to, the enslaved? (We only have to think of contemporary capitalist economies to see how this might work). What if the enslavement did not involve obvious abuse, forced labour or inhumane practices, nor even the overt denial of choice? The ethical issue of slavery might then become much more difficult to resolve. It might require much more serious contemplation if all that was left were questions about the value of "humanness." Questions of enslavement might then prompt us to question whether, and to what extent, human agency and human freedom outweigh a long, healthy and productive life.

Butler herself is an African American whose citizenship rests upon a history of slavery.[12] Like colonisation, slavery has been a prominent theme in science fiction. In the main, these stories have been written by white male authors,[13] so the inclusion of something like a "meta slavery" theme by a black woman is significant. Equally significant is the ethical complexity she injects in a situation in which the subjects, beginning with Lilith, are given a choice. To see the trilogy as an overt critique of slavery and interbreeding is far too simplistic. What Butler produces is a series of ethical conundrums, enlivened by the moral differences between humans and Oankali, by the ostensible freedom of choice given the humans and the fact that the Oankali offer the humans a chance of genetic survival. By reducing the obvious moral depravity of slavery as humans have practiced it, the deeper ethical issue of making use of other species can be addressed. In particular, the concept of free will becomes central, but while humans are allowed choice and are not overtly forced into breeding, the principle of individual desire applies. No matter how beneficial the Oankali relationship is to humans, the narrative suggests that the desire for freedom will always dominate human consciousness. This then becomes the ground on which the ethics of colonialism, and the use of the colonised, however benign, may be approached.

This issue of survival couples meta-slavery with the theme of miscegenation. Lilith is not intended to serve the Oankali, except in one respect. They can produce everything they need, except essential genetic variation supplied by other species. But her perception of this is central to the meta-slavery theme:

> She was intended to live and reproduce, not to die. Experimental animal, parent to domestic animals? Or … nearly extinct animal, part of a captive breeding program? Human biologists had done that before the war—used a few captive members of an endangered animal species to breed more for the wild population. Was that what she was headed for? (60)

Although she is never fully resigned to her choice, and feels something of a traitor to humanity, she becomes more accepting after a Mars colony is established where human "resisters"—healed and made fertile by the Oankali—may continue to build up a human population. Although Lilith has chosen to mate with the Oankali, and to lead other humans to Earth, her relationships are deeply conflicted:

> Yet she did think of Ahajas as a friend—Ahajas, Dichaan, Nikanj … But what was she to them? A tool? A pleasurable perversion? An accepted member of the household? Accepted as what? Round and round. It would have been easier not to care. Down on Earth, it would not matter. The Oankali used her relentlessly for their own purposes, and she worried about what they thought of her. (179)

Lilith and her human mates, Joseph (who dies but whose seed is used to produce her first child) and Tino, have several children but the nature of their conception means that they are all "construct" and eventually she even gives birth to ooloi children, something the Oankali communal mind regards as both unprecedented and dangerous. An underlying genetic and social transformation is taking place that raises ethical questions of its own. Nevertheless, the ethical conflict centres on Lilith and her feelings about the survival of the human race. Despite having freely chosen to lead other humans to earth and to enable the Oankali to mate with those who are willing, she still feels torn.

Ethics of Cloning and Interbreeding

The trilogy leaves one question unasked: what, after all, is the value of such a volcanic human existence, given the moral flaws that are displayed so obviously by the various resister villages? What is unethical, after all, about cloning? Why is it so important that the human race continue in its present form? We don't see arguments against the social and moral transformation of the human species, nor its generally improved health and extended life span, which are in a sense, genetic transformations. Only in the most rabid nativist quarters do we see arguments against racial interbreeding. So, what are the ethical objections to cloning and interbreeding? These are questions the trilogy raises, but only answers with the human desire for independence. When Lilith gives birth to a child with dead Joseph's sperm, facilitated by her ooloi, Nikanj, her response is predictably anthropocentric:

> "But it won't be human," she whispered. "It will be a thing. A monster."
>
> "You shouldn't begin to lie to yourself. It's a deadly habit. [says Nikhanj] The child will be yours and Joseph's. Ahajas' and Dichaan's. And because I've mixed it, shaped it, seen that it will be beautiful and without deadly conflicts, it will be mine. It will be my first child, Lilith. First to be born, at least. Ahajas is also pregnant."
>
> "Ahajas?" When had it found the time? It had been everywhere.
>
> "Yes. You and Joseph are parents to her child as well." It used its free sensory arm to turn her head to face it. "The child that comes from your body will look like you and Joseph." (247)

What the Oankali offer is not just long life, freedom from disease and the ability to procreate, but continuation of the human species, though in mutated form. One of the resister villages leaves a book behind in which they document their choice(s):

> Iriarte said, gesturing [at the book] … "The Oankali invited them again to join the trade villages, and they voted to do it. To have Oankali mates and kids. They say, 'Part of what we are will continue. Part of what we are will go to the stars someday. That seems better than sitting here, rotting alive or dying and leaving nothing. How can it be a sin for the people to continue?'" …
>
> Damek studied the book … "But there are two writers. One says 'We're joining the Oankali. Our blood will continue.' But the other one says the

> Oankali should be killed—that to join with them is against God. I'm not sure, but I think one group went to join the Oankali and another went to kill the Oankali. God knows what happened." (339)

Here are two radically different choices. The hard work of trying to build a life in a resister village in the Amazon jungle is not worth the grinding effort and a decision made by one group demonstrates that the desire for survival is strong but not clear-cut. Equally strong is the desire for rejection, isolation and revenge.

The ethical problems around cloning centre on the concept of producing children with particular (improved) characteristics. Ronald Dworkin, for example, argues that there is nothing wrong with the ambition "to make the lives of future generations of human beings longer and more full of talent and hence achievement." In fact, he maintains, the principle of ethical individualism makes such efforts obligatory.[14] Jürgen Habermas on the other hand suggests that cloning is an assault on children's autonomy.[15] Pointing them toward particular life choices may violate their right to choose. But Butler enters this argument at a much deeper level. The prospect of the Oankali producing a different, hybrid species questions the very notion of humanness itself. Yet it may not be the imputed attack on the integrity of the human species that is the problem but rather the removal of choice. What are we then to make of the fact that the Oankali do give humans a choice, a choice that some accept since "Part of what we are will continue"?

Is Oankali Culture Ethical?

These are some of the subtle ways in which Butler complicates the ethics of empire. By bringing together two species that are morally asymmetrical, the Oankali are far more observant of, indeed genetically compelled to obey, their concept of ethics than humans theirs. Humans are hierarchical, violent, dangerous, terra-centric and anthropocentric; the Oankali curious, acquisitive, life generating, nomadic and continually open to difference and change. The Oankali offer an indisputably better life to the human survivors of a destroyed world. So how do we relate the ethical conundrums in *Xenogenesis* with the ethics of empire? Does the imputed moral superiority of the Oankali echo the sentiments of Rohrbach above, that primitive peoples "have no right which that the white man could respect"?[16] How do we resolve the "benefits of colonialism" fallacy in a situation that is so radically beneficial to the colonised? The resolution must be through principles that supervene the potential benefits of colonisation. This appeal to principles rather than results follows the position of Kant's deontological theory in *Groundwork of the Metaphysics of Morals*, which is based on duties, obligations and rights rather than on consequences of actions.[17] While the consequences of Oankali actions are beneficial to humans they fail other principles that might even be held to apply to a different species: two of these are the principles of relational injustice and responsibility to otherness.

Relational injustice occurs when an individual is placed in an *unjustifiably* lower status in relation to another within a relationship.[18] Under this principle, any form of colonisation is wrong because it institutes relational inequality as a matter of course, preventing individuals from relating to each other as communities. Such a relationship, which occurs between groups and individuals, cannot be rectified, at least in principle, by what might be called *distributive* justice, the distribution of beneficial goods, whether, physical or psychological.

Relational injustice assumes that there is something intrinsically valuable in equal relations between individuals in a particular social situation. According to Wong, if we were asked to choose between a world where all relations are deeply egalitarian and grounded upon mutual respect and compassion, and a world where all relations are asymmetrical, with a clearly arranged status order and hierarchy and given that the benefits are equal we would intuitively choose the first (despite the Oankali's perception of humans' hierarchical nature). We could concede that the Oankali treatment of humans is kind and ethical in their own terms, since they are driven to heal and support life. Yet the world they control is relationally unjust. Why then, can't the relational injustice be offset by the benefits of Oankali dominance?

> We intrinsically feel that there are certain items that money cannot buy—for instance, relational attributes such as *genuine love, compassion, or dignity and respect by other individuals*. These are dimensions that exist independently of material benefits or individual welfare, in that they necessarily involve interactions between two or more individuals.[19]

Yet even here the trilogy makes the situation far from clear. The Oankali don't prevent human mates from loving, but once mated any physical contact becomes repulsive to them (except the touching of hair, which is dead). The sexual euphoria that binds them to the Oankali family is always mediated by an ooloi. The Oankali may offer a sense of dignity in giving choice to the humans, but the choice available is one of only three: mate with the Oankali, stay on a doomed earth prone to disease and unable to procreate or travel to Mars where a human culture can be built up. The relational injustice in this, even granted that the state of the earth is the humans' fault, lies in the removal of dignity and meaningful agency, even though Oankali ethics demands that they prevent humans from riding their Contradiction to the obliteration of the species.

The significance of human desire here lies in the somewhat surprising choice by some humans to struggle to build a human society on Mars under enormously difficult circumstances. The benefits to the humans are to continue a recognisably human life and to procreate the human race on the planet. The desire for independence and agency outweighs the material and genetic benefits available to anyone mating with the Oankali. Once established on Mars, there can be no relational injustice except that generated by the human Contradiction—the desire for hierarchy. We may wonder whether relational injustice is part of the human condition, like their fear of difference, but in the case of the colony on Mars it cannot be determined by colonialism because they occupy an empty planet.

Responsibility to otherness is a principle to which the Oankali seem at first to be committed. While one of the characteristics of humans is the fear of difference, the Oankali value it greatly. When explaining this to her construct son Akin, Lilith says:

> "Humans persecute their different ones, yet they need them to give themselves definition and status. Oankali seek difference and collect it. They need it to keep themselves from stagnation and overspecialization. If you don't understand this, you will. You'll probably find both tendencies surfacing in your own behavior." And she had put her hand on his hair. "When you feel a conflict, try to go the Oankali way. Embrace difference." (329)

The Oankali embrace of difference seems to be a markedly superior ethical choice than humans' fear of it. But craving difference might well be the antithesis of responsibility to otherness, since it is an attitude that underlies their nomadic colonisation of various worlds to prevent their stagnation. Responsibility to otherness

> refers to a need to be attentive to that which lies beyond the margins of our identity, our concepts, and our projects—that which is "other" to me or us. Especially important here is the appreciation of how the very processes of constructing identity, developing concepts, and conceiving projects necessarily generate "others."[20]

So, for all their apparent moral superiority, the Oankali's perception of otherness is not to be responsible to it but to join with it and thus make difference the "same." Yet even when we offer this stricture we can see that this principle is a profoundly human position. Otherness might include race, class and gender, but can the responsibility to otherness include a species so completely genetically other as the Oankali?

"Responsibility" in historical colonialism has been limited to the "white man's burden." As Daniel J. Castellano notes, Teddy Roosevelt had opined that "the great powers of the world had a twofold responsibility to suppress 'savagery and barbarism' and 'to help those who are struggling toward civilization.'"[21] This general thesis has been the alibi of all imperial expansion—the establishment of the British race for the benefit of the world, the need to raise those caught in darkness into the light. Much science fiction production focuses on what Ursula K. Le Guin calls "the White Man's Burden all over again" and what Gregory Benford describes as the "Galactic Empire motif": the concept of a human empire of many planets, scattered across the stars.[22] Benford suggests that this plot is a

> common, unimaginative indulgence of science fiction. There are generally no true aliens in such epics, only a retreading of our own history. This underlying structure is so common in science fiction, even now, that it is difficult to know whether we should attribute it to simple lack of imagination or to a deep unconscious need to return repeatedly to the problem.[23]

In the light of the prevalence of imperial motives in science fiction, the impact of the Oankali on human society offers a different, but no less critical view of the ethics of empire. Ultimately the Oankali feel no responsibility to otherness. Their craving for difference exists only to acquire otherness, through genetic mutation. Despite having no empire, the universe is their oyster, their genetic survival, their only consideration.

Transcultural Ethics

Ethical theory, whether human or inter-species, always assumes, at least implicitly, that ethics are determined by the unenforceable obligations of the powerful. "The obligations are unenforceable precisely because of the other's lack of power."[24] But given that the Oankali carry out their gene trade in compliance with their sense of ethics, and given that construct humans share the characteristics of human and Oankali we might conceive of a participatory option, in which the ethical considerations that arise at the meeting of cultures are a function of reciprocity and transformation, that is, can be *made to be* a function of transformation through the agency of appropriation. A distinctly postcolonial question is: can ethics be transcultural? That is, can they occupy an interstitial space, a contact zone in which two different species can be mutually transformed or at least transformable?

Although coined in the 1940s by Fernando Ortiz and transported into literary studies in the 1970s by Angel Rama, the term was popularised in postcolonial studies by Mary Louise Pratt who referred to it as "a phenomenon of the contact zone." These are social spaces where "disparate cultures meet, clash and grapple with each other, often in highly asymmetrical relations of dominance and subordination—like colonialism, slavery, or their aftermaths as they are lived out across the globe today."[25] But importantly, the contact zone is the space where *both* the powerful and the colonised experience a mutual transformation.

As with everything else concerning the Oankali, such a contact zone is at base genetic, occurring at the level of species hybridity. When talking about construct humans born through the intermediary of an ooloi,

> "They must be given more Human characteristics than Oankali-born construct males," Dichaan answered. "Otherwise, they could not survive inside their Human mothers. And since they must be so Human and still male, and eventually fertile, they must come dangerously close to fully Human males in some ways. They bear more of the Human Contradiction than any other people." The Human Contradiction again. The Contradiction, it was more often called among Oankali. Intelligence and hierarchical behavior. It was fascinating, seductive, and lethal. It had brought Humans to their final war. (442)

But there is a transcultural or trans-species space opened up by Akin, the one whose failure to bond with his sibling has sent him out among the humans. As Dichan, Akin's Oankali father, says to Tino, his human one, "He's teaching us what a Human-born male can be. There are still so few like him because we're too unsure to form a consensus—" (423). Akin is the means by which the Oankali might learn something about humans, such as their inexplicable desire for independence. "The resisters don't seem very complex—except biologically." "Yet they resist. They would rather die than come here and live easy, pain-free lives with you" (423). This is impossible for the Oankali to understand because their life, their way of being, is one of constant genetic exploration, change and multiplicity. So, while the fundamental misunderstanding between the two species is ontological it is at heart an ethical difference.

Because Akin is denied the essential bonding with his sibling he comes to live in a space between Oankali and human where he can empathise with both and this empathy has profound ethical consequences

> Who among the Oankali was speaking for the interests of resister Humans? Who had seriously considered that it might not be enough to let Humans choose either union with the Oankali or sterile lives free of the Oankali? Trade-village Humans said it, but they were so flawed, so genetically contradictory that they were often not listened to.
>
> He did not have their flaw. He had been assembled within the body of an ooloi. He was Oankali enough to be listened to by other Oankali and Human enough to know that resister Humans were being treated with cruelty and condescension. (404)

It is Akin who thinks of a solution for human resisters to continue human life. Earth is destined to die as it becomes consumed by the Oankali, but he manages to convince them to establish a colony on Mars where humans might continue to live limited though healthy and fertile lives, something that is not possible on Earth. This had a healing emotional impact on his parents Tino and Lilith:

> In one way, the Mars colony freed both my Human parents to find what pleasure they could find in their lives. In another it hadn't helped at all. They still feel guilt, feel as though they've deserted their people for aliens, as though they still suspect that they are the betrayers the resisters accused them of being. No Human could see the genetic conflict that made them such a volcanic species—so certain to destroy themselves. Thus, perhaps no Human completely believed it. (562)

When Akin communicates with the Oankali consensus through the Akjai, those Oankali who have not attempted to merge with humans, we get a clear picture of something like a transcultural ethical space, and a glimpse of what Oankali ethics might entail. The Akjai points out that the Human-born constructs "had had to learn the Oankali understanding of life as a thing of inexpressible value. A thing beyond trade" (470). Someone asks: "Could Humans be given back their independent lives and allowed to ride their Contradiction to their deaths? To give them back their independent existence, their fertility, their own territory was to help them breed a new population only to destroy it a second time" (470). This, of course, for the Oankali, would be completely unethical. "We've given them what we can of the things they value—long life, freedom from disease, freedom to live as they wish. We can't help them create more life only to destroy it" (470).

It is up to Akin to bridge the gap between Oankali and Human ethics. "Look at the Human-born among you," he told them. "If your flesh knows you've done all you can for Humanity, their flesh should know as mine does that you've done almost nothing. Their flesh should know that resister Humans must survive as a separate, self-sufficient species. Their flesh should know that Humanity must live!" (470–471). This is both a transcultural (trans-species) moment and a utopian moment. By appealing to the human born amongst the Oankali he hopes to offer a different life—on Mars—for the resister humans who would rather have freedom than long life. Remarkably, the Akjai, who has supported Akin in this consultation with the Oankali communal mind agrees with Akin. "All people who know what it is to end should be allowed to continue if they can continue" (471). This is a truly transcultural moment. The Akjai, knowing the ethical problems of allowing humans to "ride their Contradiction to their deaths" concedes the importance of independence to humans, even if it is fatal.

This transcultural space challenges the ethical principle of unenforceable obligations. The Oankali have an obligation to sustain life, the Humans an obligation to be free—a fundamentally asymmetrical relationship, enforceable on neither side. But it also offers a space of engagement that compromises the principle of relational injustice because it is a relational space. While the Oankali have the power in the relationship Akin manages to form an ethical compromise in which resister humans are given the chance to continue human society on Mars while the Oankali fulfil their obligations by ensuring the Humans will be disease free, fertile and given sole rights to the Mars colony. Ethical relationality is embodied in hybrid constructs such as Akin, who enact in themselves a degree of transcultural equality.

Conclusion

By conceiving a world in which a morally and intellectually advanced species interacts with a disappearing human race, Octavia Butler has provided a setting for a contemplation of the ethical dilemmas of interspecies interaction that might be simply dismissed in the context of historical imperialism. Most scholars would agree that the nineteenth-century imperialist claims to superiority have no ethical standing and in fact are now regarded by many as ludicrous. But by putting the situation into a science fiction context, Butler forces us to consider the ethical ramifications of the Oankali empire. Humans have the Contradiction of intelligence and hierarchy, they are aggressive, fearful, hate difference and in every way are inferior to the Oankali who are offering them continued life, health and fertility, with the freedom to choose. But the choice is a limited one and though the Oankali value genetic difference they are not responsible to otherness. Ultimately, the apparently benign rule and the generous benefits of their contact with humans comes up against principles that outweigh consequences, however beneficial. The principles of relational injustice and responsibility to otherness operate regardless of the Oankali's saving the human race. But perhaps most significant, the trilogy reveals how a transcultural contact zone, something that has become fundamental to postcolonial studies, may be seen to apply to the ethical relationship between two intelligent species. This is something that reaches into both postcolonial studies and science fiction—the prospect of a mutually agreed ethics.

Notes

1. Elements of this chapter have been adapted from Ashcroft, *Utopianism.*
2. Langer, *Postcolonialism*, 1.
3. Langer, *Postcolonialism*, 4.
4. Biggar, "Ethics and Empire."
5. "Ethics and Empire: An Open Letter from Oxford Scholars."
6. Azikiwe, "Ethics," 290.
7. Azikiwe, "Ethics," 291.
8. Appadurai, "Empire of Ethics."
9. Hoy, *Critical Resistance*, 184.
10. Page numbers provided in-text refer to the Kindle *Lilith's Brood* omnibus edition of the *Xenogenesis* trilogy, comprising *Dawn* (1987), *Adulthood Rites* (1988) and *Imago* (1989). See Butler, *Lilith's Brood.*
11. Mehan, "Teaching," 166.

12. In the first half of the nineteenth century, the largest single industry in the United States, measured in terms of both market capital and employment, was the enslavement (and the breeding for enslavement) of human beings. Over the course of the period, the industry became concentrated to the point where fewer than 4000 families (roughly 0.1 percent of the households in the nation) owned about a quarter of this "human capital," and another 390,000 (call it the 9.9 percent, give or take a few points) owned all of the rest.
13. Robert A. Heinlein ("Logic of Empire" [1941], *Farnham's Freehold* [1964], *Citizen of the Galaxy* [1957]); Thomas M. Disch (*Mankind Under the Leash* [1966]); Poul Anderson ("Margin of Profit" [1956], "The Master Key" [1964]); and Gene Wolfe ("How the Whip Came Back" [1970]). One particularly relevant example is the "Gor" series, including *Slave Girl of Gor* (1977), by John Norman, in which independent women from Earth become submissive, even eager, slaves.
14. Dworkin, *Sovereign Virtue*, 452.
15. Habermas, *Future of Human Nature*.
16. Azikiwe, "Ethics," 290.
17. See Guyer, *Kant's* Groundwork; Henry, *Kant's* Groundwork; Ferguson, "Kant."
18. Wong, "On Relational Injustice."
19. Wong, "On Relational Injustice."
20. White, "Heidegger," 81.
21. Castellano, "Rooseveltian Imperialism."
22. Le Guin, "American SF," 84; Benford, "Aliens and Knowability," 55.
23. Benford, "Aliens and Knowability," 55.
24. Hoy, *Critical Resistance*, 184.
25. Pratt, *Imperial Eyes*, 4.

Works Cited

Appadurai, Arjun. "Empire of Ethics: Studying UK's Colonial Past Through an Ethical Lens Legitimises a Slippery Slope." *Scroll.in*. May 28, 2018. https://scroll.in/article/866242/empire-of-ethics-studying-uks-colonial-past-through-an-ethical-lens-legitimises-a-slippery-slope.

Ashcroft, Bill. *Utopianism in Postcolonial Literatures*. Abingdon, Oxon: Routledge, 2017.

Azikiwe, Ben N. "Ethics of Colonial Imperialism." *The Journal of Negro History* 16, no. 3 (1931): 287–308.

Benford, Gregory. "Aliens and Knowability: A Scientist's Perspective." In *Bridges to Science Fiction*, edited by George E. Slusser, George R. Guffey and Mark Rose, 53–63. Carbondale, IL: Southern Illinois University Press, 1980.

Biggar, Nigel. "Ethics and Empire." The McDonald Centre. Accessed August 2, 2018. http://www.mcdonaldcentre.org.uk/ethics-and-empire.

Butler, Octavia E. *Lilith's Brood: The Complete Xenogenesis Trilogy*. Open Road Media, 2012. Kindle.

Castellano, Daniel J. "Rooseveltian Imperialism." *Repository of Arcane Knowledge*. 2012. http://www.arcaneknowledge.org/histpoli/roosevelt.htm.

Dworkin, Ronald. *Sovereign Virtue: The Theory and Practice of Equality*. Cambridge, MA: Harvard University Press, 2000.

"Ethics and Empire: An Open Letter from Oxford Scholars". *The Conversation*. December 20, 2017. https://theconversation.com/ethics-and-empire-an-open-letter-from-oxford-scholars-89333.

Ferguson, Benjamin. "Kant on Duty in the Groundwork." *Res Publica* 18, no. 4 (2012): 303–319.

Guyer, Paul (ed.). *Kant's* Groundwork of the Metaphysics of Morals*: Critical Essays*. Lanham, MD: Rowman & Littlefield, 2000.

Habermas, Jürgen. *The Future of Human Nature*. Cambridge, UK: Polity Press, 2003.

Henry, Allison E. *Kant's* Groundwork for the Metaphysics of Morals*: A Commentary*. Oxford: Oxford University Press, 2011.

Hoy, David Couzens. *Critical Resistance: from Poststructuralism to Post-Critique*. Cambridge, MA: MIT Press, 2004.

Langer, Jessica. *Postcolonialism and Science Fiction*. Basingstoke: Palgrave Macmillan, 2011.

Le Guin, Ursula K. "American SF and the Other." In *The Language of the Night: Essays on Fantasy and Science Fiction*, edited by Susan Wood, 83–85. London: The Women's Press, 1989.

Mehan, Uppinder. "Teaching Postcolonial Science Fiction." In *Teaching Science Fiction*, edited by Andy Sawyer and Peter Wright, 162–178. Basingstoke: Palgrave Macmillan, 2011.

Pratt, Mary Louise. *Imperial Eyes: Travel Writing and Transculturation*. London: Routledge, 1992.

White, Stephen. "Heidegger and the Difficulties of a Postmodern Ethics and Politics." *Political Theory* 18, no. 1 (1990): 80–103.

Wong, Brian. "On Relational Injustice: Could Colonialism Have Been Wrong Even if it Had Introduced More Benefits Than Harms?" University of Oxford. March 1, 2018. http://blog.practicalethics.ox.ac.uk/2018/03/oxford-uehiro-prize-in-practical-ethics-on-relational-injustice-could-colonialism-have-been-wrong-even-if-it-had-introduced-more-benefits-than-harms/.

[illegible] "The M[illegible] Centre." [illegible]
2018. http://[illegible]
[illegible]
[illegible] 2012. [illegible]
[illegible]
2012. http://[illegible]
[illegible]
Cambridge, MA: [illegible] Press, 2008.
[illegible]
[illegible] 26, 2014. https://[illegible]
[illegible]
[illegible]
[illegible]
[illegible] Indiana University Press, 2011.
[illegible]
[illegible] MIT Press, 2008.
[illegible]
[illegible] 2014.
[illegible]
[illegible]
[illegible]
[illegible]
Basingstoke: Macmillan, 2011.
[illegible]
[illegible]
[illegible]
[illegible]
[illegible]
[illegible] https://[illegible] 2018 [illegible]
[illegible]
[illegible]

CHAPTER 9

The Postcolonial Cyborg in Amitav Ghosh's *The Calcutta Chromosome*

Nudrat Kamal

Introduction

Science fiction has had a long and vexed relationship with colonialism and empire. Themes of discovering strange lands and conquering "alien" people have been a staple in the genre since its very beginning, and science fiction has often been critiqued for deploying and furthering imperial fantasies that formed the basis of the Global North's colonial (and neocolonial) project. According to Adam Roberts, science fiction serves as the "dark subconscious to the thinking mind of Imperialism," the seedy underbelly hidden beneath the rationalistic veneer given to colonial and neocolonial ideas.[1] In her work *Science Fiction and Empire,* Patricia Kerslake argues that "the theme of empire … is so ingrained in SF that to discuss empire in SF is also to investigate the fundamental purposes and attributes of the genre itself."[2] For many scholars exploring the links between colonialism and science fiction, the question is not whether science fiction engages with colonialism, but rather, how much and to what extent. According to John Rieder, science fiction as a genre emerged at the

N. Kamal (✉)
Department of Social Sciences and Liberal Arts, Institute of Business Administration, Karachi, Pakistan

Z. Kendal et al. (eds.), *Ethical Futures and Global Science Fiction*, Studies in Global Science Fiction,
https://doi.org/10.1007/978-3-030-27893-9_9

height of the imperial project, at a time when "colonialism made space into time ... and gave the globe a geography not just of climates and cultures but of stages of human development that could confront and evaluate one another."[3] He argues that many of the motifs of science fiction "represent ideological ways of grasping the social consequences of colonialism."[4] Therefore, for Rieder, the generic structure of science fiction and colonialism are "unthinkable" without each other, and he argues that the origins of science fiction must be contextualised as a product of imperialist culture, beginning in late nineteenth-century British and French fantasies of global conquest and then emerging in the "new" imperialist cultures of Germany, Russia, the United States and Japan in the twentieth century.[5]

Given this complicity between science fiction and empire, it is therefore all the more significant that Amitav Ghosh, an author who is a key figure in postcolonial literature, would choose science fiction as a genre for his 1996 novel *The Calcutta Chromosome: A Novel of Fevers, Delirium and Discovery*—his first (and, to date, only) work of science fiction.[6] The novel, which won the Arthur C. Clark Award for Best Science Fiction Novel in 1997, deftly utilises and then subverts science fiction motifs, conventions and tropes to deconstruct and move beyond the simplistic binary oppositions between the Global North and Global South that underpin colonial ideologies. This chapter will argue that Ghosh, by exploring postcolonial themes through a science fiction lens, offers a new form of knowing and being as an alternative both to Western, positivistic formulations of science and to visions of the future that derive from such formulations of science. This new form of knowing and being finally culminates in Ghosh's conception of a "postcolonial" cyborg. In doing so, this chapter will argue that *The Calcutta Chromosome* can be seen as articulating the concerns of a nascent but growing subgenre of science fiction, namely, postcolonial science fiction—a subgenre that is increasingly engaged with creating new visions of "postcolonialism" that take into account the complexity of problems facing diverse global communities in the wake of imperialism and neoimperialism's expansion and consolidation of global capitalism. According to Istvan Cicsery-Ronay Jr., despite SF's historical complicity with the encompassing and hegemonic "imaginary world-model of Empire," the genre does not necessarily need to mechanistically replicate imperialist ideological structures.[7] He holds that science fiction may also, in its deployment of the globalising models of Europe, provide the means for us to unearth the ideological underpinnings of global capital, the unique manifestations of globalisation in unique national and postcolonial cultures and the processes whereby these cultures exist alongside and engage this new polymorphous global habitus. Postcolonial science fiction, therefore, with its critique and subversion of science fiction's complicity with empire, is uniquely suited to imagining new visions of future that are based on modes of relation between people who are less exploitative and therefore more ethical. Ghosh's novel deploys what is termed by Malisa Kurtz as "a strategy of postcolonial experimentation" towards envisioning new "postcolonialisms that will materialise into more ethical practice," which, she argues, is a particular tool of postcolonial science fiction.[8] Kurtz contends that postcolonial SF "performs the important task of deconstructing the exploitative logic of colonialism and capitalism while simultaneously constructing alternative spaces of resistance by privileging a non-imperial ethics of relationality."[9] In Kurtz's view, this process of deconstructing and reconstructing the SF genre allows "postcolonialism" to be conceptualised as an ethical attitude as opposed to a historical time period or oppositional stance that risks reiterating centre/periphery binaries. Kurtz's conceptualisation draws on Graham Huggan's definition of postcolonialism as "an ensemble of loosely connected oppositional practices ... and by an aesthetic of largely textualised, partly localized resistance."[10] Kurtz argues that postcolonial science fiction's most significant feature is "a shared ethical commitment to mapping the commonalities and differences in experiences of empire." Therefore, she views postcolonial SF "as simultaneously a subgenre and a process, strategy, or mode of relation established between people committed to imagining less exploitative futures."[11] Kurtz's conceptualisation of postcolonial science fiction is a particularly useful lens through which Ghosh's novel can be read because she views postcolonial science fiction as constructivist, offering alternative epistemological frameworks that allow us to understand our relationship to the future beyond colonial paradigms. As the chapter will argue, Ghosh's imagining of an alternative, postcolonial future in *The Calcutta Chromosome* is similarly constructivist, articulating an ethics of relationality that attempt to think beyond Global North/Global South binaries.

The Calcutta Chromosome defies categorisation and resists any easy summary, being equal parts science fiction thriller, medical mystery and postcolonial historical novel. Beginning in New York at an undisclosed time in the future, we meet Antar, an Egyptian immigrant spending his days in his apartment working for a global megacorporation, the International Water Council, with the help of a super-computer called Ava. While doing inventory, Ava shows Antar the trace of an identity card that has been lost in the virtual system and after Ava reconstructs it, Antar realises that it belongs to his former colleague, Murugan, who disappeared while on leave in Calcutta back in 1996. Murugan had been researching the Nobel prize–winning scientist Ronald Ross, whose ground-breaking discovery that malaria is transmitted via mosquito bites had been made in Calcutta 100 years earlier, when Calcutta was part of the British Raj in the Indian subcontinent. Murugan had always insisted that Ross's discovery had been manipulated by an Indian "counter-science" group that had "systematically interfered with [his] experiments to push malaria research in certain directions while leading it away from others."[12]

As Antar tries to piece together the mystery of Murugan's disappearance, we go back and forth between the present (which is our future), Murugan's investigations in Calcutta in 1996 and Ross's time in Calcutta in 1898. Through many computer-mediated encounters with characters both from the past and the present, from Egypt, India, America and Britain, Antar realises that this counter-science cult is led by subaltern figures who are in the margins of both colonial and native society. These include Mangala, a lower-caste, illiterate scavenger woman and Ross's favourite servant, and a lower-caste orphaned villager known variously as Lutchman, Lakshman and Laakhan. According to Murugan, the group's counter-science ideology is based on the hypothesis that "to know something is to change it."[13] He explains:

> Maybe this other team started with the idea that knowledge is self-contradictory; maybe they believed that to know something is to change it, therefore in knowing something, you've already changed what you think you know so you don't really know it at all: you only know its history. Maybe they thought that knowledge couldn't begin without acknowledging the impossibility of knowledge.[14]

Working on this premise, Mangala and her followers figure out that "if you wanted to create a specific kind of change, of mutation, one of the ways in which you could get there is by allowing things to be known."[15] The counter-science group wants to utilise and intervene in the existing research on the malaria virus so that it can affect a mutation in individual chromosomes of the malaria virus in order to unlock the secret of immortality based on a chromosomal process of "interpersonal transference." This chromosomal mutation in the virus, called the Calcutta Chromosome by Murugan, allows personality traits (or the "soul") to be transferred from one person to another. To effect this mutation they want to nudge

Ross into making his breakthrough in malaria science, setting into motion events that encompass the centuries spanning between them and Antar's present (and beyond).

The Calcutta Chromosome as Postcolonial Science Fiction

In *The Calcutta Chromosome*, the coming together of genres as seemingly divergent in their stylistic and thematic modes as science fiction and postcolonial literature seems unlikely at first glance. Therefore, before moving on to a discussion of the novel itself, it is important to address what may be an obvious question: how can a genre that has been accused of participation in and complicity with the very fantasies of empire that made postcolonial literature relevant in the first place be used to expound on the themes of the latter? To understand the relationship of postcolonialism to science fiction, the complicity between "traditional" science fiction and empire have to be grappled with. Here, my use of the term "traditional science fiction" refers to the canonical SF that, according to Rieder, found its original expression in the late-nineteenth and early-to-mid-twentieth-century imperialist ideologies of Europe.[16] As Csicsery-Ronay argues, this canonical SF emerges from the juncture of three conditions: "the technological expansion that drove real imperialism, the need felt by national audiences for literary-cultural mediation as their societies were transformed from historical nations into hegemons, and the fantastic model of achieved technoscientific Empire."[17] Csicsery-Ronay's idea of the "fantastic model of achieved technoscientific Empire" has particular valence in the contemporary, postcolonial, neo-imperial context within which postcolonial science fiction, and Ghosh's novel in particular, is situated. For Csicsery-Ronay, science fiction "has been driven by a desire for the imaginary transformation of imperialism into Empire, viewed not primarily in terms of political and economic contests ... but as a technological regime that affects and ensures the global control system of de-nationalized communications."[18] It is precisely this technological regime of global control that Ghosh challenges and finds an alternative for in his novel. While Csicsery-Ronay's analysis of science fiction's complicity with empire emphasises the imperialist dimensions of technological expansion, Rieder argues that the relationship between science fiction and colonialism is most evident in two frameworks that the genre utilises: the colonial gaze and the ideology of progress that underlies science fiction narratives. Rieder's study, accordingly, analyses how "science fiction lives and breathes in the atmosphere of colonial history and its discourses, how it reflects or contributes to the ideological production of ideas about the shape of history, and how it might, in varying degrees, enact a struggle over humankind's ability to reshape it."[19] According to Rieder, science fiction emerged at the height of the colonial project, which he describes as "the entire process by which European economy and culture penetrated and transformed the non-European world."[20]

The term "postcolonial science fiction" itself emerged amidst critiques of the relationship between science fiction and colonialism, particularly in scholarship in the late 1990s and early 2000s. As a genre, postcolonial science fiction was only consolidated in 2004 with the publication of Nalo Hopkinson and Uppinder Mehan's *So Long Been Dreaming: Postcolonial Science Fiction and Fantasy*, an anthology of science fiction, speculative fiction and fantasy by people of colour.[21] As Mehan argues in his afterword to the anthology, the visions of the future articulated by postcolonial science fiction "are both a questioning of colonial/imperialist practices and conceptions of the native or the colonized, and an attempt to represent the complexities of identity that terms such as 'native' and 'colonized' tend to simplify."[22] This hybrid genre is one that utilises science fiction tropes in a radically different way. According to Jessica Langer, instead of completely moving away from science fiction's colonial tropes of discovery and conquest, postcolonial science fiction instead "hybridizes them, parodies them and/or mimics them against the grain in a play of Bhabhaian masquerade."[23] In other words, it subverts "traditional" science fiction's tilt towards perpetuating colonial tropes by critiquing it in different ways. For instance, as Mehan says, "one of the key strategies employed by these writers is to radically shift the perspective of the narrator from the supposed rightful heir of contemporary technologically advanced cultures to those of us whose cultures have had their technology destroyed and stunted."[24] Postcolonial science fiction can therefore be defined as that particular kind of science fiction that acknowledges and then subverts, in different ways, the genre's genealogical and ideological debt to colonialism, and in doing so articulates a vision for the future which is more ethical and less exploitative.

Despite, but also perhaps because of, science fiction's origin of complicity in colonialism, many of the genre's concerns, themes and motifs can in fact provide a powerful means of exploring postcolonial themes. On closer inspection, it becomes clear that science fiction and postcolonial literature have several things in common. According to Andy Sawyer in his foreword to *Science Fiction, Imperialism and the Third World*, "the tools of sf, its speculative drive and its ability to distort language, are among the most powerful weapons available" to radical and anti-colonial thinkers.[25] By using these tools of the genre, postcolonial science fiction has to consciously write from outside the traditional strands of Western science fiction while acknowledging those strands as progenitors, which is what Ghosh does in his novel. Two foundational signifiers of traditional science fiction are the idea of the alien and the alien space, and these signifiers are closely linked to imperialist and colonialist ideologies. As Langer argues, "the figure of the alien—extraterrestrial, technological, human-hybrid or otherwise—and the figure of the far-away planet ripe for the taking are deep and abiding twin signifiers in science fiction, are perhaps even the central myths of the genre."[26] She goes on to explain that these central signifiers of science fiction—what she calls the use of the Stranger and the Strange Land—are in fact "the very same twin myths of colonialism."[27] The very myths of science fiction that are the most closely intertwined with colonialism can thus be used to undertake a critique of colonialist ideology. This is what postcolonial science fiction attempts to do in its narrative conceptualisation of the identity of the Other, which challenges colonial discourses, assumptions and stereotypes.

This impulse might accord with "traditional" science fiction as well; if we use Ray Bradbury's definition, the genre is, or should be, about "hating the way things are, wanting to make things different."[28] This questioning of the status quo, or critique of the way human society is structured, is central to postcolonial literature as well, which uncovers the ways in which centre-periphery binaries affect, in material as well as ideological ways, colonised or previously colonised societies. Similar to science fiction, postcolonial literature also seeks to imagine ways to make such societies different, and their relationship to the Global North more equitable. Furthermore, science fiction, despite its problematic relationship with imperial fantasies, nevertheless is deeply engaged with questions of difference and Otherness. As Roberts points out, the genre "has always had sympathies with the marginal and the different," and bringing to light marginalised people's forgotten and erased stories is one of the most important objectives of postcolonial literature.[29] Science fiction's exploration of Otherness and marginality, and its ethical potential in imagining more equitable futures, lends itself well to the postcolonial ethos of relating to other people in non-exploitative ways. Ghosh illustrates this perfectly by making the most important players in *The Calcutta Chromosome* individuals belonging to marginalised communities who exist on the fringes of society—women, immigrants, lower-caste people.

Another generic trope of science fiction that is particularly suited to the concerns of postcolonialism is its discursive interest in history. In science fiction, even though the future might be the stage dressing, what is often played out on this stage is human history and our ambivalent and complex relationship with disturbing aspects of this history. Whether it is war, genocide, or nuclear destruction, science fiction routinely explores the unpleasant aspects of actual human history. This is particularly evident in another subgenre of science fiction called alternate history, in which "what if" scenarios are explored in depth.[30] A deep engagement with history is in fact instrumental in articulating visions of the future, and science fiction that attempts to imagine futures without confronting the past will be limited in its perspective. As Susie O'Brien argues, there is a "ludicrous quality to future speculations that fail to take into account the historical violence that undergirds the present and, by extension, the place from which speculation unfolds."[31] This questioning of the way history is constructed and remembered is something science fiction deeply shares with postcolonial literature, which also contests and challenges imperial history in its different formulations. Addressing the erasure of histories of colonial exploitation and oppression is a central concern of postcolonialism, and postcolonial science fiction can therefore utilise science fiction's generic engagement with human history in order to articulate this concern. The interrogation of, and intervention into, actual history is a significant theme in *The Calcutta Chromosome*, as Ghosh reimagines the malariological discovery of Ronald Ross, a real-life historical figure whose time in Calcutta is chronicled in his *Memoirs*.[32] Ghosh draws heavily from this source, underlining the novel's commitment to engaging with the colonial history of scientific progress. In fact, Lutchman, Ross's assistant who is an important figure in the counter-science group in the novel, appears in the story similar to how Ross himself describes in his *Memoirs*: Ross wrote that he learnt about the difference between species of mosquitoes from Lutchman, and so Ghosh's elaboration of this in the novel is his way of fictionalising Ross's memoirs. In this way, *The Calcutta Chromosome* is a blend of alternate history and an attempt to read a text (in this case, Ross's *Memoirs*) from a perspective different from that of its author. By employing traditional science fiction's motif of interrogating history through the utilisation of different, marginalised viewpoints, postcolonial science fiction can more dynamically engage with the violences of a colonial past. As Kurtz argues, "postcolonial science fiction reminds us that the global North and South not only have different pasts and presents, but if we continue to ignore these fundamental differences, they will have different futures as well."[33]

Postcolonial science fiction, then, not only engages with many of the generic foundations of traditional science fiction but also overtly questions many of the assumptions that underpin it. Of these assumptions, the most significant for postcolonial science fiction are, according to Kurtz, "the genre's colonial gaze, the appeal to an ideology of progress, focus on the future of the construction of an assumed cosmopolitan future, and an implicit faith in technological solutions or the inclination towards techno-optimism."[34] As the rest of the chapter will show, *The Calcutta Chromosome* is particularly emblematic of postcolonial science fiction because it directly confronts these assumptions and, as the next section will argue, a confrontation with these concerns is fundamentally ethical.

Science Fiction and Postcolonial Ethics

Given traditional science fiction's ideological debt to colonialism, a central concern of postcolonial science fiction is ethical: how does postcolonial science fiction avoid replicating the structure of the colonial gaze? How can new visions of postcolonialism be created through this subgenre of science fiction, which can also be viewed as a process or strategy for more ethical praxis? Here, I am referring to a postcolonial ethos as is elaborated upon by Simone Bignall, who argues that the process of postcolonisation requires "a particular concept of transformative agency that enables one consciously to enact a postcolonial ethic of relation to others, and to engage a collaborative politics of material transformation in order to construct postcolonial institutions and communities of practice with others."[35] By foregrounding this concern of enacting a postcolonial ethics of relationality, postcolonial science fiction can articulate what Bignall calls "careful forms of sociability," which represents the ethical drive that characterises this emerging subgenre.[36]

For Kurtz, there are particular benefits that emerge from viewing postcolonial science fiction as "a mode of relation established between people committed to imagining less exploitative futures."[37] She argues that viewing postcolonial science fiction as being premised upon this shared ethical commitment is "central to reimagining the truly productive potential of postcolonial sf."[38] This will enable us to utilise postcolonial science fiction as a critical lens through which continuations of contemporary neocolonial frameworks can be interrogated. Kurtz elaborates on this potential of the subgenre by arguing that "rather than seeing the genre as yet one more way capitalism appropriates and resells difference by marketing the margins, postcolonial sf can be seen as the manifestation of a collective ethos, or desire, to imagine lines of flight from the colonial logic that binds us."[39] This ethical potential of science fiction is echoed by Ericka Hoagland and Reema Sarwal, who contend that visionaries of science fiction such as Wells and Bradbury are invested in an ethical project of revealing the stresses and contradictions internal to imperialist histories. Hoagland and Sarwal argue that the "rewriting/revising of history and the recovery of the subaltern subject, integral components to postcolonial studies, are mirrors of science fiction's complex relationship with history and the haunting presence of aliens and others like Bradbury's Martians."[40] Similarly, Kurtz argues that postcolonial studies can in fact benefit from "thinking through the lens of science fiction, where creative projects function as ethical experiments towards mapping out the possibilities of transnational affiliation."[41] Thus, there is an overlapping interest in both science fiction and postcolonial literature in exploring the ethics of Otherness in the wake of empire that is particularly meaningful in this chapter's analysis of *The Calcutta Chromosome*. Accordingly, this chapter will be particularly cognisant of the ways in which Ghosh is articulating a postcolonial ethos in the novel's exploration of a new form of knowledge and existence. For Kurtz, postcolonial science fiction produced from diverse national contexts is an expression of "a transnational desire to understand such questions as: what do we need to do so that tomorrow is not characterised by the violence against others we exhibit today? Or, more specifically, how can we create new visions of 'postcolonialism' that will materialize into more ethical practice?"[42] Ghosh's novel is concerned with a similar imagining of a more ethical mode of relating that human beings can develop, a mode which attempts to undo the violence that undergirds relationships under empire.

When Western Science Met the Colonised

Along with traditional science fiction's perpetuation of imperialist casting of the alien Other as barbaric and less civilised—and therefore deserving of being conquered and colonised—is the furthering of a connected but distinct ideology. This is the trope of casting the conquering heroes as being on the side of seemingly objective and infallible science and the conquered Others as being mired in superstition and irrationality. This science fiction trope is closely associated with the Orientalist binary opposition of the Euro-American coloniser as rational, scientific and logical and its colonised counterpart as irrational, spiritual and religious. In *The Calcutta Chromosome*, Ghosh places this tension between rationality and spirituality at the centre of the novel, attempting to move beyond this simplistic dichotomy by questioning whether the two are really so incompatible after all. The novel's Indian counter-science group seamlessly combines Western methods of scientific inquiry with Indian or, more specifically, Hindu concepts of reincarnation without any contradiction between the two. As Murugan tells Antar, the goal of the group is as much grounded in the physical, biological and scientific as it is in the spiritual and transcendent. He says, "What these guys were after was much bigger; they were after the biggest prize of all, the biggest fucking ball game any human being has ever thought of: the ultimate transcendence of nature … immortality."[43] But in the next breath, he describes the process in scientific terms:

> What I'm really talking about is a technology for interpersonal transference … How much do you think you'd pay for that kind of technology, Ant? Just think, a fresh start: when your body fails you, you leave it, you migrate—you or at least a matching symptomology of your self. You begin all over again, another body, another beginning. Just think: no mistakes, a fresh start. What would you give for that, Ant: a technology that lets you improve on yourself in your next incarnation?[44]

By framing immortality in scientific terms such as "technology," "interpersonal transference" and "matching symptomology," Ghosh is employing an idea that is common in science fiction: the notion of trading in an old body for a new and better one. Thus, he explores the tension between science and religion both as a general rational-spiritual issue and as a specific postcolonial issue, but importantly, he does so in a way that challenges Orientalist binary oppositions. After all, Mangala and her counter-science group use conventional Western tools of science to further their transcendental goals. For instance, the following passage shows a minor character secretly observing Mangala during one of her rituals:

> First the assistant went up to the woman, Mangala, still regally ensconced on her divan, and touched his forehead to her feet. Then in the manner of a courtier or acolyte he whispered some word of advice in her ear. She nodded in agreement and took the clean slides from him. Reaching for the birdcages she allowed her hand to rest upon each of the birds in turn, as though she were trying to ascertain something. Then she seemed to come to a decision; she reached into a cage, and took one of the shivering birds into her lap. She folded her hands over it and her mouth began to move as though muttering a prayer. Then suddenly a scalpel appeared in her right hand; she held the bird away from her and with a single flick of her wrist beheaded the dying pigeon. Once the flow of blood had lessened, she picked up the clean slides, smeared them across the severed neck, and handed them to the assistant.[45]

As can be seen here, the imagery most associated with Mangala and her rituals is replete with test tubes, scalpels, slides and other forms of microscopic study. Moreover, the iconic talisman for Mangala and her cult is a clay figurine that combines the images of a pigeon and a primitive microscope with a "simple, semicircular mound, crudely modeled and featureless except for two large, stylized eyes, painted in stark blacks and white."[46] Again, the combination of tools of Western science with Eastern iconic deities is significant in showing the science-religion hybrid the novel is exploring. As Suchitra Mathur argues, "Mangala's counter science, then, is clearly not antiscience; it does not reject completely the methods (experimentation) and processes (vivisection) of modern science. Instead, it appropriates them for its own purposes."[47] By using Western scientific tools and mingling science with mysticism, Mangala's counter-science group is not the opposite of Western science as an Orientalist idea of irrational superstition, but a hybrid of both and thus capable of moving beyond the limits of conventional Western science. According to James H. Thrall, the counter-science proposed by Ghosh "is a form of rational inquiry of an entirely different order, Eastern rather than Western in orientation, with goals commensurately more vast, and independent from limited perspectives that can accomplish only what is perceived as possible."[48] Thus, Ghosh's exploration of the rational-spiritual dichotomy is explicitly postcolonial in nature, suggesting a solution that hybridises the East and the West instead of placing them in opposition to one another.

Along with a complication of Orientalist binary oppositions, Ghosh also highlights Western scientific discourse's interaction with and complicity in colonisation. As Ross himself acknowledges in his memoirs, his discovery in Calcutta was helped by his Indian assistants, whose full names he did not bother to learn and who received none of the credit in his Nobel Prize, and he tested his hypotheses on Indian people who remained a nameless horde. Ghosh rewrites this highly biased account of Western scientific discovery. Instead of Ross being a genius who worked alone, he is an almost clueless pawn in a game orchestrated by the very people he hardly even sees, nudging him along every step of his discovery:

> "Eureka," [Ross] says to his diary, "the problem is solved."
>
> "Whew!" says Lutchman, skimming the sweat off his face. "Thought he'd never get it."[49]

Ghosh is giving validity to colonised India's local systems of knowledge, which according to Langer is a feature of postcolonial science fiction: "postcolonial science fiction foregrounds the concept that indigenous and other colonized systems of knowledge are not only valid but are, at times, more scientifically valid than is Western scientific thought."[50] Indeed, Mangala and her assistants are already aware of the connection between malaria and mosquitoes before Ross even thinks to make the discovery—they are so far ahead in their research that they are even aware that certain strands of malaria can be used to cure syphilis. That locals in India had this knowledge before Ross made the discovery is historically accurate. In an interview Ghosh gave to Paul Kincaid, he explains:

> Ross deserves a great deal of credit for this because his work was indeed a very elegant piece of research, but in effect much that he 'proved' was already well known amongst common folk in India and Africa. Ross's *Memoirs* clearly show that he used folk knowledge in advancing his work. His real achievement then, lay in translating folk knowledge into the language of science.[51]

Thus, Ghosh complicates the notion that it was the West's superior scientific and reasoning skills that were the cause of its material domination of the world, and instead argues that local systems of knowledge were equally legitimate in their attempts to know scientific truths, even if this knowledge was incomprehensible to Western scientific discourse.

Silence Is Power

For readers belonging to previously colonised cultures, science fiction's thrilling tales about conquering foreign lands is not merely fiction, and when confronted with these works they would more likely see themselves in the colonised aliens than the conquering heroes. As Nalo Hopkinson says, "one of the most familiar memes of science fiction is that of going to foreign countries and colonizing the natives … and for many of us, that's not a thrilling story; it's non-fiction and we are on the wrong side of the strange-looking ship that appears out of nowhere."[52] One task of postcolonial science fiction, therefore, is to present such tales from the so-called "wrong side of the strange-looking ship," and this is what Ghosh does in *The Calcutta Chromosome*. All of the players that move the plot forward are the marginalised and the silenced, the ones whose accounts have historically been missing, especially in the archives of Western scientific development. The truth about Mangala and Lutchman is uncovered, not just by Antar and Murugan, but also by Sonali Das, a middle-aged Bollywood actress, and Urmila, a young Indian journalist who is struggling against third world patriarchy. In Antar's timeline, this narrative centering of the marginalized is represented by his immigrant friends, as well as Tara, a young Indian immigrant working as a nanny for rich New Yorkers. All of the characters, in other words, are non-Western, people who are peripheral to a West-centric world order.

But Ghosh goes further than merely giving non-European protagonists agency in furthering their own narratives. In an interesting twist, he makes their very marginality the source of their power. This is especially the case with Mangala, Lutchman and the counter-science group, whose very modus operandi is silence, invisibility and secrecy. The very fact of being marginal to colonial society is what gives them their power. As Murugan explains:

> Thinking of it in the abstract, wouldn't you say that the first principle of a functioning counter-science would have to be secrecy? The way I see it, it wouldn't just have to be secretive about *what* it did (it couldn't hope to beat the scientists at that game anyway); it would also have to be secretive *in* what it did. It would have to use secrecy as a technique or procedure.[53]

The counter-science group draws strength from its marginality, not just in the abstract way of operating in the shadows, but also by drawing literal power from their silence. "Fact is we're dealing with a crowd for whom silence is a religion," Murugan explains later.[54] By making silence an essential feature of the counter-science group's power over both Ross and scientific discovery in general, Ghosh is engaging with Gayatri Spivak's concept of the subalternity of colonially oppressed groups being dependent on their being silenced, which she explains in her seminal work "Can the Subaltern Speak?"[55] Ghosh subverts Spivak's contention that the subaltern is unable to speak by rendering their very silence a form of communication. This concept of silence being literally powerful is clearest in the words of Phulboni, a Bengali author in the novel who turns out to be a member of the counter-science cult, who says: "Mistaken are those who imagine that silence is without life; that it is inanimate, without either spirit or voice. It is not: indeed the Word is to this silence what the shadow is to the foreshadowed, what the veil is to the eyes, what the mind is to truth, what language is to life."[56]

Furthermore, the fact that Mangala exists on the fringes of society is what enables her to be open to the seemingly impossible discoveries she makes during her scientific research. Precisely because she is so far removed from the centre of conventional Western science, she is able to stumble upon the discovery of personality transmutation. As Murugan explains, Mangala is "completely out of the loop, scientifically speaking," and that is what makes her receptive to this unconventional discovery which is "exactly the kind of entity that would be hardest for a conventional scientist to accept."[57] Mangala's existence on the fringes of society forces her to work "in a different way; a way so different it wouldn't make sense to anyone who's properly trained."[58] At a later point, Murugan says that Mangala "wasn't hampered by the sort of stuff that might slow down someone who was conventionally trained: she wasn't carrying a shit-load of theory in her head, she didn't have to write papers or construct proofs ... she didn't care about formal classifications."[59] By "proper" and "conventional" training, Ghosh is referring to the West's ideology of delimiting legitimate and illegitimate forms of science, and by making Mangala succeed in her "illegitimate" form of scientific discovery, he is challenging the notion that only Western, positivistic forms of scientific discourse can lead to the truth. According to Diane Nelson, the novel "offers a powerful alternative, a 'how-to guide for postcolonial new humans,' that figures the possibility of a new mode of being and knowledge in the contemporary world."[60] Ghosh conceives of a scenario in which native systems of knowledge combine with and subvert Western science to discover a new form of existence. Moreover, by exploring new forms of being, Ghosh is at the same time writing in the vein of traditional science fiction, which works on the premise of pushing conventional scientific discourse to its limit and then exploring the unlikely events that result.

The Postcolonial Cyborg

These new forms of knowing and being explored in *The Calcutta Chromosome* come together in Ghosh's conceptualisation of a post-human, postcolonial cyborg. The key to the technology of interpersonal transference (i.e., the Calcutta Chromosome) resides in the function of the vector, which in the novel evolves from carriers such as mosquitoes, to pigeons, and then finally, during Antar's time, to the computer. The vector transports the "soul" from host to host, body to body, ensuring that Mangala is reincarnated from her old body to that of an elderly Armenian woman called Mrs. Aratounian in 1996 Calcutta, then to Urmila's body, and finally to the body of Antar's neighbour, the Indian immigrant Tara. As the vector evolves from organic (mosquitoes, pigeons) to technological (the computer Ava), it becomes clear that by harnessing cyberspace, Mangala and her followers have become much more global in their search for new carriers and hosts to develop ongoing experiments.

Ava, based on a human-computer interface (HCI), acts as a catalyst to further the events that Mangala has set in motion, functioning like the mosquito in the transmission of the malarial virus and the Calcutta Chromosome. Just as the mosquito has "recombinatory powers" that channel and redistribute the malarial virus, Ava has the ability to transpose information and data, in the end allowing Antar to merge with it to become part of the counter-science cult and a carrier in their reincarnation experiments. The novel ends with Antar getting initiated into the cult with the help of Mangala and the other members, via the medium of Ava:

> He felt a cool soft touch upon his shoulder and his hand flew up to take off the SimVis headgear. But now there was a restraining hand upon his wrist, and a voice in his ear, Tara's voice, whispering: "Keep watching; we're here; we're all with you."
>
> There were voices everywhere now, in his room, in his head, in his ears, it was as though a crowd of people were in the room with him. They were saying: "We're with you; you're not alone; we'll help you across."[61]

As Antar migrates into the world of Ava's virtual consciousness to join the others through Ava's SimVis headgear, computer and human become integrated in the counter-science group's quest to further the growth of biological and organic life, producing a sort of cyborg that is different from the conventional machinery-and-software-based "post-human" cyborgs usually found in science fiction. According to Donna Haraway, cyborg ontology implies the end of the "organic" as we have understood it in the past; rather, it signals a post-human age.[62] But instead of signalling the end of organic life as we know it, the new cyborg technology in *The Calcutta Chromosome* acts as an instrument for human evolution. According to Christopher Shinn, the cyborg technology in the novel "facilitates a significant reversion to organic life. Rather than the end of the 'organic', Ghosh's novel ultimately reaffirms the importance of the organic in a 'post-human' world."[63] For Ghosh, the post-human world is not one in which technology controls nature and organic life, but one where nature perpetuates itself and evolves through the help of technology.

The postcolonial post-human that Ghosh presents at the culmination of the novel is created through the establishment of connections: Western science mingling with Eastern spirituality; humans interacting with other biological organisms, such as mosquitoes; and colonised people connecting with the kinds of technology that were part of the colonising project. The relationship between colonised people and the technology which was instrumental in colonising them is particularly significant in the novel. Ghosh draws a connection between the use of older technologies, such as railroads, and new forms of technology that help the neo-imperial project, such as global communications networks that extend the reach of multinational corporations. Indeed, in all the three timelines that Ghosh presents, the colonised people are shown to be using the tools of the colonisers for their own ends. In addition to Mangala and her counter-science group using Western methods and tools of scientific inquiry, railways figure into the events significantly as well. The railway system was one of the most important technologies that the British employed to further their power and control in India, and it is significant that both Mangala and Lutchman are found by Ross in a railway station. He explains: "I found her where I find all my bearers and assistants: at the new railway station. ... That's the place to go if you need a willing worker: always said so—it's full of people looking for a job and a roof over their heads."[64] But, Ghosh argues, the railway isn't a one-way technology that the coloniser can use to control the colonised. Instead, it is a tool that the colonised can take into their own hands. This is signified by Lutchman leading a minor character, who is about to discover the counter-science group's secret, to the railway station and causing him to almost get killed by a train: "a locomotive is bearing down on Grigson, snorting clouds of steam. He panics and starts running between the tracks; he's set to become roadkill. But in the last half second he manages to jump: the fenders miss him by a fraction of an inch."[65] Other characters who come close to finding the truth about Mangala suffer similar fates, getting killed in mysterious accidents on the railway tracks.

If railways signify colonial control in the novel, the advanced computer technology employed by the International Water Council, the megacorporation Antar works for (and is monitored by), signifies a tool for control in the neocolonial order—an order where corporate empires have replaced colonial ones, where Antar and other non-Western subjects' labour is yoked to the global machinations of corporate enterprise. Advanced computer technology is how the International Water Council manages to keep its employees under control. At the beginning of the novel, as Antar is working with Ava, he is constantly aware that Ava is watching him through a camera and tracking his performance. He recalls a time when he decided to read a book while on duty, and how Ava recognised his inattention and warnings suddenly came blaring on the screen. The consequence was that, "at the end of the week, he received a notice from his employer, the International Water Council, telling him that his pay had been docked because of 'declining productivity', warning him that a further decline could entail a reduction in his retirement benefits."[66] In the neocolonial world, corporate empires keep people under control via surveillance and computer technology, which is why it is significant that Mangala and the counter-science group evolve their methods and tools into something appropriate for this new world order. This is what makes the "postcolonial" cyborg Ghosh posits against this neocolonial empire so meaningful—this cyborg is based on a postcolonial connectivity, between people around the world in different positions of power and oppression, between humans and technology, between science and spirituality.

In the novel, the cyborg's mode of being and knowledge has a degree of autonomy because it exists in an in-between space, neither completely outside the gambit of the dominant socio-political structure, nor completely controlled by it. This is shown by the physical spaces occupied by the counter-scientists—the outhouses, anterooms, ramshackle houses under construction, and private apartments—where the actual work of "interpersonal transference" takes place. According to Mathur, "these are fringe spaces characterized by the relative invisibility that is accorded to the marginalized, and as such, allow the freedom that accompanies such liminality… [the counter-science group] is the Other of modern science, the double that it dismisses (not demonizes or represses), and as such, it co-exists with modern science in relative autonomy."[67] The fact that for Ghosh, Mangala and the counter-science group can exist alongside modern Western science is key, because it represents a new mode of knowledge and existence that is not inherently trapped within colonial binaries of the Global North and the Global South, but can move beyond it. This mode of knowledge and existence, one which has the potential to transcend the oppressive and exploitative colonial and neocolonial binaries, offers a possible ethical blueprint and suggests a truly postcolonial ethos.

Conclusion

In *The Calcutta Chromosome*, Ghosh brings science fiction tropes and conventions together with postcolonial themes in a way that makes the novel a valuable contribution to both genres of literature. Through Mangala and her counter-science group, he offers a new form of knowing and being which is an alternative to Western, positivistic formulations of science but without upholding simplistic colonial binaries. Ghosh expands upon this postcolonial cyborg to offer a future that not only pushes the boundaries of our traditional conception of what is possible, as science fiction at its best tends to do, but also disentangles itself from the fraught and restrictive society predicated on oppressive colonial power structures. As Kurtz argues, "postcolonial sf does not just critique colonialism and neocolonial structures, it also seeks to enact alternatives. … Separately, the terms 'postcolonial' and 'science fiction' are problematic and difficult to define, but together they may create a new site of knowledge that asks us to think simultaneously about the continuum of history and what links past violences (postcolonial) to future alternatives (science fiction)."[68] In *The Calcutta Chromosome,* Ghosh does precisely this, creating an alternative future and a new form of knowledge that is cognisant of, and in fact grows out of, past colonial violences, while offering a better form of being than the one that is entrapped by colonial and neocolonial power structures. In doing so, the novel allows us to explore the potential of postcolonial science fiction as a genre, expanding the possibilities both of science fiction and the ways in which postcoloniality is configured.

Notes

1. Roberts, *Science Fiction*, 66.
2. Kerslake, *Science Fiction and Empire*, 191.
3. Rieder, *Colonialism*, 6.
4. Rieder, *Colonialism*, 20.
5. Rieder, *Colonialism*, 3.
6. Ghosh, *Calcutta Chromosome.*
7. Csicsery-Ronay, "Science Fiction and Empire."
8. Kurtz, "Globalization," 28.
9. Kurtz, "Globalization," 36.
10. Huggan, *Postcolonial Exotic*, 6.
11. Kurtz, "Globalization," 4.
12. Ghosh, *Calcutta Chromosome*, 36.
13. Ghosh, *Calcutta Chromosome*, 104.
14. Ghosh, *Calcutta Chromosome*, 78.
15. Ghosh, *Calcutta Chromosome*, 217.
16. Rieder, *Colonialism.*
17. Csicsery-Ronay, "Science Fiction and Empire," 231.
18. Csicsery-Ronay, "Science Fiction and Empire," 232.
19. Rieder, *Colonialism*, 3.
20. Rieder, *Colonialism*, 25.
21. Hopkinson and Uppinder, *So Long Been Dreaming.*
22. Mehan, "Final Thoughts," 269.
23. Langer, *Postcolonialism*, 3–4.
24. Mehan, "Final Thoughts," 270.
25. Sawyer, "Foreword," 1.
26. Langer, *Postcolonialism*, 3.
27. Langer, *Postcolonialism*, 3.
28. Bradbury, "No News," 169.
29. Roberts, *Science Fiction*, 69.
30. An important example of this subgenre is Philip K. Dick's *The Man in the High Castle* (1962), which explores a different outcome for World War II.
31. O'Brien, "'We Thought,'" 339.
32. Ross, *Memoirs.*
33. Kurtz, "Globalization," 28.
34. Kurtz, "Globalization," 28.
35. Bignall, *Postcolonial Agency*, 3.
36. Bignall, *Postcolonial Agency*, 220.
37. Kurtz, "Globalization," 28.
38. Kurtz, "Globalization," 217.
39. Kurtz, "Globalization," 217.

40. Hoagland and Sarwal, "Introduction," 9–10.
41. Kurtz, "Globalization," 28.
42. Kurtz, "Globalization," 28.
43. Ghosh, *Calcutta Chromosome*, 79.
44. Ghosh, *Calcutta Chromosome*, 79–80.
45. Ghosh, *Calcutta Chromosome*, 170–171.
46. Ghosh, *Calcutta Chromosome*, 44.
47. Mathur, "Caught," 131.
48. Thrall, "Postcolonial Science Fiction?" 298.
49. Ghosh, *Calcutta Chromosome*, 78.
50. Langer, *Postcolonialism and Science Fiction*, 8.
51. Ghosh, "Reprint."
52. Hopkinson, "Introduction," 5.
53. Ghosh, *Calcutta Chromosome*, 50.
54. Ghosh, *Calcutta Chromosome*, 149.
55. Spivak, "Can the Subaltern Speak?"
56. Ghosh, *Calcutta Chromosome*, 30.
57. Ghosh, *Calcutta Chromosome*, 250–251.
58. Ghosh, *Calcutta Chromosome*, 105.
59. Ghosh, *Calcutta Chromosome*, 245.
60. Nelson, "Social Science Fiction," 250.
61. Ghosh, *Calcutta Chromosome*, 310.
62. Haraway, *Simians*.
63. Shinn, "On Machines," 147.
64. Ghosh, *Calcutta Chromosome*, 154.
65. Ghosh, *Calcutta Chromosome*, 143.
66. Ghosh, *Calcutta Chromosome*, 19.
67. Mathur, "Caught," 133.
68. Kurtz, "Globalization," 36.

Works Cited

Bignall, Simone. *Postcolonial Agency: Critique and Constructivism*. Edinburgh: Edinburgh University Press, 2010.

Bradbury, Ray. "No News, or What Killed the Dog?" In *Quicker than the Eye*, 141–150. New York: Avon, 1996.

Csicsery-Ronay, Istvan, Jr. "Science Fiction and Empire." *Science Fiction Studies* 90 (2003): 231–245.

Ghosh, Amitav. *The Calcutta Chromosome: A Novel of Fevers, Delirium and Discovery*. 1995. New York: Avon, 1997.

———. "Reprint: An Interview with Amitav Ghosh." By Paul Kincaid. *The Dark Labyrinth*, August 27, 2013, https://ttdlabyrinth.wordpress.com/2013/08/27/reprint-an-interview-with-amitav-ghosh/

Haraway, Donna J. *Simians, Cyborgs, and Women*. New York: Routledge, 1991

Hoagland, Ericka, and Reema Sarwal. "Introduction." In *Science Fiction, Imperialism and the Third World: Essays on Postcolonial Literature and Film*, ed. Ericka Hoagland and Reema Sarwal, 5–19. Jefferson, NC: McFarland, 2010.

Hopkinson, Nalo, and Uppinder Mehan, eds. *So Long Been Dreaming: Postcolonial Science Fiction and Fantasy*. Vancouver: Arsenal Pulp Press, 2004.

Hopkinson, Nalo. "Introduction." In *So Long Been Dreaming: Postcolonial Science Fiction and Fantasy*, edited by Nalo Hopkinson and Uppinder Mehan. Vancouver: Arsenal Pulp Press, 2004.

Huggan Graham, *The Postcolonial Exotic: Marketing the Margins*, New York: Routledge, 2001.

Kerslake, Patricia. *Science Fiction and Empire*. Liverpool: Liverpool University Press, 2007.

Kurtz, Malisa. "Globalization, Postcolonialism, and Science Fiction: Nomadic Transgressions." PhD diss., Broch University, 2016. http://hdl.handle.net/10464/9291.

Langer, Jessica. *Postcolonialism and Science Fiction*. Houndmills, Basingstoke Hampshire: Palgrave Macmillan, 2011.

Mathur, Suchitra. "Caught Between the Goddess and the Cyborg: Third-World Women and the Politics of Science in Three Works of Indian Science Fiction." *The Journal of Commonwealth Literature* 39, no. 3 (2004): 119–138.

Mehan, Uppinder. "Final Thoughts." In *So Long Been Dreaming: Postcolonial Science Fiction and Fantasy*, edited by Nalo Hopkinson and Uppinder Mehan. Vancouver: Arsenal Pulp Press, 2004.

Nelson, Diane M. "A Social Science Fiction of Fevers, Delirium and Discovery: The Calcutta Chromosome, the Colonial Laboratory, and the Postcolonial New Human." *Science Fiction Studies* 30, no. 2 (2003): 246–266.

O'Brien, Susie. "'We Thought the World Was Makeable': Scenario Planning and Postcolonial Fiction." *Globalizations* 13, no. 13 (2015): 329–344.

Rieder, John. *Colonialism and the Emergence of Science Fiction*. Middletown: Wesleyan University Press, 2008.

Roberts, Adam. *Science Fiction*. London: Routledge, 2000.

Ross, Ronald. *Memoirs: With a Full Account of the Great Malaria Problem and Its Solution*. London: John Murray, 1923.

Sawyer, Andy. "Foreword," *Science Fiction, Imperialism and the Third World: Essays on Postcolonial Literature and Film*, ed. Ericka Hoagland and Reema Sarwal, 1–4. Jefferson, NC: McFarland, 2010.

Shinn, Christopher. "On Machines and Mosquitoes: Neuroscience, Bodies, and Cyborgs in Amitav Ghosh's *The Calcutta Chromosome*." *MELUS* 33, no. 4.1 (2008): 145–166.

Spivak, Gayatri Chakravorty. "Can the Subaltern Speak?" In *Marxism and the Interpretation of Culture*, edited by Cary Nelson and Lawrence Grossberg, 271–313. Chicago: University of Illinois Press, 1988.

Thrall, James H. "Postcolonial Science Fiction? Science, Religion and the Transformation of Genre in Amitav Ghosh's *The Calcutta Chromosome*." *Literature and Theology* 23, no. 2 (2009): 289–302.

CHAPTER 10

Wagering the Future: Split Collectives and Decolonial Praxis in Assia Djebar's *Ombre sultane* and Nalo Hopkinson's *Midnight Robber*

Lara Choksey

Assia Djebar's *Ombre sultane* (*A Sister to Scheherazade*) (1987) and Nalo Hopkinson's *Midnight Robber* (2000) bookend a phase of post-Cold War globalism in which Afrofuturism and African futurism stage and subvert the fracture between Pan-African decoloniality and globalised neoliberal finance regimes. They register this through their depiction of the continuation of gender relations based on national development and global production. Both texts work at undoing these divisions by subverting the developmental paradigm under which national growth is expected to occur and by seeking local commons across national and cultural divisions. They invoke and push against realist literary forms that adhere to chronological time: a fiction of development and the conditions for what Walter Rodney has called the paradox of underdevelopment.[1] This paradox is

L. Choksey (✉)
Wellcome Centre for Cultures and Environments of Health,
University of Exeter, Exeter, UK

Z. Kendal et al. (eds.), *Ethical Futures and Global Science Fiction*,
Studies in Global Science Fiction,
https://doi.org/10.1007/978-3-030-27893-9_10

symptomatic both of the neo-imperial institution of a globalised neoliberal finance system, and the *longue durée* of capitalist exploitation in the Caribbean and Africa, and the figure of "woman" is central to sustaining its imaginative reproduction.

The exclusion of women from political decision-making is symptomatic of the encounter between two constitutions of time: the chronological time of decolonial political transition and the anachronism of decolonial praxis. Working under a broad definition of decolonial speculative fiction as "part of a broader recuperation of SFF by writers from colonized territories (former or current), often reclaiming mythologies and indigenous history appropriated by colonial and imperial anthropology, and incorporating modern narratives of oppression," I read these texts as offering new forms of African futurist collectivity (both continental and diasporic) through tropes of science fiction, fantasy and horror, in which women shape its terms, imagining new political spaces in which to think the future.[2] Foregrounding the exclusion of women subjects from the work of political transition in alternative lifeworlds, these texts ask urgent questions about ethical praxis at the intersection of gender and political participation, towards decolonial futures.

While the various parts played by women in anti-colonial movements have been recognised in anti-colonial historiographies (not least in Algeria, the setting of *Ombre sultane*), the presence of women as political actors after liberation is more obscure. Horace Campbell has argued that the history of Pan-Africanism has been told through the individual histories of great men, and Rhoda Reddock concurs that women's work has not received the same recognition.[3] This absence has been co-opted by the rhetoric of Western liberal democracy as a signifier of underdevelopment, and a justification for neo-imperial Euro-US interventions, both hard and soft (i.e. through war and international governance, and development/cultural programmes). A tendency to focus on the actions of individuals or even collectives with declared aims—making these subjects the prime movers of historical transformation—privileges a model that centres not only individuals, but declarations and manifestoes, while occluding the long-term cultural change needed to overturn exclusionary and violent practices continually activated in the present. These practices are not only products of colonialism, but also enable the productive dynamics of neo-imperialism. The activities characterised as politics are reduced to the governance of the new nation, and in these texts, these activities fall to men.

Djebar and Hopkinson foreground women actors excluded from participating in, organising or historicising trajectories of rebellion and resistance and disconnected from political lineages. These characters play out decolonial strategies in ways that either resist incorporation into heroic, masculinist narratives of anti-colonial history or whose effects fall outside the remit of women's emancipation by the standards of Western liberal feminisms. Rather, they undertake strategies which involve an on-going questioning of what is relevant to their survival and care, reconfiguring what counts as women's politics. The interweaving of strategy and narrative is a site of speculation, in that it departs from the chronological temporality of literary realism: how to imagine motherhood, for example, outside the continuum of oppression and (re)production to which it has been relegated by anti-capitalist feminist critique, while simultaneously acknowledging the way that liberal democratic nations have been complicit in policing and restricting the reproductive rights of non-western women, both domestically and abroad. In these texts, speculative practices among women begin with the necessity of stopping hypermasculine violence, portrayed as a result of unacknowledged post-colonial trauma and embedded in the inherited structures of colonial governance and geography. This initial imperative leads to speculations on new spaces for collective living and opens onto different possibilities for gendering.

Pan-Africanism, Afrofuturism and African Futurism

The time period between the texts' publication dates—1987–2000—was characterised by a diminishment of anti-colonial international solidarity, corresponding to the establishment of neoliberalism proper and the fall of the Soviet Union. This period also saw the reinvigoration of debates around Pan-Africanism, primarily led by diasporic Black intellectuals—W. E. B. Du Bois in the US and Henry Sylvester-Williams in the UK—at the turn of the twentieth century. In the era of decolonisation, the movement envisioned transnational commonality between African peoples—those displaced by the transatlantic slave trade and those who remained in continental Africa, as well as with anti-colonial movements across the world. While Pan-Africanism initially tied the cultural to the political, these two spheres of activity became loosened from each other along the lines of competitive nationalism: a shared pan-African culture would be more complicated than resurrecting or inventing national heritages for the sake of nation-building, as Frantz Fanon had foreseen.[4] The demands of competing on the international stage to become nations among nations undermined Marcus Garvey's Pan-African call in 1921 "to unite these 400 million [African] people toward the one common purpose of bettering their condition."[5] As the twentieth century went on, Pan-Africanism came into conflict with nationalist prerogatives to develop post-colonial African nations as competing players in the world-system.

The difference between decolonisation and decoloniality is significant here: whereas decolonisation involves the realm of public affairs, and takes place in law and government, decoloniality—or decolonial praxis—involves changing the realm of social organisation, or culture. The former involves regime change, the forging of new laws and the international recognition of new political leadership. The latter is based in the everyday practices that move the formerly colonised territory out of the oppressive policies under which its population laboured for hundreds of years, based on the management of gender, sexuality, racialisation and productivity. This might involve the re-inscription of pre-colonial forms of sociality and cultural practice, but also necessitates the creation of new ones.

Parallel to this difference between decolonisation and decoloniality is the difference between Pan-Africanism as a political and ideological movement, and Afrofuturism and African futurism as a series of intertextual cultural suggestions. Pan-Africanism began as a set of political demands, eventually subsumed by African nationalisms, and thereby restricted in the kinds of future visions it can propose in an official, unified capacity. Historically, decolonisation has involved the legal and governmental transformation of the public sphere from colonial to indigenous rulers, but its programmes often do not involve the transformation of social organisation. In this lacuna, Afrofuturist and African futurist texts imagine transformation, and I read this as a form of praxis: decolonial praxis conveys change, but cannot ensure it, and is speculative work. In a discussion of *Ombre sultane*, Kasereka Kavwahirehi argues, "L'écrivain trace un chemin de liberté, mieux de libération" ("The writer traces a path of liberty, better than liberation"), a formulation that attends to the continuous activity of emancipation, distinct from the finite realisation of an independent state.[6] Djebar and Hopkinson speculate on the decolonial relevance of women's practices—their cultivation of new habits, towards different ethical sensibilities—and how these practices might, in Elleke Boehmer's words, "change the subjects that have dominated the nationalist texts," in the process of building African futures.[7] The characters in these texts encounter the future as a wager, speculating on it, rather than trying to guarantee an outcome or fix upon a solution.

It is important to note differences between Afrofuturism and African futurism, as such differences register the distinction between African American and African cultural histories and their different connections to imperialism, the transatlantic slave trade, and colonisation. If Afrofuturism "addresses African American concerns in the context of twentieth century technoculture," African futurism performs a different kind of cultural work, which is the principal focus of my discussion in this chapter.[8] While *Midnight Robber*'s Afrofuturism involves—to cite Hopkinson—"what words a largely African diasporic culture might build" in neo-colonial futurity, Djebar's speculations are based in continental Northern Africa, concerned with reformulations of national sovereignty in post-colonial Algeria, 25 years after the Algerian War of Independence.[9] In reading the texts together, I draw out their neo- and decolonial themes alongside processes of gendering, taking both continental and diasporic African science fiction as a "temporary, flexible, non-monolithic, and, above all, strategic identity," in Mark Bould's words.[10]

If the *bildungsroman* centres on individual (male) development, often initiated by moving its protagonist from rural inexperience to metropolitan cultivation along the lines of class mobility, *Midnight Robber* reverses this journey and jumbles its subjects—place, character, intertext, system and historical facts—into an archipelago of different influences. Incongruously placed, a constellation of historical and cultural references—Maroon resistance, interplanetary travel, digital magic, Pan-African heroes, indigenous bird-folk and folklore characters coming to life—offer ways out of a politics and public sphere that determine (re)productive function and inhibit collective self-determination. *Ombre sultane* imagines a post-colonial Algeria that can make space for women's political decision-making, which would deliver on the promises of the revolution with regard to women's rights. Djebar bends the conventional ghost story, giving voice to a female character in search of a function after giving up the role of wife and mother and expressing the spectrality of this undetermined subject-position in relation to national interests.

Djebar and Hopkinson question the immediacy of political time, showing instead how distant and disjoined strategies can produce alternative ethical sensibilities. Weaving a politics of care into the speculative *bildungsroman* and quasi-ghost story, these texts work at undoing the knots of centralised dependency that keep decolonised nations and peoples both tied to the paradox of underdevelopment and split off from each other. Praxis among women happens in a temporality at odds with modern developmental paradigms. For characters functioning under the sign of "woman" in these texts, the present is an opaque surface that cannot be fully grasped or managed: they must come at it in jolts and at angles, occasionally meeting others—by chance, risking loss—with whom they might gather and collect, for the sake of imagining different futures.

Dependency and Splitting

Both texts direct speculative praxis through their characters' encounters with the prospect of motherhood. Tan-Tan (in *Midnight Robber*) and Hajila (in *Ombre sultane*) conceive children in abusive relationships with men, and both are forced to make decisions about how to deal with the situation. In a discussion of feminist Afrofuturism, Tricia Rose calls for "an investigation of maternal, matriarchal African narratives about the power of reproduction," rather than feminist rewritings of science fiction simply involving giving female characters guns; she urges, instead, "radical feminist models of pregnancy and motherhood" and being open to the possibility of "creating universes of feminist children," rather than denying the power and creativity of reproduction according to a fiction of motherhood created by male needs.[11] Alondra Nelson notes the centrality of motherhood to challenging the patriarchal underpinnings of liberation: "In the zeal for a liberatory detour, it is the mothers who are uncertain."[12] Would it be possible to unbind motherhood from patriarchy and challenge the way that, as Silvia Federici has described, "Capitalism must control the work of reproduction, as it is a central aspect of the process of accumulation"?[13] Neither Hajila nor Tan-Tan desires to have a child, but both are required to deal—practically and painfully—with the reality of pregnancy. This prompts different kinds of speculation to the oblique fantasies generated by neoliberal economics, which are directed towards empty future space-units which need not ever exist or be realised in reality. Speculating on futures with or without children (Tan-Tan has the child, while Hajila does not) involves engaging in practical and difficult ways with what motherhood means, and imagining different frameworks of care which either involve more distributed care responsibilities or which attend to the possibility of choosing not to have a child. In both, these speculations are imbricated in feminisms that are not reducible to women's movements in the Euro-US.

For these characters, motherhood means being tied to obligations of domestic care and experiences of male violence. Eva Feder Kittay's notion of "dependency work" offers a way of describing work that involves caring for others in private spaces—domestic, maternal, non-medical, emotional. For Kittay, this work is sustained by a one-way flow of labour and overwhelmingly imbricates women subjects as its primary administrators. Kittay shows how this kind of work is often, though not always, dependent on the privacy and invisibility of domestic spaces. In this work, gender functions as a principle of labour division. In Michael Fine's words, "gender operates as a social process that brings the 'reproductive bodily distinction' into play, rendering care a sphere of feminine activity."[14] This is allied to an undervaluation of the work of dependency—work that is carried out in response to the vulnerability of another, involving the negation of vulnerability for the person carrying out the work. Theorists of social reproduction feminism (SRF) have addressed the lacunae in Marx's, Althusser's and Bourdieu's uses of social reproduction, to show how gender relations are "always concretely interconnected in the ongoing maintenance and reproduction of an overall capitalist social formation."[15] Dependency work is crucial not only to the reproduction of labour power, but also to the reproduction of *conditions* of labour in which "woman" as sign plays a central role. It is not a criterion of the global development index; its processes, structures and restrictions do not feature explicitly in what constitutes a "developed nation". Nonetheless, capitalism needs dependency work as a central form of unpaid labour to keep production moving.

In the texts, this flow of labour is sustained through male violence, which takes place out of public sight, in the home, while simultaneously being "seen" by the state—in the sense that this violence sustains a system of unpaid domestic labour: in *Ombre sultane*, Hajila is beaten by her husband, who slashes her arm with a broken beer bottle, in the kitchen; in *Midnight Robber*, Tan-Tan is raped, repeatedly, by her father Antonio, in her bedroom. This violence takes place in a context where women are required to supplement lacunae of national or global welfare with taking care of men. In both texts, this is an inherited duty: Hajila has inherited it from Isma, the second wife to leave their husband, and Tan-Tan has inherited it from her mother, Ione. Hajila and Tan-Tan both take care of men who have been abandoned by their previous carers, not maternal, but sexual. They are filling in after Isma and Ione refused to continue. The men excuse their violence this way: as replacements, they are not as valuable. Reddock argues, "Post-colonial masculinities often expressed their subordination to European masculinities by a hyper-masculinity represented in violence and domination over women, as well as against less powerful men."[16] In this structure, the women are not afforded subjectivity, but are interchangeable. They do not exist in the developmental paradigm that characterises male experience, around which modern constructions of masculinity are organised. Instead, they are restricted to an experience of time that unfolds according to their ability to care for, or to reproduce and reproduce for, male subjects in their respective homes. There is no space for female self-determination outside the logic of childhood, adolescence and menopause.

Ombre sultane, set mostly in Algiers, was published three years after the Algerian government passed the Family Code of 1984. The promise of socialist solutions to gender inequality made by Le Front de libération national (FLN) premised a new Algerian woman able to stand alongside her brothers, father, husband and son to fight colonial rule. Yet this new figure was forced to disappear after liberation, required instead to substitute non-existent reparations with a retreat into the family home to provide care, relieving the state of accounting for national unemployment figures on the global development index. This gave men jurisdiction over the domestic sphere, granting male authority over marriage, divorce and child custody. It also spoke to a longer history in post-independent Algerian governance, in which the networks of women freedom fighters central to the success of the Algerian revolution were not institutionalised after liberation. The state forgot the central role women played, instead instituting an economic development programme that failed to differentiate women's needs (and the inheritance of gendered divisions of labour) from men's, assimilating them instead with the national agenda. It also forgot the FLN's invocation of *Ijtihad* (critical thinking), which advocates both freedom of movement and freedom of education for women. The Family Code was a way for the state to deal with an economic crisis stemming in part from dependency on foreign loans, while trying to keep the Muslim Brothers out of power. Removing women from the realm of work and making them the responsibility of their fathers and husbands, rather than the state, would take women out of unemployment statistics and deflect political support for the Muslim Brothers. As Marnia Lazreg has noted, this "flight into Islam" was a way for the government to retain and augment political power at home and economic validity internationally.[17]

The husband in *Ombre sultane* works for the government. This detail links the character directly to national governance, the bearing of the political on the domestic and the global on the national and local. To reproduce itself, the Global North needs to maintain structures of dependency; it needs a Global South. In *Ombre sultane*, this logic of projected dependency also structures the domestic, displaced onto the differentiation between male and female via materially different kinds of labour. The absence of repair in the post-colonial state is expected to be supplemented by feminine care. The novel is structured by the absence of reparation, and not just economic: the absenting of colonial violence from international memory by the failure of the colonial rulers to repair; the failure of women to heal these wounds of war and occupation—their own and those of the men they are charged with caring for; and the absence of public measures against domestic abuse.

Isma narrates the novel, at first standing invisibly in the corner of a city apartment, watching the new wife of her nameless ex-husband. For the first few chapters, it is uncertain whether Isma is dead or alive, and her musings seem to suggest that she is a ghost. Was she murdered? And why is she watching this other woman, calling her the familiar "tu," as if she knows her intimately or as if a child? It emerges that she is not dead, but absent: she left some time ago, the second wife who replaced the first, who also never returned. Isma found Hajila in the slums surrounding the city, illiterate and from a poor family, to be her replacement: "je t'entravais, toi, innocente" ("you, the innocent victim whom I enslaved") (10; 1).[18] The placement of "*toi*" in the centre of the sentence gestures to Hajila's containment, split off from her own innocence (she did not deserve this future) and trapped in someone else's determination of her place. The interface of intimacy/infantilisation stands out here: Hajila is in all respects described as less than Isma. Isma's selection of her replacement is coded as a performance of metropolitan benevolence: despite the abuse, surely anything would be better than living in the slums. Hajila, the Arab-Berber woman, is given a chance of a better life in the francophone city, while in reality is made to do the kinds of dependency work that the educated metropolitan woman refuses.

Kavwahirehi argues that the force of liberation in *Ombre sultane* is bound up in the "power of the colonial language, which has marginalised Arabic, and, following this, the power and logos of masculinity (phallogocentrism) which keeps the Arab-Berber woman in a minority position, like a child."[19] The wives are arranged in degrees of proximity to colonial language: the first wife is French; Isma's father sent her to boarding school in France; Hajila is illiterate, and her marriage is her family's way out of a shantytown. Hajila struggles to make sense of her new world, where her stepchildren speak French—a language she doesn't know—and her husband barely speaks to her. She begins wandering the city, unveiled. When their husband drinks too much beer—beer from the national brewery—he comes back to beat Hajila, calling out Isma's name. He is beating her for leaving; he is beating her because he cannot keep order. The broken beer bottles the husband uses to slash his wife's arm reflect a situation in which the absence of reparations—both economic and affective—produce trauma-related addiction, a symbol of economic liberation from colonial used to inflict violence on women. Hajila is caught between the ghost of an absent wife and her husband's memory of her. In this scene, she realises her position outside any logic of care: "Tu attends stupéfaite, plaquée contre le chambranle. Quelle est cette étrangère qui revient, par sa voix? Tu recules." ("You flatten yourself against the window-frame, waiting, stupefied. Who is this strange woman who reappears, conjured up by his voice? You recoil.") (119; 84, translation modified). Embodying the ghost of another woman, Hajila is stupefied—struck senseless—by her lack of ground in the place assigned to her. There is no space for her. Isma's absence—standing in for the larger absence of national reparation—dominates the room. Neither of them exists. A Cixousian play of writing, absence and presence marks Djebar's quasi-ghost story: Isma is writing herself into presence, to describe her absence—which has now been transmitted to Hajila (already made absent from national politics by the linguistic dominance of French and the second-class status of Arab-Berber subjects)—as well as the bigger absence which structures all three lives.[20]

Midnight Robber also depicts the psychic splitting generated by this forced embodiment of other absent women. Hopkinson's narrative, in contrast, moves through a web of anti-colonial resistance, bulldozing metaphor. If speculative fiction displaces the function of metaphor by twisting its relevance, then Hopkinson achieves this in part by confounding any sense of chronology in decoloniality. Hopkinson uses science fiction and fantasy tropes to subvert the *bildungsroman* form and to show how this form cannot keep up with the leaps Tan-Tan is forced to take across time and space. The two planets on which the narrative takes place are bound to neo-colonial rule: different sites of colonisation imagined as different planets. The difference between Toussaint and New Half-Way Tree—one "civilised," the other "rough"—echoes the uneven development of African and Caribbean nations, whose speed and forms are determined by corporate (dis)interests. Tan-Tan is born on the planet of Toussaint, a planet colonised by the Marryshow Corporation, "impregnate[d] … with the seed of Granny Nanny," the digital communications system that operates through telepathy (2).[21] New Half-Way Tree is a low-tech world where so-called Nation Worlds "exile their undesirables where they were stripped of the sixth sense that was Granny Nanny," meaning that they are disconnected from participation in the digitised (inter)world information system (247). Tan-Tan is from an affluent family, the daughter of the Mayor of Toussaint, Antonio Habib, and Ione, a beautiful woman who does not love her. After Antonio kills his wife's lover in a fight, he flees, taking Tan-Tan with him to New Half-Way Tree, "the mirror planet of Toussaint," "how Toussaint planet did look before," a planet not (yet) transformed into a satellite nation for corporate resource extraction (2). This peripheral place does not yet have a stake in the (inter)global economy, where indigenous communities live in the Bush and negotiate small trade operations with "Tall People." Tan-Tan's movement from the semi-periphery of Toussaint to the periphery of New Half-Way Tree traces the unevenness that has prevented sustained Pan-African unity between diaspora and continent.

In her reading of the novel, Giselle Anatol identifies "the intersection of geography and gender ideology … apparent in the language that Hopkinson uses to describe the colonisation of Toussaint."[22] She argues, "the land is construed as a woman's body, and one is reminded of the initial gendering of the so-called 'New' World territories that stretched before the European imperial gaze starting in the late fifteenth century."[23] Following this, the territorial split between planets invokes temporalities of uneven development, wherein resource-rich parts of the world are exploited and kept in a state of underdevelopment for the sake of developing resource-poor nations. As in *Ombre sultane,* this relation is then displaced onto the exploitation of beings-gendered-female. The land of Toussaint is valuable in a way that the land of New-Half Way Tree is not, and it is here that different kinds of violence begin to emerge. Antonio's murder of his wife's lover Quashee is an act of male rivalry, and revenging the pleasures of his wife that are being denied to him; having found out about their affair, he no longer appreciates "the perfection of his grounds: every tree healthy, every blade of grass green and fat and juicy" (12–13). The anger over the betrayal is laced into a sense of proprietorship and in the way the cultivation of his land—and its demonstration of abundance—corresponds to his status. On Toussaint, Tan-Tan is a precious object, "the beautiful thing he had made, this one daughter, this chocolate girl," described in the vocabulary of a luxury commodity, "with the cocoa-butter skin soft like fowl breast feathers," which he loves to stroke (13).

On New-Half Way Tree, unable to uphold this fiction of ownership and being able to touch anything he wants, he begins seizing this property violently, taking this "beautiful thing" and trying to destroy it. He is not being watched: Granny Nanny's surveillance does not reach here, and he begins assaulting Tan-Tan's body, physically and sexually. She becomes pregnant, twice, the first time getting an abortion with the help of the local shopkeeper's wife (who says nothing), and the second time, unable to because she has sought refuge in the Bush after killing Antonio. Hopkinson avoids pathologising Antonio; the figuration of Tan-Tan's body through the diminution of his territorial assets is rendered as a structural correspondence, shown in how Tan-Tan's body loses value (to him) according to her proximity to colonial/corporate spoils. Tan-Tan's adolescence is erased/replaced by her father's need to use her body to erase/replace her mother's absence, and her body is made into a conduit for his loss, no longer a luxury item but a necessity, a tool, Antonio's instrument of self-preservation through which he can continue to establish his personhood through property. Experiencing the "special thing" that Antonio wants to show her for the first time, "something more horrible than she'd ever dreamt possible," she feels them both split in two: Daddy is now two daddies, and "she felt her own self split trying to accommodate them both" (140). Good Antonio smiles at Good Tan-Tan, while Bad Antonio hurts Bad Tan-Tan's bad body. Tan-Tan is forced to split herself to accommodate the pain of unsolicited sexual violence, and the possibility of a third space—a place of interpretation that would allow her to imagine her own place—is annihilated. Her reaction to Antonio leaving her alone after her first abortion is registered by Bad Tan-Tan as a loss "because he stopped loving her," and accompanied by Good Tan-Tan's fear that he will come back (148). When he says, "I never want to hurt you," she is "confused and angry"; she cannot reconcile the contradictory messages of pain and love (149). She is unable to find a timeline on which to live out her childhood and adolescence and come into adulthood, but is forcibly moved between these phases according to her father's needs.

The exchange of daughter for mother is characteristic of the kind of substitutions involved in dependency work. In the following section, I'll explore how both Djebar and Hopkinson counter these substitutions by staking out a place for these characters in speculative temporalities, wagering the present on futures that may or may not arrive. This writing of time allows the recognition of the kinds of splitting discussed above: the segregation of women from each other in public, and through this, the prevention of women's collectivity; the splitting of self to accommodate the cognitive dissonance and trauma of domestic abuse; and the splitting of women from their own labours of care.

Care as Consensus: Split Collectives

In these texts, care is not figured as giving sanctuary or as rehabilitation. Sanctuary and rehabilitation imply direct action in response to events that have already happened, and that may still be ongoing. Rather, care happens *around* these events, before, during, after; it is constituted by pre-emptive listening, an attending to a situation, in the knowledge of the structural conditioning of violence. This happening of care can be described as a "split collective," signifying collectivity without an immediately identifiable identity, not restricted to forms of direct communication, and arising out of the need to cross gaps in space and time generated by the division of peoples by colonial bordering. A split collective is not based on shared presence in space and time; it enables a form of politics to become visible that negotiates the exclusions of those restricted from appearing and speaking in public.

Both narratives reject the institution of a modern, public political sphere as a place for working out decoloniality; rather, they explore private or semi-public spaces as sites in which this care is needed, which declarations and manifestoes—public proclamations—can only go so far in addressing. In *Ombre sultane,* Isma's spectral narration is crucial, but unverifiable: it is evidence of events witnessed only by the walls of the family home, which cannot be proved. She cannot provide sanctuary because she is not really present, but can only attend to the process of hypermasculine violence unfolding and, somehow, try to interrupt it. In *Midnight Robber*, this listening is passed from Granny Nanny—a networked version of the Queen of the Maroons, who provided sanctuary to self-emancipating slaves in Jamaica in the eighteenth century—to Tan-Tan, who adapts a masque, the Robber Queen, to care for strangers, based on her own experiences: "Me tell you, don't hurt your son no more. Me will know. Me, Tan-Tan, the Robber Queen" (246). Myths and ghosts listen to the present, waiting for an opportunity to materialise, a movement that will shift mythology to ontology. This is an uncertain politics, which cannot make claims to realist participatory decision-making; nonetheless, it works upon what Sylvia Wynter, following Larry Neal, calls "a post-Western aesthetics based on a new system of ideas" to abolish a globalised value-system whose reproduction (re)produces violating and violated subjects in the periphery and semi-periphery.[24] This makes space for a different kind of consensus that does not have to involve character-to-character negotiations, but which is derived from the situation at hand, often at a distance.

This mode of care is not dependency work. Drawing on Kittay, María Puig de la Bellacasa suggests an understanding of care as "disruptive thought": "care is not about fusion; it can be about the right distance."[25] Care can be an ambivalent, speculative venture, "vital in interweaving a web of life, expressing a key theme in feminist ethics, an emphasis on interconnection and interdependency," as distinguished from dependency cultivated by modern industrialised societies.[26] Distance is key to the characterisation of collectivity in these texts: characters do not have to be physically or even temporally proximate to care, and for that care to have an effect in future worlds. The work of care does not have to be causal or generative of radical change. Rather, care in these texts emerges as, following Bellacasa, "the concrete work of maintenance, with ethical and affective implications, and as a vital politics in interdependent worlds."[27] Maintenance is carried out in the hope of something more worldly than global, not restricted to proving existence on a global stage.

Ombre sultane is based on a globally circulated myth that became interchangeable with eighteenth- and nineteenth-century European imaginaries of the East, "a western commodity appropriated for orientalist, critical and artistic purposes ever since its first European translation into French"[28]: Scheherazade, who told 1001 tales to save her life from the Sultan's vengeance of another woman's betrayal, and the lives of the women that would come after her, by drawing her younger sister Dunyazad into the bedroom to elicit stories, which she leaves deliberately unfinished. A temporary collective subverts the Sultan's attempt to isolate her from help. Yomi Olusegun-Joseph highlights how Djebar's re-writing of this myth acknowledges the onus the narrative places on both Scheherazade and Dunyazad to remain alive and to stay awake: "what if the sister had relaxed her guard?"[29] Djebar instead centres the collective experience of bodily vulnerability as a condition of self-determination. Woman's deliverance in the novel "lies in the celebration of her own body, and the motivation for this celebration much come through a determined drive that would liberate her from object to subject."[30] This shared experience makes possible a "hybrid effort" of Western and Arab-Islamic feminisms, rather than the former shaping the terms of the latter.[31] This resists the humanitarian benevolence of third-wave Euro-US feminism, in which female emancipation is figured as individual social mobility in male-dominated spheres, whose framework largely wants to save women of colour and women outside Europe and has excluded the particularity of their concerns and interests, along with transsexual and non-binary critiques feminisms. This is especially important given that this wave of feminism was already being incorporated into Euro-US foreign policy as a criterion of civilisation and development, used "to shore up the image of imperialist justice," as Dean Spade and Craig Willse have argued.[32] Instead, what Olusegun-Joseph calls "an Algerian woman's brand of feminism" across socio-economic boundaries becomes relevant.[33]

Djebar plays with the spectrality of the Algerian woman who disappeared after national liberation, making both Isma and Hajila into ghosts. "Une ombre-sœur? Les sœurs n'existent-elles que dans les prisons?" Isma asks. ("A phantom-sister? Do sisters only exist in prisons?") (116; 82, translation modified). This shared status as ghost breaks the narrative out of its assumed genre when we realise Isma is still alive, transforming the mythology of un-dead life into a description of ontological insecurity. It is only by haunting and being haunted by each other that they begin to reach consensus and share agency. They meet only twice in person, after the husband has begun to sexually and physically abuse Hajila. Isma tells Hajila to meet her at the *hammam*, where she gives her a spare key to the family apartment, so that she still has a way out. The *hammam* is not a Western space, but again—like Scheherazade—a heavily orientalised and unilateral image of an Ottoman "East," connected to female sexuality and the Sultan's *harem*; Djebar claims this space back from myth, and reinscribes it as a site of collectivity, out of public view:

> Chaque nuit, le bain maure, qui sert de dortoir aux ruraux de passage, devient un harem inversé, perméable—comme si, dans la dissolution des sueurs, des odeurs, des peaux mortes, cette prison liquide devenait lieu de renaissance nocturne. De transfusion. Là s'effectuent les passages de symbole, là jaillissent les éclairs de connivence; et leurs frôlements tremblés. (198)
>
> (Every night the Turkish bath serves as a dormitory for country-folk in transit and so becomes a harem in reverse, accessible to all—as if, in the melting-pot of sweat, odours, and dead skin, this liquid prison becomes a place of nocturnal rebirth. And of transfusion. Here, women can communicate by signs; here, a split-second glance, a barely perceptible touch, will seal their secret collusion.) (148)

The *hammam* is constructed as a place for the merging of female experience, a site of gathering between women, where they can move freely. This is a place for shedding skins, where transfusion into other shapes and forms can occur. Verbal communication no longer creates a barrier to the interplay of signs between subjects. Djebar calls the language of the body "the fourth language, for all females," which prevents the complete incarceration of "woman" into patriarchy, so that the body can "seek some unknown shore as destination for its messages of love."[34] This is a place where dead and dying matter can be swept away and taken in by other living bodies. This is not a place for manifestos and deal-making, but less direct forms of exchange that circumvent the implementation of solutions. It facilitates the transfer of a key to the apartment, so that Hajila might make another escape: "Que tu gardes cet enfant dans ton ventre ou que tu le rejettes, c'est à toi d'en decider!" ("It's up to you to decide whether you keep the child you're carrying, or whether you get rid of it") (204; 153). "Garder" is a verb of protection, but also of bordering: to protect the infant or to protect herself against it. "Garder" becomes a verb of embodied resistance, expressing Hajila's agency and choice.

In this moment, care is the articulation of distance; body is separated from Hajila's duty as woman. Suspended in this distance afforded by the bathing-house meeting, Hajila can become a subject. The transmission of a language, *in* language, of a border not previously acknowledged also pronounces the splitting of labour from care, while acknowledging the relevance of this splitting. The fourth language of the body comes in to relieve the dissonance of bodies split off from subjective experience. This is a kind of politics and an act of solidarity, which emerges out of the absence of any holistic and assured sense of what "woman" has been made to mean. It is a moment of consensus in which a commonality is laid out, temporarily, and (almost) anonymously: the body is a border that registers what passes through, speaks its own language and has a place of its own. Djebar subverts the appropriation of Arab-Islamic tropes into Western imaginaries, which in this case would reduce this body to an object in need of saving and direct intervention. The *hammam* is crucial as a site of liberation, tied to pre-colonial history, and a space in which decolonised imaginary of embodied experience emerges.

This embodiment depends on enacting myths. In *Midnight Robber*, the mythologies held in the Granny Nanny network—activated as virtual representations in the minds of Toussaint residents—take physical form on New Half-Way Tree. The text invokes a network of African, African-Caribbean and African-American references, shaping its lifeworlds around these aesthetics. The refusal throughout *Midnight Robber* to settle on a specific, discrete and stable location, constructed through a restricted set of cultural referents and fixed to a singular heritage, defers the arrival of a decolonised universalism. The text's archipelago form subverts the expectation that narratives offer some kind of situated coherence. Tricksters abound, and characters from folktales live and shape the present, while on Toussaint, they are usually hidden in Granny Nanny's web as "information" to be called up retrospectively, as a remedy: Anansi the spider and the tales around him; Dry Bone the old man of death, who tries to keep Tan-Tan bound to him; the rolling calf with "eyes of red flame," which would chase anyone it catches outside late at night "where you had no business to be … till you dropped dead of fright and exhaustion" (229). It's here that Tan-Tan becomes the Robber Queen, an identity adapted from the Robber King, which helps her take on various local battles, and eventually, to escape a sentence for killing her own father. These references are not simply ways of coping with the present; they become ontological conditions for life on New Half-Way Tree. The move there from Toussaint involves suspending the chronological distinction between pre-modern and modern lifeworlds; here, they co-exist, anachronism as ontology, rather than diagnosis. Moreover, this ontology is not essence, but performance—it is creative and spontaneous, foregrounding process.

Care does not require establishing a coherent narrative of cause and effect, and consensus is not always transparent or easily translatable. Tan-Tan's early interactions with the douen community (the indigenous bird folk who live in the Bush of New Half-Way Tree, whom she perceives as primitive and strange) illuminate the difficulties of articulating a common agenda across socio-economic divisions of modernisation. When she arrives among them, on the run, she is critical of their anti-modern ways and strange eating habits, and cannot understand them. They, in turn, are suspicious and critical of her. This comes to a head in her confrontation with the douen chief's daughter, Abitefa. When Abitefa plays a trick on Tan-Tan, enjoying Tan-Tan's lack of expertise in this world, Tan-Tan snaps. She "leaped at Abitefa, dragging her to the ground," and "boxed her in her ugly mouth"; in response, Abitefa "bit Tan-Tan's hand" (227). It is a brief moment of violence, but does not carry the annihilating weight of her final, deadly encounter with Antonio. This is a struggle to reach consensus on a mode of living. Abitefa's trick is a cultural declaration, the conjuring the spirit of Anansi. While Tan-Tan knows the stories, she has not been forced to live through them. This tussle over living is a way of reaching consensus on how they will survive. This wager is not articulated through the racialised typology of coloniser and colonised, but through a resistance to the obliteration of indigenous social organisation in the name of development and progress. Tan-Tan and Abitefa have not been able to communicate through the universal language of the Nation-Worlds, but their own confrontation allows for a new understanding to emerge. After their joint banishment from the douen community because of the threat Tan-Tan poses, and Abitefa's perceived collusion with her, they survive together with little food in a hostile Bush. No one comes to save them. Crucially, Tan-Tan has learned that she can defend herself without this being fatal or destructive to others. The violence between them is structural (produced by miscommunications arising from cultural difference), but it is also overcome. They learn to trust each other, and the battle over dominance ceases.

The Robber Queen, Dry Bone and Anansi are not always happy characters; frequently, they are violent and fearsome. But they help demonstrate ways of relating to and resisting with others in a world still structured by colonial violence. From Abitefa, Tan-Tan learns the life-saving power of play and trickery. It's through Abitefa that she starts adopting the tactics of the Robber Queen, preventing violence, standing up for oppressed children and trying to break the cycle of violence. She also realises how her own performance might be distributed among women and girls as a form of resistance: "She began to notice little girls playing at Robber Queen in the settlements" (288). Myths are active, but not essential: they are part of the performance, enabling the putting-on and taking-off of guises. When Tan-Tan has her final confrontation with Janisette, her stepmother, who has been tracking her to avenge Antonio's death, she takes on this guise to defend her right to live, speaking in the third-person, on behalf of herself. Speaking to a gathered crowd in the town square, and facing the possibility of her own execution, she recites her testimony,

> *The plan for love never come to transaction.*
> *When Antonio find out, he rape she, beat she, nearly kill she.*
> *Lying under he pounding body she see the knife.*
> *And for she life she grab it and perform an execution.*
> *She kill she daddy dead. The guilt came down 'pon she head,*
> *The Robber Queen get born that day, out of excruciation.* (325)

The split collective here is Tan-Tan's consensus with her own split self. She has decided she has a right to live, and that she is worthy of defence. This is the closest she comes to a public manifesto: a story of her own rebirth through violence, this new existence explicitly tied to the attempted destruction of her body. This is also a retroactive realisation of how, killing her father, she reached consensus over what needed to be done. Anatol argues that this ending complicates—rather than resolves—not only "patriarchal, masculinist ideologies," but also "Eurocentric visions of feminism that discount cultural and historical codes, and second-wave feminist critiques that view maternity as a hindrance to woman's full potential."[35] Tan-Tan does not use this freedom to move back into the human world, but stays in the Bush, where she chooses to raise Tubman, her son, with Abitefa and Melonhead, a new "plan for love" emerging in the decision that she should keep existing. Her birth "out of excruciation" stands in here for a wider narrative, but her ability to give herself this narrative marks a possibility for a longer strategy of survival.

Concluding Remarks

I've argued that these texts are Afrofuturist and African futurist speculations about the anachronisms of decolonial liberation through the optic of gendering, which show their women protagonists wagering their futures in a present that cannot or refuses to guarantee their survival and self-determination. Djebar and Hopkinson show that working towards a robust and self-sufficient politics of care between women means placing relations outside the developmental paradigm—either by subverting genre from within or by attending to the spectral presence of certain subjects and thereby departing from the model of fraternal alliance that characterises participatory democracy in the West. They keep care open-ended, refusing its abstraction into a fixed set of practices. Through this, they create porous spaces with the potential for other kinds of influence, not restricted to the imperative to reproduce a national inheritance. Saying this, neither give up the problem of territory in the context of decolonial world-building: the problem of nation—and its attending exclusions—is left unresolved. The unforeseen accident, the chance meeting, the door left unlocked: these are the figures through which women's collectivity takes place. Distance is key: these are always wagers for an uncertain future, necessitating a leap into possible annihilation. Practices of decoloniality are shared between characters (person, nonhuman, eco-system, network), invoking ethical sensibilities for cultivating long-term, habit-forming and habit-remembering change. While care has been historically bound to feminised labour, it functions in these texts to motivate strategies of consensus, moving away from the model of dependency/supplementation imposed by neo-colonial globalism on decolonial nations, which keeps gendered violence activated and relevant. Suspending both the obligation of dependency work and the dominance of the nation-as-family, these texts explore what might be done with the energy left over: what new futures become possible when women do not have to do this work, and what new forms of sociality—freed from the violence of gendering—might emerge?

Notes

1. Rodney, *How Europe Underdeveloped Africa*, 30.
2. Choksey, "Decolonial."
3. Campbell, "Pan Africanism," 286; Reddock, "Gender Equality," 257.
4. See Fanon, "On National Culture," 166–199.
5. Garvey, "If You Believe."
6. Kavwahirehi, "*Ombre sultane*," 52. My translation.
7. Boehmer, *Stories of Women*, 94.
8. Dery, "Black to the Future," 180. As Alondra Nelson points out, while Dery coined "Afrofuturism" in 1993, its currents and tropes date back to post-Reconstruction speculative fiction by Charles Chesnutt and W. E. B. DuBois. See Nelson, "Introduction," 14.
9. Aylott, "Filling."
10. Bould, "African Science Fiction," 11.
11. Dery, "Black to the Future," 218, 221.
12. Nelson, "Feminist Critique," 1.
13. Čakardić and Souvlis, "Feminism."
14. Fine, "Dependency Work," 149.
15. Ferguson, "Social Reproduction."
16. Reddock, "Gender Equality," 256.
17. Lazreg, "Gender," 767.
18. References to *Ombre sultane* and *A Sister to Scheherazade* will be cited with in-text page numbers for the French original and the English translation, respectively. Djebar, *Ombre sultane*; Djebar, *Sister to Scheherazade*.
19. Kavwahirehi, "*Ombre sultane*," 55. My translation.
20. Cixous, "Coming to Writing."
21. References to *Midnight Robber* will be cited with in-text page numbers. Hopkinson, *Midnight Robber*.
22. Anatol, "Maternal Discourses," 112.
23. Anatol, "Maternal Discourses," 112.

24. Wynter, "On How We Mistook," 114.
25. Puig de la Bellacasa, *Matters of Care*, 1, 6.
26. Puig de la Bellacasa, *Matters of Care*, 5.
27. Puig de la Bellacasa, *Matters of Care*, 5.
28. Olusegun-Joseph, "Different," 227.
29. Olusegun-Joseph, "Different," 143.
30. Olusegun-Joseph, "Different," 229.
31. Olusegun-Joseph, "Different," 229.
32. Spade and Willse, "Sex," 24.
33. Olusegun-Joseph, "Different," 232.
34. Olusegun-Joseph, "Different," 234.
35. Anatol, "Maternal Discourses," 115.

Works Cited

Anatol, Giselle. "Maternal Discourses in Nalo Hopkinson's *Midnight Robber*." *African American Review* 40, no. 1 (2006): 111–124.

Aylott, Chris. "Filling the Sky with Islands: An Interview with Nalo Hopkinson." SPACE.com. Jan 11, 2000. www.space.com/sciencefiction/books/hopkinson_intv_000110.html.

Boehmer, Elleke. *Stories of Women: Gender and Narrative in the Postcolonial Nation*. Manchester: Manchester University Press, 2005.

Bould, Mark. "African Science Fiction 101." *SFRA Review* 311 (2015): 11–18.

Čakardić, Ankica, and Souvlis, George. "Feminism and Social Reproduction: An Interview with Silvia Federici." *Salvage*. January 19, 2017. http://salvage.zone/online-exclusive/feminism-and-social-reproduction-an-interview-with-silvia-federici/.

Campbell, Horace. "Pan Africanism and African Liberation." In *Imagining Home: Class, Culture and Nationalism in the African Diaspora*, edited by S. J. Lemelle and R. D. G. Kelley, 285–307. London: Verso, 1994.

Choksey, Lara. "Decolonial Speculative Fiction and Fantasy." *Global Social Theory*. October 2018. https://globalsocialtheory.org/topics/decolonial-speculative-fiction-and-fantasy/.

Cixous, Hélène. "Coming to Writing." In *Coming to Writing and Other Essays*, edited by Deborah Jenson, translated by Sarah Cornell, Deborah Jenson, Ann Liddle and Susan Sellars, 1–58. Cambridge: Harvard University Press, 1992 [1986].

Dery, Mark. "Black to the Future: Interviews with Samuel R. Delany, Greg Tate, and Tricia Rose," *Flame Wars: The Discourse of Cyberculture*, edited by Mark Dery. Durham, NC: Duke University Press, 1994.

Djebar, Assia. *Ombre sultane*. Paris: Albin Michel, 2006 [1987].

———. *A Sister to Scheherazade*, translated by Dorothy S. Blair. Portsmouth, NH: Heinemann, 1993.

Fanon, Frantz. "On National Culture." *The Wretched of the Earth*, trans. Constance Farrington, 166–199. London: Penguin 2001 [1961].

Ferguson, Susan. "Social Reproduction: What's the Big Idea?" *Pluto Books*, 2017. Accessed November 29, 2018. https://www.plutobooks.com/blog/social-reproduction-theory-ferguson/.

Fine, Michael. "Dependency Work: A Critical Exploration of Kittay's Perspective on Care as a Relationship of Power," *Health Sociology Review* 14, no. 2 (2005), 146–160.

Garvey, Marcus. "If You Believe the Negro Has a Soul." History Matters. Audio, recorded 1921. Accessed May 5, 2018. http://historymatters.gmu.edu/d/5124.

Hopkinson, Nalo. *Midnight Robber*. New York: Warner, 2000.

Kavwahirehi, Kesereka. "*Ombre sultane* d'Assia Djebar et les 'Forces de la littérature'." *Études Littéraires: Algérie à plus d'une langue* 33, no. 3 (2001): 51–64.

Lazreg, Marnia. "Gender and Politics in Algeria: Unraveling the Religious Paradigm." *Signs* 15, no. 4 (1990): 755–780.

Nelson, Alondra. "A Feminist Critique of Afrofuturism." *University of Bayreuth*. 2003. Accessed May 5, 2018. http://www.bayreuth-academy.uni-bayreuth.de/resources/ANelson-_Bayreuth-Proposal_.pdf.

———. "Introduction: Future Texts." *Afrofuturism: Social Text* 20, no. 2 (2002): 1–16.

Olusegun-Joseph, Yomi. "Different from Her Sister's Voice: (Re)configured Womanhood in Assia Djebar's *A Sister to Scheherazade*." *Journal of Postcolonial Writing* 54, no. 2 (2018): 226–238.

Puig de la Bellacasa, María. *Matters of Care: Speculative Ethics in More than Human Worlds*. Minneapolis: University of Minnesota Press, 2017.

Reddock, Rhoda. "Gender Equality, Pan-Africanism and the Diaspora." *International Journal of African Renaissance Studies—Multi-, Inter- and Transdisciplinarity* 2, no. 2 (2007): 255–267.

Rodney, Walter. *How Europe Underdeveloped Africa*. Dar-es-Salaam: Bogle-L'Ouverture Publications, London and Tanzanian Publishing House, 1973.

Spade, Dean, and Craig Willse. "Sex, Gender, and War in an Age of Multicultural Imperialism." *QED: A Journal in GLBTQ Worldmaking* 1, no. 1 (2014): 5–29.

Wynter, Sylvia. "On How We Mistook the Map for the Territory, and Reimprisoned Ourselves in Our Unbearable Wrongness of Being, of *Desêtre*: Black Studies Toward the Human Project." In *Not Only the Master's Tools: African-American Studies in Theory and Practice*, edited by Lewis R. Gordon and Jane Anna Gordon, 107–118. Boulder, CO: Paradigm Publishers, 2006.

PART IV

Ethics and Global Politics

CHAPTER 11

Rewriting France's Future: From Louis-Sébastien Mercier's Pre-Revolutionary Projections to Michel Houellebecq's Islamic Agendas via Secular State Ethics

Jacqueline Dutton

Introduction

Futuristic fiction found its niche when Louis-Sébastien Mercier's premonitory novel *L'An 2440: Rêve s'il en fut jamais* (*Memoirs of the Year Two Thousand Five Hundred*) (1771) became a worldwide bestseller in the late eighteenth century.[1] Published in an era of royalist repression, this subversive critique of the French monarchy was initially banned for attempting to rewrite France's future in a more positive, liberal, revolutionary light. As the first example of a "uchronia"—a utopia set in a known place transformed by future time[2]—Mercier's text exposed the contemporary malaise of French society and proposed an alternative projection of Paris, ameliorated through democratic institutions, economic cooperation and religious recognition. It was a positive vision for France's future, heralding

J. Dutton (✉)
University of Melbourne, Melbourne, VIC, Australia

Z. Kendal et al. (eds.), *Ethical Futures and Global Science Fiction*, Studies in Global Science Fiction,
https://doi.org/10.1007/978-3-030-27893-9_11

the transformational acts of the 1789 Revolution, the Declaration of Human Rights, the abolition of slavery, and many other enlightened ideals.

Since Mercier's foundational work, futuristic projection has become an enabling literary device for science fiction and many of its avatars—from uchronias to alternative history futures to cli-fi. Despite the positive beginnings of this generic trend, futuristic fiction was very quickly transformed into a negative paradigm, paving the way for dystopian predictions in many cultural traditions. Although the most-cited examples of early dystopias may be Russian—Yevgeny Zamyatin's *Мы* (*We*) (1920)—or British—H.G. Wells's *The Time Machine* (1895), Aldous Huxley's *Brave New World* (1932), George Orwell's *Nineteen Eighty-Four* (1949)—French futuristic fiction demonstrated a notable declinism that began in the immediate wake of the Revolution. In particular, the city of light, as Paris was known in the nineteenth century, was horribly disfigured through literary and filmic futuristic projections to become a mostly nightmarish city of dark. In 2015, this tendency was ironically revised by Michel Houellebecq, with another bestselling futuristic novel, *Soumission* (*Submission*), set in Paris in 2022.[3] Like his eighteenth-century predecessor, Houellebecq critiques contemporary French society and presents an alternative projection of an Islamicised Paris that appears positive for the protagonist but may of course be interpreted as negative by (many) others.

How and why did French futuristic fiction move from Louis-Sébastien Mercier's positive pre-revolutionary projections to descend into dystopia throughout the nineteenth and twentieth centuries and end up in an ambiguous Islamic agenda in the twenty-first century? This study will seek to provide an explanation for this phenomenon, if not a definitive answer to the question, through reference to France's stance on secularism. Beginning with a deeper analysis of Mercier's novel, followed by an overview of speculative narratives in French from the nineteenth and twentieth centuries, it will demonstrate the declinist tendencies of French futuristic fiction set in Paris. The second section of the study will trace the parallel development of secular ethics propounded via the French Revolution and its aftermath to question whether secularism can actually accommodate positive utopian futures for France. By considering, in the final section, the greater challenges to French secularism witnessed in twenty-first-century politics and society, and the religious agendas that dominate more recent French novels, it may be possible to discern a new discourse on religion in futuristic fiction that, despite its dystopian overtones, could reorient writing on France's future.

Louis-Sébastien Mercier's Pre-revolutionary Projections

In the decades leading up to the French Revolution, Mercier moved between literary circles that included leading enlightenment philosophers, such as Rousseau, Diderot, Crébillon fils and Condorcet, as well as more popular social commentators like the gastronome and theatre critic Grimod de la Reynière and renowned erotic writer and utopian satirist Rétif de la Bretonne.[4] Both groups exercised their influences on the young Mercier's thinking and writing, who experimented with heroic poetry and academic debates, and gained some notoriety with a Rousseauist novel titled *L'Homme sauvage* (1767) (The Wild Man). This story of North American Indians whose Arcadian idyll is shattered by Spanish conquerors earnt him the unflattering title of "le singe de Jean-Jacques" (Jean-Jacques's Ape).[5] His subsequent plays—around 60 in total—and treatises on the theatre were not particularly memorable; several plays were banned for anticlerical, antimilitarist and antimonarchical themes and not performed until after the Revolution. The anonymous publication of *L'An 2440* in 1771 in Amsterdam made Mercier an immediate European best-seller, with around 18,000 copies in circulation by the end of 1772—copies that were unauthorised, translated, banned and smuggled past French and Spanish authorities.[6] New expanded editions in 1786, then 1798, added to the novel's renown and distribution, as did the publication (and banning) of his 12-volume critical ethnographic study of his home city in the last years of the monarchic rule, *Le Tableau de Paris* (Portrait of Paris) (1781–1788). After a period of favour with the zealous Jacobins, including friendships with Robespierre, Marat and Camille Desmoulins, and election to deputy for the Department of Seine-et-Oise in the National Convention in 1792, Mercier turned to the more moderate Girondins, voted against the execution of Louis XVI, was imprisoned and almost lost his head—saved only by Robespierre's own demise.[7] His final tribute to Paris demonstrated a more jaded, satirical view on the city left behind by the Revolution in *Le Nouveau Paris* (The New Paris) (1797).

Mercier's groundbreaking uchronia, *L'An 2440*, read by American Presidents George Washington and Thomas Jefferson as well as around 63,000 other contemporaries during his lifetime, was clearly his passport to fame.[8] In this text, he expressed his anticlerical, antimilitarist and antimonarchical stance by presenting a positive vision of future Paris, while criticising the contemporary church, armed forces and royal reign. Enrico

Rufi argues that Mercier's writings and readings place him as a pioneer of secular thought in French life and politics, but in this early novel, secularism is not yet suggested and religion still has an important role in improving the quality of life and society in the year 2440.[9]

To summarise the plot, the original 1771 text begins with a conversation between the narrator and an Englishman about the deficiencies of French society in the eighteenth century. Upon falling asleep that night, the narrator's dreams transport him to Paris in the year 2440 where his 700-year-old self (the same age that Mercier would have been) discovers the remarkable changes that have improved Paris beyond belief. With a philosopher-guide, he tours the city, witnessing positive progress in the political, educational, legal, literary, social and religious systems of the time. Forty-four chapters recount in detail various features ranging from government to the contents of libraries—Rousseau, Diderot and Voltaire, care of the sick, food and foodways, the Prince's role as Publican of free dining houses, art, women, home life and writing. The journey ends in the ruins of Versailles, when the narrator meets a weeping, contrite Louis XIV, who has been brought back to life by divine justice to contemplate his "deplorable work," before a snake bites the dreamer on the neck and wakes him up.

It is not until the fifteenth chapter titled "Théologie et Jurisprudence" ("Theology and Jurisprudence") that religion is alluded to for the first time, as an object of study and discourse. The narrator declares: "Heureux mortels: vous n'avez donc plus de théologiens" (85) ("Happy mortals! You have then no theologians among you?") (65). The philosopher-guide explains that there is no need to dispute the nature or attributes of the Supreme Being, as their only desire is to praise and adore him in silence. A few chapters later, the absence of religious orders is likewise rationalised as an unnatural cluster of people supported by others, whose lives would be better spent in society than in sacred prisons (101–104) (86–90). Even the pope has been dethroned, and Italy is enlightened, like France: "Ce royaume tient aujourd'hui son rang et porte une physionomie vive et parlante, après avoir été emmaillotée pendant plus de dix-sept siècles dans les haillons ridicules et superstitieux qui lui coupaient la parole et lui gênaient la respiration" (104) ("That kingdom now holds its proper rank, bears a lively and expressive aspect, after having been wrapped up, for more than seventeen centuries, in ridiculous and superstitious rags, which stopped its breath, and deprived it of all power of utterance") (92). These clerical reforms were peaceful and happy, the work of philosophy, and a great success: religion has remained intact and "without doubt" (92). The most holy are not publicly revered as saints, but humbly help those in need, with labour or other assistance. The Temple of God is unadorned, but a priest burns incense, hymns are sung, a pastor preaches, silent meditation is observed, at the same hour of the morning every day. The narrator observes that future religious practice seems aligned with the original faith of Enoch, Elias and Adam (104). This chapter on "Le Temple" is one of the longest in the novel (109–117) (97–110), and the following chapters on the "Le Prélat" (The Prelate) and "Communion des deux infinis" (The Communion of the Two Infinities) continue to reinforce the ideals of a natural religion that remains relevant to the people.

Mercier's positive vision of a future Paris projects the demise of a tyrannical monarchy, the abolition of slavery, the destruction of the Bastille and the dismantling of the Catholic church—so it is perhaps no wonder that he claimed credit for pre-empting the Revolution or even fuelling its fires.[10] While the novel does offer a portrait of progress based on reason, science and knowledge, rather than a millenarian prediction, it does not present a society founded on secular ethics. Clerical power structures are dismantled, perhaps foreshadowing a turn towards secularism, but deist religion remains part of the pleasure of everyday life in 2440 and serves to unite the community in worship of the Supreme Being. It is significant that the first uchronia frames a future ideal France as "eutopia" (positive utopia) in its pre-revolutionary projection, that is, before the advent of secularism as a state-sanctioned ethical system. However, the positive uchronia is quickly undermined by dystopian examples of future France in the nineteenth and twentieth centuries, and Paris is particularly ravaged by time.

City of Light: City of Dark

The origins of Paris as *ville lumière* (city of light) have been variously attributed to seventeenth-century police chief Gilbert Nicolas de la Reynie who ordered streets to be lit with lanterns, flames and candles in response to Louis XIV's directive to clear up crime; eighteenth-century Enlightenment philosophers and scholars who gathered in Paris salons and cafés during the *Siècle des Lumières*; and nineteenth-century gas-lighting invented by Philippe Le Bon, which illuminated Baron Haussman's redesigned Paris with 56,000 street lights (1853–1870). However, throughout this time, the city of light certainly had its shadows. As David Garrioch confirms, pre-revolutionary Paris described by Rousseau in *La Nouvelle Héloise* and Mercier in *Le Tableau de Paris* was a city of extremes—wealth and poverty, beauty and filth.[11] The Revolution of 1789 brought new challenges to the rapidly growing urban centre, making it both a scene of Terror in the 1790s and a site for renewal in the nineteenth century.

During this period, utopian socialists such as Henri de Saint-Simon, Charles Fourier, Etienne Cabet and Victor Considerant proposed positive reforms for practical social experiments, sometimes couched in religious terms, though faith was more anchored in science and progress than Catholicism. Auguste Comte's positivism and "Religion of Humanity" maintained a spiritual emphasis despite the erosion of the role of organised religion in French political life. Utopian literature ranged from Gérard de Nerval's esoteric portraits to Victor Hugo's anticlerical moral tales to George Sand's rustic novels, expressing less optimistic hopes for post-revolutionary France than the writings of their socio-politically oriented counterparts.

Nineteenth-century French futuristic fiction was even more pessimistic, from Jean-Baptiste de Grainville's *Le Dernier homme* (*The Last Man*) (1805) to Camille Flammarion's *La Fin du monde* (*Omega: The Last Days of the World*) (1894). It is unsurprising then that uchronias written during this time were dystopian, including Emile Souvestre's *Le Monde tel qu'il sera* (*The World as It Shall Be*) (1846), which inverts the power paradigm in the year 3000, placing Polynesia at the centre and France/Europe at the periphery, and Charles Renouvier's *Uchronie: l'utopie dans l'histoire* (Uchronia: Utopia in History) (1876), with his alternative history future that actually coined the term "uchronia." Focusing on two examples of uchronias set in Paris, it is possible to determine the negative connotations that the city—and by extension France—manifests in the minds of futurists.

Jules Verne's *Paris au XXe siècle* (*Paris in the Twentieth Century*) (1863), published posthumously in 1994, shows how life in Paris might be lived around 100 years hence, and is more "romanesque" than Mercier's time travelling journey.[12] His descriptions of the technological transformation of 1960s Paris into a city of electricity, science and engineering are framed by a narrative that follows an alienated young poet—Michel—to his eventual demise and death. In this future Paris, industrialisation and commercial exploitation dominate, personified by the protagonist's banker uncle and his successful "eminently practical" family (45). Education has been reduced to fulfil only the essential needs of society—sciences to progress the economy, and "living" languages to promote speedy communication for business. Verne also pre-empts the invention of the fax machine, praised for its capacity to link France with America in the space of a second (61). A business developer's motto crystallises these ideals saying that construction and instruction are really the same thing, instruction being just another form of construction which is slightly less solid (28). As in many of Verne's other works of the period, including *De la Terre à la lune* (*From the Earth to the Moon*) (1865), *Vingt mille lieues sous la mer* (*Twenty Thousand Leagues Under the Sea*) (1869) and *L'île mystérieuse* (*The Mysterious Island*) (1874), science is held up as a new form of religion, though in *Paris au XXe siècle,* it is enslaved to productivity, which renders it nefarious. Capitalist greed encourages war, and Verne implies that the 1863 American war of secession is a sign that such conflicts can indeed be motivated by financial rather than political causes. In fact, Paris in the twentieth century is fundamentally inspired by the American dream of progress, which although admired by Verne in other texts, is challenged by this pessimistic tale of hopeless love and death for those who cannot accept the new norms of capitalist society. Their complaints that France has lost its superiority, that French women have lost their charm and become like hard, shrewd American women (119), sets these social rejects apart from their successful capitalist peers who embrace an internationalised and commercialised existence in twentieth-century Paris. The city of light has clearly become, in spite of its electrification, a city of dark due to the unbridled rampage of economic progress.

In contrast, Alfred Franklin's *Les Ruines de Paris en 4875* (The Ruins of Paris in 4875) (1875) is a kind of archaeological work examining the relics of a once-great society now degraded and banal. It is more openly satirical, projecting a future re-discovery of Paris by an exploratory expedition from Noumea.[13] In a typical utopian inversion, the New Caledonian capital has become the heart of civilisation following a violent cataclysm that devastated the entire "vieux monde" (7) ("old world"). Using the model of an alternative future (as opposed to alternative history),[14] Franklin describes the findings of archaeologists, scientists and historians, which include interpreting the "mairie" ("town hall") and Marianne as signs of a religious cult to the goddess "Marianne République" and the Egyptian relics near the Seine as evidence that Parisian writing and culture was based on hieroglyphs. There are many amusing assumptions caricaturing contemporary Parisian practices, including an analysis of "ces barbares vêtus de peaux de bêtes" ("these barbarians clothed in animal skins") (10) as pleasure-seeking, frivolous, impressionable people with pride being their most obvious flaw, along with a complete inability to govern their city, as demonstrated by their obsession with celebrating the overthrow of each successive ruler (10–12). Post-Revolutionary Paris is thereby depicted as ridiculous, not only in its ironic, secular worship of the Republic but also in its political instability, accompanied by constant, violent upheavals.

Twentieth-century authors were not much more hopeful for uchronian Paris, which had become, essentially, a dark city. The philosophical novel *Sur la pierre blanche* (1905) (*The White Stone*) by Anatole France is one of the rare examples that does present a more positive future for the "United States of Europe" in the year 2270, represented as both a socialist utopia and a secular entity. Religion is considered unimportant, having little purchase on power since the separation of the Church and the State.[15] Human relations are freed from "des barbaries légales et des terreurs théologiques" ("legal barbarics and theological terrors"), but there are signs that humanity is not entirely happy in this socialist future. As Wells, Zamyatin, Huxley and Orwell took over the mainstream dystopian literary scene, French uchronias made their way into the science fiction novels of Barjavel and became post-apocalyptic in the films of Jean-Luc Godard (*Alphaville* 1965) and Chris Marker (*La Jetée* 1962). Robert Merle's novel *Malevil* (1972) reprises the religious theme as the threat of evil is explicitly theocratic in his text: religion is a source of oppression and there is an anticlerical undercurrent. Uchronias based on alternative history futures flourished in the twentieth century, and into the twenty-first, proposing the victories of Napoleon (*La Victoire de la Grande Armée* [The Victory of the Great Army] by former President Valéry Giscard d'Estaing, 2010), Hitler (*L'Altermonde* [The Otherworld] series by Jean-Claude Albert-Weil, 1996–2004), and Jean-Marie Le Pen, who represented the far-right political party, the Front National, in the second round of the Presidential elections in 2002 (*Les Cent jours* [The 100 Days] by Guy Konopnicki, 2002). It was also in the early years of the twenty-first century that French secular state ethics were called into question, and alternative futures (not histories) began to appear in French literary uchronias again. To understand how and why secularism came under scrutiny during this period, it is necessary to acknowledge the historical evolution of France from a Catholic to a secular state.

French Secular State Ethics

France's current stance on secularism is a strict separation of church and state, also known as *laïcité*. This means the complete removal of God, religion and overt religious symbols from the public sphere, including the state-run education system and the public service. However, religious belief, practices, symbols and education can be pursued in the private sphere—there is no religious censorship as long as the practices are not public or government sanctioned. A number of Catholic private schools do exist in France, and a handful of Jewish and Muslim ones, where religious practices and education are permitted, in spite of the fact that teachers are paid by the state if they follow the national curriculum. Secular education is a cornerstone of France's policy on *laïcité*.

Canadian philosopher Charles Taylor defines French *laïcité* as one of three principal strands of secularism, with each category determined according to the relationship between publicly espoused and privately practised religion.[16] Essentially, these categories could be termed as official, practical and optional. The French case is official, in that it does not necessarily reduce religious belief in the population, but the state interdiction effectively replaces religious discourse with a secular one, thereby offering an alternative "Republican" belief system. A second category is practical secularism where the population may be equally, more, or less religious than in an official secular state, but there is a steady decline in the number of people who declare themselves to be religious. This is the case in many European Union countries and the United Kingdom—even though the Head of State, Queen Elizabeth II, is also the Head of the Church of England. A practically secular society can therefore be observed in an officially religious state. In the optional version of secularism, there is no official stance on a particular religion by the state—religious belief is an option, one among many others, for both the population and the state. In this scenario, the religiosity of the people (and its leaders) can be high and yet the public discourse is not focused on particular religious beliefs—the society as a whole is religiously plural.

There are many theories as to the reasons for France's strict secular ethics, dating their source in the Enlightenment when dogma and clericalism were challenged by reason and science. Historian Peter Gay's influential studies proposing that the rise of secularism (or modern paganism) was due to opposition to religion during the Enlightenment have since been contested, and there is now general consensus that although anticlericalism certainly coloured visions of authority and power during this period, secularism came later.[17] The final chapter on "Religion" in Rousseau's *Contrat social* (*The Social Contract*) (1762) clearly influenced thinking on religion, with his theories of the unorganised religion of "man" focused on morality and deism, and the organised religion of the "citizen," more hierarchical and dogmatic, linked to the state politics. However, his own conversion from Protestant to Catholic belonged to another category—a kind of "world" religion that potentially rivalled the power of the state if not specifically aligned.

The Revolution is also seen as a touchpoint for secularism, with the Declaration of Human Rights opening the way for universal understanding, regardless of race, religion, age or gender, and yet these rights were deemed to be "natural, inalienable, and *sacred*," thereby retaining the religious terminology.[18] The Constitution of 1791 proclaimed the right to free worship and civil marriage, and under Napoleon's Consulate then Empire, the links between church and state were re-established according to the 1801 Concordat recognising certain cults and providing salaries for their clergy. In 1806, the Imperial University was created to administrate the entire educational system and manage the learning of citizens according to the new centralised regime. During the first 20 years following the Revolution, the fundamental institutions of human rights, marriage and education were, therefore, secularised to an extent, and in 1882, the Jules Ferry Law confirmed public, free and secular education as obligatory for all, but *laïcité* was not legally formalised until the Law of Separation of Church and State in 1905, putting an end to public funding of any religious cults.

This *fin-de-siècle* reinforcement and early-twentieth-century legislation of secular ethics present a paradox when contrasted to the concurrent affirmation of religious practice and power, illustrated by the construction of 24 new churches in Paris and a steady rise in pilgrimages to Lourdes.[19] Currents and counter-currents flowed more or less freely, and the world wars saw champions of both *laïcité* and Catholicism leading political discourse, but the conservative Catholic values of the Vichy regime certainly cemented a secular backlash in post-war years. In fact, the term "laic" only actually entered the French Constitution in 1946, under the Fourth Republic headed by interim leader (and Catholic) General Charles de Gaulle: "Article 1: La France est une République indivisible, laïque, démocratique et sociale" (Article 1: France is an indivisible, laic, democratic and social Republic). Secularism remained the principle of the "Trente Glorieuses," the 30 years of rebuilding and prosperity following World War II, and underpinned anti-bourgeois May 68 movements for students' and workers' rights. The new religious movements of the 1960s and 1970s therefore represented a double distancing from the mainstream, rejecting not just Catholicism but also state secularism, which James

A. Beckford posits as a trigger for French dystopian narratives depicting alternative belief systems in the 1980s and 1990s.[20]

During the same period, the children of Islamic migrants from North Africa and elsewhere began entering French schools wearing veils. When asked by teachers to remove these religious symbols, the young women frequently refused to do so, eventually provoking the Minister of Education Lionel Jospin to submit a question to the Conseil d'Etat in 1989, which returned a directive encouraging dialogue rather than a strict ban. With the change to a conservative coalition government in 1993, Minister of Education François Bayrou issued a circular in 1994 supporting schools wishing to ban religious symbols, thereby strengthening the state's commitment to secular ethics in education. After several heavily mediatised cases, and upon the advice of the Commission of Experts on secularism in the French Republic constituted in 2003 and led by Bernard Stasi, a law was passed by French parliament in 2004 forbidding visible religious symbols in schools.[21] Perfectly timed to coincide with the centenary of the 1905 Law of the Separation of Church and State, secularism was socially reinforced through the educational system, enshrined in the legal system via a new law and celebrated in the political system with public discourses, publications and awareness campaigns during 2005.[22]

Intercultural violence exploded in the Parisian *banlieue* (suburbs) with extended riots in October–November 2005, after two adolescents of Arab origin were electrocuted and a third injured fleeing police persecution. Was this the beginning of a new regime of terror in France? It was reported as such in the media, the threat of delinquency and unassimilated migrants being to blame for all manner of malaise in French cities and even rural villages. Nicolas Sarkozy, Minister for the Interior during the riots, was virulent in his attack on the "*racaille*" (scum) and, when he was elected President in 2007, reinforced anti-immigration policies and increased police powers. The global obsession with Islamic terrorism after the Twin Towers bombing on September 11, 2001, justified the need for greater security in France, and secularism became the "saving grace" for banning religious garb on the sports field and at the beach, culminating in the 2010 law against wearing a full *burqa* and *niqab*. France has been the unfortunate victim of several terrorist attacks, ranging from the shootings in the Jewish school in Montauban in 2012 to the devastating Charlie Hebdo and Saint Denis attacks in January 2015, the Paris massacre in the Bataclan in November 2015 and the Bastille Day deaths in Nice in 2016. Free speech and secularism have been targeted by Islamic fundamentalist individuals and groups as symbols of the decadent West and impediments to harmonious intercultural relations.

By no coincidence, it was during the last decade or so (2005–2015) that secularism took on more "sacred" qualities. Kim Knott, Amélie Barras and others have theorised this apparently antithetical concept of the sacred secular, demonstrating that the consensual politics around definitions and applications of secularism, the non-negotiable, non-accommodating ideology attached to *laïcité*, and its mobilisation to redefine rules of behaviour in certain spaces point to a process of sacralisation that pits secularism against religion, and especially Islam.[23] Renewed scholarly interest in secular studies and ethics also attests to its growing importance in international relations, as well as almost every other area of humanities or social science.

Jean Baubérot is one of the key consensual voices for the French government in the renaissance of secularism, described in *Le Monde des idées* as the "great thinker of *laïcité*" who is "renewing the conceptual framework of *laïcité*."[24] Historian of religions and secularism, he was appointed to the first Chair dedicated to *laïcité* in 1991 and four years later became the first director of the research group on sociology of religions and secularism at the prestigious Ecole Pratique des Hautes Etudes and the Centre National de Recherche Scientifique (EPHE-CNRS). Baubérot's authority on the subject is unparalleled, answering questions on the government website "La Documentation française" and signing a contract with them in 2015 to write a 96-page easily accessible pamphlet (with Micheline Milot) for wide distribution, costing just 5 euros 90, entitled *Parlons laïcité en 30 questions* (Let's talk about secularism in 30 questions) (2017). While this presentation format is a standard one for French civic education pamphlets and educational tools, it is somewhat ironic, or perhaps indicative of the times, that Houellebecq's 2015 novel introduces a publication entitled *Dix questions sur l'Islam* (Ten Questions on Islam) that assists the protagonist to take the final steps towards his religious conversion (259).

Michel Houellebecq's Islamic Agendas

Houellebecq's *Soumission* was released on the morning of January 7, 2015, and by midday, the international press had linked it to one of the most tragic events in recent French history. The author's face was on the cover of the French weekly satirical newspaper *Charlie Hebdo*, and 12 employees of the paper had been killed in a terrorist attack at the paper's office in Paris's 11th arrondissement by Islamic extremists Cherif and Said Kouachi.

The upsurge of support for the victims, their families and freedom of speech represented by the paper, and by extension France itself, took shape in a slogan, logo and hashtag "Je suis Charlie" that entered mainstream culture as a meme and was the subject of academic conferences and special issues of journals.[25] This apparently random confluence of events was emphasised by the novel's socio-political narrative: the rise of the Muslim Brotherhood political party in France and the appointment of its candidate, Mohammed Ben Abbes, as President in 2022, following a coalition arrangement with the centre-left political parties Union for a Popular Movement (UMP), Union of Democrats and Independents (UDI) and Socialist Party (PS), led by François Bayrou, in order to overcome Marine Le Pen's electoral victory. It is this collective, public narrative that frames the intimate, individual story of the protagonist François's intellectual and religious journey. A moderately successful professor at the University of Paris III—Sorbonne Nouvelle, specialised in the work of *fin-de-siècle* decadent author Joris-Karl Huysmans who famously converted to Catholicism, François turns away from the depressing, meaningless existence he leads to embrace the Islamic way of life and faith. His religious conversion seems ostensibly a means to keep his job but also implies a pathway leading to a more meaningful spiritual life.

In several press reviews of Houellebecq's *Soumission*, journalists cited a mostly forgotten French novel *Le Camp des saints* (*The Camp of the Saints*) (1973) by Jean Raspail as an extreme racialised precursor to this new novel.[26] *Le Camp des saints* then surfaced again as a surprising political reference made by Steve Bannon, prior to becoming US President Donald Trump's Chief Strategist for eight months from January to August 2017. Bannon had referred to *The Camp of the Saints* in three public speeches between October 2015 and April 2016 to emphasise his anti-immigrant views[27] and subsequently to justify his controversial ban on travellers from seven Muslim-majority countries.[28] Even the French did not make an immediate connection of his reference to Jean Raspail's 1973 novel, which draws its title from the *Apocalypse According to Saint John*, Revelation chapter 20, verse 9. The French edition had, however, been republished in 2011, in its eighth edition, with updated epigraphs and contemporary quotes, selling 20,000 copies in two months, making it the number one novel on Amazon's bestseller list in France, and of course, its appeal was not lost on far-right politicians, including Marine Le Pen.[29] It first appeared in English translation in 1975 and was reissued in 1983, and then again in 2001, with the support of American heiress Cordelia Scaife May and the anti-immigration campaigner, retired ophthalmologist John Tanton.[30]

The *Camp of the Saints* style invasion that Bannon mentions in his speeches comes from Raspail's portrait of a world transformed by an Indian leader who sails to France with 800,000 desperate refugees. It is lack of leadership and foresight by weak, secular, French and European authorities that allows the Indians to flood into the country, unleashing a "third-world takeover" of "western Christian civilisation" by non-white peoples from all over the globe, but especially India and Pakistan, clearly linking the novel to the recent Indo-Pakistani war (1971) and projecting its apocalyptic aftermath. This phenomenon is known as "The Great Replacement." The "Saints" to which the title refers include Calgues, the Catholic hero, who is like a medieval Crusader, lamenting the fall of Constantinople and praising the foundation of the KKK. The novel is clearly a call to arms, to take back control of western power, to dismiss impotent secularist ethics and reclaim the former glory of Catholic France.

The idea of the "Great Replacement," successfully exploited by Raspail, and famously theorised by Renaud Camus in his 2011 essay of the same title, has spawned a new wave of post-millennial uchronic novels, including Patrick Besson's *La Mémoire de Clara* (Clara's Memory) (2014), Jean Rolin's *Les Evénements* (The Events) (2014) and Julien Suaudeau's *Dawa* (Dawa) (2014). In tandem, several novels have posited the success of the far-right as a future political scenario as in Marc Villemain's *Et je dirai au monde toute la haine qu'il m'inspire* (And I Will Tell the World of All the Hate it Inspires in Me) (2006), Karim Amellal's *Bleu Blanc Noir* (Blue White Black) (2016), Eric Pessan's *La Nuit du second tour* (The Night of the Second Round) (2017) and the very popular political comic series *La Présidente* (The President) (2015) by historian François Durpaire and illustrator Farid Boudjellal. These texts attest to a trend that neither begins nor ends in our current era, but is certainly inspired by contemporary discourse and predictions.

Houellebecq is well-known for his provocative prose, which has often been labelled as dystopian,[31] nihilistic,[32] among other things. In his recent and timely book *Without God: Michel Houellebecq and Materialist Horror* (2016), Louis Betty concludes that the author is a "deeply and unavoidably religious writer," foregrounding his complicated relationship with religion as a key to understanding his worldview and work.[33] In spite of—or perhaps because of—his contradictory, ambiguous and ambivalent positions on major social and political issues, Houellebecq is one of the widest read contemporary French authors both in France and in translation around the world, and the *Charlie Hebdo* terrorist attacks added to the notoriety (and sales) of *Soumission*. This novel is a uchronia in the sense that it presents an imagined society set in the near future in Paris (2022). As in several other recent French uchronias, there is much more emphasis on the protagonist's dissatisfaction with the future status quo, with descriptions of an inherently dystopian society filling most of the pages and the prospect of a better option for the future being introduced only towards the end of the novel. In *Soumission*, Parisian society is therefore not clearly distinguishable from the Paris of 2015. The main differences can be observed in terms of the novel's principal themes of education and religion,[34] which are of course important markers of the tensions between secular ethics and Islamic agendas in the author's futuristic Paris.

Houellebecq's own education was principally undertaken in small primary and secondary schools in the countryside near Paris before undertaking preparation for entry into the elite *grandes-écoles* system at the Lycée Chaptal in Paris. He was accepted into the prestigious tertiary level Institut National Agronomique Paris-Grignon and graduated as an agronomist. He did not study in the university system which is depicted in the novel. At the end of *Soumission*, he acknowledges his debt to critic and lecturer Agathe Novak-Lechevalier from the University of Paris—Nanterre for providing insights into the working of these French institutions.

In terms of religion, Houellebecq is not Islamic, nor is his mother despite rumours to the contrary.[35] In 2001 following the publication of *Plateforme* (Platform) in which a tourist resort in Thailand is bombed by terrorists, he made a much-cited statement that "La religion la plus con, c'est quand même l'islam." ("Really, the most stupid religion is Islam.")[36] This declaration was nuanced in 2015, with his admission that he is probably Islamophobic, having a *fear* of Islam due to terrorism associated with it, rather than a *hatred* of the religion itself.[37] He was not raised according to a particular belief system, saying in an interview with Novak-Lechevalier: "j'ai été élevé par des gens déchristianisés, mais déchristianisés depuis tellement longtemps qu'ils n'étaient même plus anti-cléricaux ... et il est très difficile à moi-même de m'expliquer pourquoi je suis allé au catéchisme comme enfant" ("I was brought up by dechristianised people, but dechristianised for such a long time that there weren't even anti-clerical any more ... and it's very difficult for me to explain why I went to catechism as a child").[38] Attracted by Catholicism while living in a small French village where there was nothing else to do, Houellebecq was disappointed to find that the metaphysical questions of good and evil were treated in a humanitarian rather than in an intellectually stimulating way. He even thought about getting baptised at one stage, but discovered Pascal at the age of 15 and left religion behind.[39] Defining himself previously as atheist, but more agnostic now, he says that he will never embrace religion.[40] Houellebecq is far more influenced by science, science fiction (including Verne), and "Pascal's wager."[41]

The teleological approach to ethics that Pascal's wager implies—essentially betting on belief if the odds look good for improving the chances of happiness and it doesn't impede too many of life's pleasures—is an appropriate framework for considering the protagonist François's proposed conversion to Islam in *Soumission*.[42] It is a highly effective *dénouement* offering a glimmer of hope to this dystopian portrait of contemporary French society and definitively undermining secularism's role as an indisputable pillar of French identity. Implicit and explicit critiques of secularism pervade the entire text though, with special focus on education and religion.

Though François declares himself "aussi politisé qu'une serviette de toilette" (50) ("as political as a bath towel") (35), he realises that the current political climate is about to change him and his career as an educator. Having spent his teaching years at the Sorbonne Nouvelle seducing students and socialising rarely with few selected colleagues, the first shocks appear via these two pressure points. His affair with the lovely Jewish student Myriam ends when she decides to leave France for Israel, as her family is terrified of persecution under either Marine Le Pen's "Indigènes de France" ("Indigenous French") or Ben Abbes's Muslim Brotherhood (102–104). Chances of seducing other students seem unlikely to increase under the new political system either:

> Lorsque je retournai à la fac pour assurer mes cours, j'eus, pour la première fois, la sensation qu'il pouvait se passer quelque chose; que le système politique dans lequel je m'étais, depuis mon enfance, habitué à vivre, et qui depuis pas mal de temps se fissurait visiblement, pouvait éclater d'un seul coup. Je ne sais pas exactement ce qui me donna cette impression. … Peut-être aussi la démarche des filles en burqa, plus assurée et plus lente que d'ordinaire, elles avançaient de front par trois dans les couloirs, sans raser les murs, comme si elles étaient déjà maîtresses du terrain. (78)

> (When I went in to teach my class, I finally felt that something might happen, that the political system I'd grown up with, which had been showing cracks for so long, might suddenly explode. I don't know exactly where the feeling came from. … It may also have been the way the girls in burkas carried themselves. They moved slowly and with new confidence, walking down the very middle of the hallway, three by three, as if they were already in charge). (56)

François also begins to seek out the company and advice of his colleague Marie-Françoise, whose husband Alain Tanneur of the political police (DGSI) provides in-depth political analysis of the situation. Tanneur describes the probable negotiations between the Socialists and the Muslim Brotherhood to establish a coalition against the "Indigènes de France," judging that the coalition shouldn't have difficulties agreeing on matters of finance or security, maybe a few issues with foreign policy, but their real sticking point is education:

> La vraie difficulté, le point d'achoppement des négociations, c'est l'Education nationale. L'intérêt pour l'éducation est une vieille tradition socialiste, et le milieu enseignant est le seul qui n'ait jamais abandonné le Parti socialiste. ... La Fraternité musulmane est un parti spécial, vous savez. ... Pour eux l'essentiel c'est la démographie et l'éducation; la sous-population qui dispose du meilleur taux de reproduction, et qui parvient à transmettre ses valeurs, triomphe; à leurs yeux c'est aussi simple que ça, l'économie, la géopolitique même ne sont que de la poudre aux yeux: celui qui contrôle les enfants contrôle le futur; point final. Alors le seul point capital, le seul point sur lequel ils veulent absolument avoir satisfaction, c'est l'éducation des enfants. (81–82)
>
> (The real difficulty, the sticking point, is education. Support for education is an old Socialist tradition, and teachers are the one profession that has stood by the party, right to the end. ... The Muslim Brotherhood is an unusual party, you know. ... To start with, the economy is not their main concern. What they care about is birth rate and education. To them it's simple—whichever segment of the population has the highest birth rate, and does the best job of transmitting its values, wins. If you control the children, you control the future. So the one area in which they absolutely insist of having their way is the education of children). (54)

Marie-Françoise adds that it will be the same situation for universities, in particular, the Sorbonne, which will become one of the richest universities in the world, financed by the Saudi Arabians (84). As the Socialists eventually accede to the demands of the Muslim Brotherhood, the strict secular domain of education is undermined and Islamicised in the Paris of 2022. State secular ethics are therefore dead. The impact of this shift on François is significant: he is informed by Robert Rediger, the new Islamic President of Paris-Sorbonne, that the new statutes of his university do not allow non-Islamic staff to continue working there, and so François is retired with a substantial pension (178). However, the door is left open, with many enticements visible in the frame, to be enjoyed if François chooses to convert to Islam.

Houellebecq's narrative has prepared François for this moment by forcing him to interrogate his own ideas on secularism and religion. Firstly, there is the conversation with Tanneur, who correctly predicts Ben Abbes's antisecular tactics. Instead of denigrating Catholics as *dhimmis,* second-class citizens, or pronouncing Catholicism dead, Tanneur suggests the new President will raise them to higher status, so as to better discredit state secularism:

> En ce qui concerne la France, je suis absolument persuadé – je suis prêt à prendre le pari – que non seulement aucune entrave ne sera apportée au culte chrétien, mais que les subventions allouées aux associations catholiques et à l'entretien des bâtiments religieux seront augmentées – il peuvent se le permettre, celles allouées aux mosquées par les pétromonarchies seront de toute façon bien plus considérables. Et, surtout, le véritable ennemi des musulmans, ce qu'ils craignent et haïssent par-dessus tout, ce n'est pas le catholicisme: c'est le sécularisme, la laïcité, le matérialisme athée. Pour eux les catholiques sont des croyants, le catholicisme est une religion du Livre; il s'agit seulement de les convaincre de faire un pas de plus, de se convertir à l'islam: voilà la vraie vision musulmane de la chrétienté, la vision originelle. (156)
>
> (In France, I promise you, they won't interfere with Christian worship—in fact, the government will increase spending for Catholic organisations and the upkeep of churches. And they'll be able to afford it, since the Gulf States will be giving so much more to the mosques. For these Muslims, the real enemy—the thing they fear and hate—isn't Catholicism. It's secularism. It's laicism. It's atheist materialism. They think of Catholics as fellow believers. Catholicism is a religion of the Book. Catholics are one step away from converting to Islam—that's the true, original Muslim vision of Christianity). (109–110)

In a sense, Tanneur's reading of the country under Ben Abbes's presidency is a reconstitution of France's pre-revolutionary religiosity where Catholicism is replaced by Islam. As secularism is no longer even a subject worth debating (115), having been debased by Marine Le Pen's derisory and desperate efforts (110), Tanneur suggests that François goes to Rocamadour to understand France's great Catholic history: "A Rocamadour, vous pourrez vraiment mesurer à quel point la chrétienté médiévale était une grande civilization" (161) ("At Rocamadour you'll see what a great civilisation medieval Christendom really was") (113). His sojourn there offers insights but no long-term inspiration: "Au bout d'une demi-heure je me relevai, définitivement déserté par l'Esprit, réduit à mon corps endommagé, périssable, et je descendis tristement les marches en direction du parking" (170) ("After half an hour, I got up, fully deserted by the Spirit, reduced to my damaged, perishable body, and I sadly descended the stairs that led to the car park") (120).

Having eliminated secularism and Catholicism as belief systems to make sense of Paris in 2022, François finds himself drifting towards the agnostic teleology of Pascal's wager, substituting the dominant religion of Islam for Pascal's Catholicism. François is systematically seduced into the fold, as it becomes clear that the former bastions of French (Catholic and/or secular) literary culture are now under Islamic management. Firstly, Bastien Lacoue from the prestigious Pléiade editions offers him the opportunity to edit the works of Huysmans, and at their meeting demonstrates his alignment with the new Islamic regime by inviting François to the reception for the reopening of the (Islamic) Sorbonne. At this event, François observes the camaraderie of his colleagues who have already converted to the new faith and receives an invitation to Sorbonne President Robert Rediger's house. Witnessing the luxurious trappings of successful alignment with the Islamic university, including a range of pretty young girls and older good cooks as wives, excellent wine like Meursault and opulent comfort in the house where Dominique Aury wrote the famous erotic novel *Histoire d'O* (*Story of O*) (241–262), François reasons, à la Pascal, that he has not much to lose by embracing the faith. As Rediger compares a man's submission to Islam to a woman's submission to a man in *Histoire d'O*, and François reads Rediger's takeaway message in the pamphlet *Dix Questions sur l'Islam*, the final seduction is underway—via the power of literature and the written word. François delves deeper into Rediger's publications, doing his research on the subject, and finds clear, convincing prose, including a fascinating article comparing his own atheist author, Huysmans, who converted to Catholicism, to René Guénon, a Catholic author who converted to Islam and who initiated the construction of the Grande Mosquée in Paris and proposed an Islamic university (275). The models are laid out before him, and a fatalistic tone conjoins François's paraphrasing of Rediger's text with Houellebecq's own projection of the decline and fall of European society:

> Il fallait se rendre à l'évidence: parvenue à un degré de décomposition répugnant, l'Europe occidentale n'était plus en état de se sauver elle-même ... elle avait perdu du terrain, elle avait dû composer avec le rationalisme, renoncer à se soumettre le pouvoir temporal, ainsi peu à peu elle s'était condamnée, et cela pourquoi? Au fond, c'était un mystère; Dieu en avait décidé ainsi. (276)

(The facts were plain: Europe had reached a point of such putrid decomposition that it could no longer save itself … it had given way, it had been forced to compromise with rationalism, it had renounced its temporal powers, and so had sealed its own doom—and why? In the end, it was a mystery; God had ordained it so.). (178)

As François finishes his preface on Huysmans for the Pleïade edition and closes the book on that chapter of his life, including all possibilities of a "return" to secularism or Catholicism, his mind and spirit open to an Islamic future as professor, polygamist and positive person: "Que ma vie intellectuelle soit terminée, c'était de plus en plus une évidence … mais je commençais à prendre conscience – et ça, c'était une vraie nouveauté – qu'il y aurait, très probablement, autre chose" (295). ("Yes, my intellectual life was finished … but I started to realise—and this was a real novelty—that life might actually have more to offer") (208). What this "other thing" could be is open to interpretation—possibly spirituality, or it might be quite simply, a different future, another option. François's conversion is imagined in the conditional tense in the final chapter of *Soumission*,[43] ending with the optimistic lines: "une nouvelle chance s'offrirait à moi; et ce serait la chance d'une deuxième vie, sans grand rapport avec la précédente. Je n'aurais rien à regretter" (299–300) ("I'd been given another chance; and it would be the chance at a second life, with very little connection to the old one. I would have nothing to mourn") (211).

Houellebecq's uchronia set in Paris 2022 only becomes positive when the French secular nation's struggles with multicultural interventions are overridden by the force of Islamic faith. This future includes education as a tool of fundamentalist Islamic ideology, the complete subjugation of women and ongoing subversion of other freedoms through economic as well as political domination by Islamic collaborations, yet it is nevertheless presented as the most promising option for the protagonist François. While the projections for Paris in *Soumission* have been interpreted as dystopian, this novel is in fact one of the very few examples depicting a future that has positive connotations for the protagonist and other members of society, amongst a morass of speculative fiction painting pessimistic futures.

Conclusion

Laïcité and secular state ethics have long been associated with the French Revolution's utopian catchcry "*liberté, égalité, fraternité*," and the ideals behind the separation of church and state were an obvious attempt to uphold these virtuous aims. Yet the examples of French futuristic fiction examined here indicate that for past 250 years, Mercier's *L'An 2440* is the only truly positive French uchronia and it pre-dates both the French Revolution and the introduction of secular ethics. Even though the evidence of legal history and political practice shows that it is a much more recent construct, secularism has overshadowed religious hope throughout nineteenth- and twentieth-century uchronian fiction. Science and progress were allowed to take over the role that religion once played, but the lack of moral codes to safeguard humanity's future turned texts by Verne, Godard, Raspail and others into dystopian nightmares. The negative nature of secular uchronian fiction is remarkable, demonstrating that something happened to the utopian imaginary when secularism was introduced to and then imposed on French society. In many ways, secular ethics like human rights and liberty, equality and fraternity should be an enabler for positive futuristic fiction, but does it actually short-circuit any possibility of the millenarian hope—or forward-looking positivity—that has been a cornerstone of utopian thinking since before Thomas More? With the rise of the Catholic Far Right in French politics and the increasing visibility of Islamic populations, intercultural tensions and terrorist attacks, there has clearly been a religious resurgence in officially secular France. Even President Emmanuel Macron has reached out to the Catholic leaders for dialogue on France's future, attracting virulent criticism from the Socialist and Communist parties who declare Macron's actions as an attack on *laïcité*.[44] Religion has returned to the core of political debate, and to the heart of uchronian fiction in France: Houllebecq's Islamic projections for a "better future" provide much food for thought. Through this brief analysis of examples of uchronia from across the centuries, it is evident that there is a new discourse on religion in futuristic fiction that, despite its perceived dystopian overtones, has already and will continue to reorient fictional writing on France's future.

Notes

1. In-text references of French quotes are from the 1999 publication of the 1771 edition: Mercier, *L'An 2440*; English quotes are from W. Hooper's translation: Mercier, *Memoirs*.
2. Alkon, *Origins of Futuristic Fiction*. See also: Ransom, "Alternate History."
3. In-text references of French quotes are from Houellebecq, *Soumission*; English quotes are from Lorin Stein's translation: Houellebecq, *Submission*.

4. Forsström, *Possible Worlds.*
5. Forsström notes that this text is likely translated and adapted from a German novel by Johann Gottlob Benjamin Pfeil. Forsström, *Possible Worlds*, 33.
6. Wilkie, "Mercier's *L'An 2440*," 16.
7. Forsström, *Possible Worlds*, 43.
8. Wilkie, "Mercier's *L'An 2440*," 16.
9. Rufi, *Le Rêve.*
10. Trousson, "Sciences et Religion en 2440."
11. Garrioch, *Making of Revolutionary Paris*, 1–14.
12. In-text references are from Verne, *Paris au XXe siècle.*
13. There has been some confusion as to the authorship of this anonymously published work, but the attribution to Léo Lespès (Timothée Trimm) (1815–1875) seems implausible given that an updated edition was published in 1908 of *Les Ruines de Paris en 4908*, and Lespès died in 1875. See: Jean de Palacio, *Le Silence du texte*, 235; Marie-France David de Palacio, *Reviviscences romaines*, 109n22. Léo Lespès wrote *Les Ruines de Paris: Chronique du Paris brûlé* (Paris: Imprimerie nationale, 1871), which adds to the confusion.
14. Alternative or alternate history is a popular sub-genre of uchronia, as noted by Alkon, *Origins of Futuristic Fiction*, 115, and a list of "what if" works can be found at http://www.uchronia.net/
15. Ludlow, "*Sur la pierre blanche*," 623.
16. Taylor, *Secular Age.*
17. Gay, *The Enlightenment.* See, for example: Barnett, *Enlightenment and Religion.*
18. French National Constituent Assembly, *Declaration*, preamble.
19. Powers, *Confronting Evil*, 5.
20. Beckford, "Laïcité," 27–40.
21. Jones, "Beneath the Veil."
22. See: "Déclaration universelle"; "Laïcité: les débats"; "De la Révolution Française."
23. Barras, "Sacred *Laïcité.*"
24. Chemin, "Jean Baubérot."
25. Devichand, "How the World."
26. For example: Leménager, "Houellebecq."
27. In October 2015, Bannon mentioned the text in an interview, saying: "It's been almost a Camp of the Saints type invasion into Central and then Western and Northern Europe"; later saying on a Breitbart podcast in January 2016: "The whole thing in Europe is all about immigration. It's a global issue today—this kind of global Camp of the Saints"; and again in April 2016: "When we first started talking about this a year ago, we called

it the Camp of the Saints … I mean, this is Camp of the Saints, isn't it?" Quoted in Blumenthal and Rieger, "This Stunningly Racist."

28. After consideration by the Supreme Court, this travel ban is now enforceable against citizens of six Muslim-majority countries. McCarthy and Laughland, "Trump Travel Ban."
29. Albertini, "L'un des livres favoris."
30. Blumenthal and Rieger, "This Stunningly Racist."
31. Chrostek, *Utopie*; Giri, "En quête"; Wesemael, *Michel Houellebecq*; Sweeney, *Michel Houellebecq*; Morrey, *Michel Houellebecq*; Betty, *Without God*.
32. Onfray, *Miroir du nihilisme*.
33. Betty, *Without God*, 47.
34. In an interesting attempt at categorising media reactions to *Soumission* according to French or English language reporting, Joseph Voignac suggests that while reactions to the novel in France focus on integration, insecurity and Islamophobia, the Anglophone press brings out the author's "real" themes: the moral and sexual misery of a world without religion, and France's place in the world. Voignac, "Michel Houellebecq."
35. Sénécal, "Michel Houellebecq."
36. Sénécal, "Michel Houellebecq."
37. Chrisafis, "Michel Houellebecq."
38. Novak-Lechevalier, "La religión."
39. Lorentzen, "Nobody."
40. Chaudey and Denis, "Michel Houellebecq"; Novak-Lechevalier, "La religión."
41. Baird, *Studies in Pascal's Ethics*, 56–84.
42. See also: Chaouat, "Houellebecq's Wager."
43. Novak-Lechevalier posits that François does not necessarily convert, given this conditional tense. Novak-Lechevalier, "Soumission."
44. Sauvaget, "Macron."

Works Cited

Albertini, Dominique. "L'un des livres favoris de Marine Le Pen décrit une apocalypse migratoire." *Libération*. September 16, 2015. http://www.liberation.fr/france/2015/09/16/le-livre-de-chevet-de-marine-le-pen-decrit-une-apocalypse-migratoire_1383026.

Alkon, Paul K. *Origins of Futuristic Fiction*. Athens: University of Georgia Press, 1987.

Baird, A. W. S. *Studies in Pascal's Ethics*. The Hague: Springer, 1975.

Barnett, S. J. *The Enlightenment and Religion: The Myths of Modernity*. Manchester: Manchester University Press, 2003.

Barras, Amélie. "Sacred *Laïcité* and the Politics of Religious Resurgence in France: Whither Religious Pluralism?" *Mediterranean Politics* 18, no. 2 (2013): 276–293.

Beckford, James A. "Laïcité, 'Dystopia' and the Reaction to New Religious Movements in France," in *Regulating Religion: Case Studies from Around the Globe*, edited by James T. Richardson, 27–40. New York: Kluwer Academic Press, 2004.

Betty, Louis. *Without God: Michel Houellebecq and Materialist Horror*. Pennsylvania: Pennsylvania State University Press, 2016.

Blumenthal, Paul, and J. M. Rieger, "This Stunningly Racist French Novel Is How Steve Bannon Explains the World," *Huffpost*, March 5, 2017, https://www.huffingtonpost.com.au/entry/steve-bannon-camp-of-the-saints-immigration_us_58b75206e4b0284854b3dc03

Chaouat, Bruno. "Houellebecq's Wager." *Los Angeles Review of Books*. May 26, 2015. http://marginalia.lareviewofbooks.org/houellebecqs-wager-by-bruno-chaouat/.

Chaudey, Marie, and Jean-Pierre Denis. "Michel Houellebecq: 'Je ne suis plus athée'." *La Vie*. January 27, 2015. http://www.lavie.fr/culture/livres/michel-houellebecq-je-ne-suis-plus-athee-27-01-2015-59984_30.php.

Chemin, Anne. "Jean Baubérot, grand penseur de la laïcité," *Le Monde.fr*, December 8, 2017, http://www.lemonde.fr/idees/article/2017/12/08/jean-bauberot-grand-penseur-de-la-laicite_5226435_3232.html.

Chrisafis, Angelique. "Michel Houellebecq—'Am I Islamophobic? Probably, Yes'." *The Guardian*. September 7, 2015. https://www.theguardian.com/books/2015/sep/06/michel-houellebecq-submission-am-i-islamophobic-probably-yes.

Chrostek, Katharina. *Utopie und Dystopie bei Michel Houellebecq*. Berlin: Peter Lang, 2011.

"De la Révolution Française à la loi de 1905: les débats du XIXe siècle". *Sénat*. February 4, 2005. https://www.senat.fr/colloques/actes_laicite/actes_laicite2.html.

"Déclaration universelle sur la laïcité au XXIe siècle". *Le Monde.fr*. December 9, 2005.http://www.lemonde.fr/idees/article_interactif/2005/12/09/declaration-universelle-sur-la-laicite-au-xxie-siecle_718769_3232.html.

Devichand, Mukul. "How the World Was Changed by the Slogan 'Je Suis Charlie'." *BBC Trending*. January 3, 2016. http://www.bbc.com/news/blogs-trending-35108339.

Forsström, Riikka. *Possible Worlds: The Idea of Happiness in the Utopian Vision of Louis-Sébastien Mercier*. Helsinki: Biblioteca Historica, 2002.

French National Constituent Assembly. *Declaration of the Rights of Man and the Citizen* (1789). http://avalon.law.yale.edu/18th_century/rightsof.asp.

Garrioch, David. *The Making of Revolutionary Paris*. Berkeley: University of California Press, 2004.

Gay, Peter. *The Enlightenment: An Interpretation.* New York: Knopf, 1966–1969.
Giri, Hemlata. "En quête d'une société idéale: la dialectique de l'utopie et de la dystopie dans *Travail* d'Emile Zola et *La Possibilité d'une île* de Michel Houellebecq." PhD diss., University of Delhi and Sorbonne Paris Cité, 2015.
Houellebecq, Michel. *Soumission.* Paris: Flammarion, 2015.
———. *Submission.* Translated by Lorin Stein. London: William Heinemann, 2015.
Jones, Nicky. "Beneath the Veil: Muslim Girls and Islamic Headscarves in Secular France." *Macquarie Law Journal* 9 (2009): 47–69.
"Laïcité: les débats, 100 ans après la loi de 1905". *La documentation Française.* December 9, 2005. http://www.ladocumentationfrancaise.fr/dossiers/d000095-laicite-les-debats-100-ans-apres-la-loi-de-1905.
Leménager, Grégoire. "Houellebecq, Besson, Rolin… 'Le Grand Remplacement', sujet de roman?" *BibliObs.* December 18, 2014. https://bibliobs.nouvelobs.com/actualites/20141218.OBS8319/houellebecq-besson-rolin-le-grand-remplacement-sujet-de-roman.html.
Lorentzen, Christian. "Nobody Will Make Us Do Yoga: A Conversation with Michel Houellebecq." *Vice.* June 12, 2017. https://www.vice.com/en_us/article/qv43a7/michel-houellebecq-interview.
Ludlow, Gregory. "*Sur la pierre blanche*: Anatole France's Vision of a European Federation." *History of European Ideas* 16, no. 4–6 (1993): 619–625.
McCarthy, Tom, and Oliver Laughland. "Trump Travel Ban: Supreme Court Allows Enforcement as Appeals Proceed." *The Guardian.* December 5, 2017. https://www.theguardian.com/us-news/2017/dec/04/donald-trump-travel-ban-on-six-mostly-muslim-countries.
Mercier, Louis Sébastien. *L'An 2440: Rêve s'il en fut jamais.* Paris: La Découverte, 1999.
———. *Memoirs of the Year Two Thousand Five Hundred.* Translated by W. Hooper. Philadelphia: Thomas Hobson, 1795.
Morrey, Douglas. *Michel Houellebecq: Humanity and Its Aftermath.* Liverpool: Liverpool University Press, 2013.
Novak-Lechevalier, Agathe. "La religión en las novelas de Houellebecq." *YouTube.* June 12, 2017. https://www.youtube.com/watch?v=i1DFEW09dvU.
———. "Soumission: la littérature comme résistance." *Libération.* March 1, 2015. http://next.liberation.fr/culture/2015/03/01/soumission-la-litterature-comme-resistance_1212088.
Onfray, Michel. *Miroir du nihilisme: Houellebecq éducateur.* Paris: Éditions Galilée, 2017.
Palacio, Jean de. *Le Silence du texte: poétique de la décadence.* Leuven: Peeters, 2003.
Palacio, Marie-France David de. *Reviviscences romaines: la latinité au miroir de l'esprit fin-de-siècle.* Berne: Peter Lang, 2005.
Powers, Scott M. *Confronting Evil: The Psychology of Secularization in Modern French Literature.* West Lafayette: Purdue University Press, 2016.

Ransom, Amy J. "Alternate History and Uchronia: Some Questions of Terminology and Genre." *Foundation* 87 (2003): 58–72.

Rufi, Enrico. *Le Rêve laïque de Louis-Sébastien Mercier entre littérature et politique.* Oxford: Voltaire Foundation, 1995.

Sauvaget, Bernadette. "Macron, un nouveau pacte avec les catholiques?" *Libération.* April 10, 2018. http://www.liberation.fr/france/2018/04/10/macron-un-nouveau-pacte-avec-les-catholiques_1642260.

Sénécal, Didier. "Michel Houellebecq." *L'Express.* September 1, 2001. https://www.lexpress.fr/culture/livre/michel-houellebecq_804761.html.

Sweeney, Carole. *Michel Houellebecq and the Literature of Despair.* New York: Bloomsbury, 2013.

Taylor, Charles. *A Secular Age.* Cambridge, MA: Harvard University Press, 2007.

Trousson, Raymond. "Sciences et Religion en 2440." *Cahiers de l'Association internationale des études françaises*, no. 58 (2006): 89–105.

Verne, Jules. *Paris au XXe siècle.* Paris: Hachette/Le Cherche-Midi Editeur, 1994.

Voignac, Joseph. "Michel Houellebecq: Le Monde Anglophone réagit à *Soumission.*" *Revue des Deux Mondes.* October 29, 2015. http://www.revuedesdeuxmondes.fr/michel-houellebecq-le-monde-anglophone-reagit-a-soumission/.

Wesemael, Sabine van. *Michel Houellebecq. Le Plaisir du texte.* Paris: L'Harmattan, 2005.

Wilkie, Everett C., Jr. "Mercier's *L'An 2440*: Its Publishing History During the Author's Lifetime." *Harvard Library Bulletin* 32, no. 1 (1984): 5–35.

CHAPTER 12

The Appearance of Dystopian Fiction in Macedonia and its Ethical Concerns

Kalina Maleska

There is no tradition of utopian or dystopian writing in Macedonian literature, and although in the Western tradition, dystopian fiction, suggesting disillusionment, pessimism and fear, has usually followed the utopian visions of a better world, Macedonian literature is marked with the appearance of dystopias without previously formulated utopian visions. This is understandable, taking into consideration that in the increasingly globalised world, dystopian writing and films are much more accessible in Macedonia today than were the utopian concepts in the past.

This study will explore the dystopian elements in three Macedonian novels and the reasons behind the appearance of these otherwise untypical elements in Macedonian literature. Certain traits of speculative fiction can be found at the end of the twentieth century in the works of Vlada Urošević. His novels are not, strictly speaking, utopian or dystopian, but they often include elements of fantasy and alternative realities. One of his novels, entitled *Дворскиот поет во апарат за летање* (The Court Poet in a Flying Machine) (1996), can be described as a precursor of dystopian

K. Maleska (✉)
Saints Cyril and Methodius University in Skopje, Skopje, North Macedonia

Z. Kendal et al. (eds.), *Ethical Futures and Global Science Fiction*, Studies in Global Science Fiction,
https://doi.org/10.1007/978-3-030-27893-9_12

fiction in Macedonian literature, as it takes place in a kind of post-apocalyptic future. The first two works that may be said to belong to dystopian fiction in the more narrow sense are Sanja Mihajlović Kostadinovska's novella *517* (2015) and Tomislav Osmanli's *Бродот. Конзархија* (The Ship. Consarchy) (2016). All three novels question the change of values that lead to an undesired or even threatening future. In his essay "The Origins of Dystopia: Wells, Huxley and Orwell," Gregory Claeys focuses on the turn from utopia towards dystopia between the end of the nineteenth century and the mid-twentieth century. He states that he uses the term *dystopian* "in the broad sense of portraying feasible negative visions of social and political development, cast principally in fictional form. By 'feasible' we imply that no extraordinary or utterly unrealistic features dominate the narrative."[1] Similarly, in the case of the three novels that will be discussed, they are all feasible in the sense that they do not include elements that are improbable but explore possible futures—which is one of the few characteristics they share. The essay will examine the reasons why dystopian fiction has appeared only recently in Macedonian literature, what specific characteristics these novels have that mark them as dystopian and how they deal with the ethical concerns of utopian literature more broadly.

Exploring the relationship between ethics and politics, these works centre on the ramifications of the political rise to power of authoritarian and corrupt regimes whose unethical actions have led their countries and citizens to total war (in Urošević's novel) or the abolition of freedom of speech and privacy (in the other two works). In this context, a recurrent ethical issue is nostalgia, whose possible negative ethical aspects are examined in cases when the governing structures manipulate the population by exploiting nostalgia for the purposes of strengthening their nationalistic claims. Similar uses of nostalgia, both in literature and in politics at large, has led thinkers to dismiss and condemn nostalgia as "reactionary, even xenophobic," as John J. Su points out in his study *Ethics and Nostalgia in the Contemporary Novel* (2005).[2] Analysing the ethical dimensions of nostalgia, Su aims to re-examine such negative attitudes towards nostalgia and emphasises that they can be countered by exploring the positive effects of nostalgia and memory in contemporary Anglophone literature. His study argues "that more utopian visions" in literature "are possible *only* through nostalgic evocations of lost or nonexistent communities."[3]

The works of Urošević, Osmali and Mihajlović Kostadinovska reflect this twofold position towards nostalgia. The obsession of the political rulers with reconstructing an allegedly glorious past has certainly led to xenophobia and totalitarian rule in the three works discussed. On the other hand, a positive understanding of nostalgia is also present, as demonstrated when nostalgia is related not to the governing structures but to the characters' rejection of the present circumstances in which they find themselves. In this context, a parallel can be drawn to the way nostalgia is discussed by Katharina Niemeyer, who notes that the beginning of the twenty-first century has seen "an increase in expressions of nostalgia."[4] According to Niemeyer, this is not just a trend; "Rather, it very often expresses or hints at something more profound, as it deals with positive or negative relations to time and space."[5] The longing for former times is a characteristic feature of Macedonian culture in general, and it expresses an attempt to escape the constant political crises and conflicts that have plagued the country since its independence in 1991, while expressing the longing for what is seen (rightly or wrongly) as a more stable past. In such a way, nostalgia in Macedonian dystopian fiction is often used as a mark, or a symbol, of the positive ethical values that had existed in the past, such as empathy, family bonding, respect between different ethnic or religious groups, free access to knowledge, close relationships with friends—traits which have been eradicated by state systems that discourage closeness between people and ban reading so as to keep the people in a state of fear, obedience and permanent control.

The Political Circumstances and Settings of the Novels

As is the case with much utopian and dystopian writing, the contexts of these three novels reflect the social and political circumstances experienced by their writers. Although none of the authors explicitly mention the direct inspiration behind their works, or what specific developments they refer to in the description of events, it can be deduced from the stories themselves what events from the political reality inspired some of the novels' descriptions and plots.

Дворскиот поет во апарат за летање was published in the 1990s, which witnessed the wars on the territory of former Yugoslavia, often referred to as the deadliest conflicts in Europe after the Second World War. These wars undoubtedly had a great impact on the novel's post-apocalyptic setting, in which civilisation as we know it has been destroyed in some kind of catastrophe. Although it is not completely clear whether the catastrophe was due to environmental reasons or to war, there are more indications in the novel to suggest the latter. Yet its constant focus on nature, with its protagonist Igor being especially connected to the plants and animals around him, resonates with the science fiction that discusses ecological catastrophes predicted in the twenty-first century. As Brian Stableford states: "The notion that the twenty-first century would be an era of unprecedented ecological crisis, highly likely to lead to a temporary or permanent collapse of civilisation, became so firmly entrenched in speculative fiction as virtually to be taken for granted."[6]

517 and *Бродот. Конзархија* were published in 2015 and 2016, respectively, in the context of Macedonia being governed by the right-wing VMRO-DPMNE political party—a government that lasted for 11 years, until the first half of 2017, and against which there have been allegations of corruption. During that time, the government made extensive efforts to represent the Macedonian identity as closely related to that of the ancient Macedonians. A major government project initiated in 2010 and called "Skopje 2014" was part of those efforts and included the construction of many monuments referring to antiquity and the period of Alexander the Great, as well as buildings that were very different in style from the architecture in Skopje before 2010. This project has assumed ethical dimensions in Macedonia, with thousands of articles dedicated to it from both ideological and financial perspectives. Rather than leaving the issue of history to historians, the government used the project to construct a fabricated, ideological narrative (often referred to as a "myth" in the media) of Macedonian identity, misrepresenting history and promoting nationalism and xenophobia among citizens. Concerning the financial aspect, research from the Balkan Investigative Reporting Network Macedonia showed that the government spent seven times more than it first announced for the project, which has repeatedly been described as outrageous in the context of the high level of poverty in Macedonia and the poor state of health care and education.[7] The project's ethical problems have become even clearer with allegations that the real cost of the project, taking into consideration the low quality of materials used in its construction, is much lower than claimed, which has been the ground for investigations of corruption. The buildings that are part of the "Skopje 2014" project are popularly called "baroque" buildings, although architects argue that the style is not really baroque. Many allusions to these buildings and monuments that completely altered the look of the city are found in the books by Mihajlović Kostadinovska and by Osmanli.

All three books are set in different kinds of futures. At first, *Дворскиот поет во апарат за летање* seems to be set in medieval times: it represents a fortification that resembles medieval castles, inhabited by a Lord (the Macedonian word "кнез" [*knez*] may actually be translated as either Lord or Duke), a Lady (or Duchess) and their subjects, including astrologists, court poets and servants. The motif of an attack on the castle, as we witness in the first chapter, by some kind of barbarians, skilful in riding horses and wearing tattered clothes, contributes to the medieval effect. This ambiguity in the setting is generally present in Urošević's works, where, as critic Kapuševska-Drakulevska explains,

> јасно можат да се издвојат две фабуларни линии изградени според принципот на контраст: една, која создава илузија на секојдневниот живот, т.е. на искуствената реалност, и друга, која алудира на ирационалното, на неверојатното што се јавува во истото тоа секојдневие и со тоа внесува нарушување.
>
> (two plot lines built upon the principle of contrast can be distinguished: one, which creates an illusion of everyday life, i.e. experienced reality, and another, which alludes to the irrational, the incredible that appears in the everyday life, thus introducing a rupture, undermining its order.)[8]

In this novel, the medieval setting seems realistic, as a description that may have come from historical accounts of the Middle Ages to suddenly surprise the reader with the introduction of futuristic elements, such as references to machinery, which posits the castle in a post-apocalyptic future.

The first suspicion that events are, in fact, taking place in the future is the alarm spread after the attack of the savages that "Папа-тема беше исчезната" ("Papa-tema is missing").[9] Papa-tema is considered by the people in the castle to be a kind of monument of olden times, but its description indicates to the reader that it may be a large antenna whose function no one in the novel knows. That immediately establishes the setting in the far future, after some kind of global devastation—a future in which technology is no longer known but only remains as scattered archaeological artefacts. Thus, the Papa-tema and other instances of technological leftovers clearly indicate that the novel is set in a post-apocalyptic future. This raises the question, asked many times in the novel, how did people in the past treat each other and the environment that had led to the destruction of civilisation, leaving just a few scattered small towns in which people deal with agriculture. The protagonist, Igor Trol, a court poet who spends his life writing poems about the beauty of the landscape, animals, plants, people, inventing metaphors that emphasise the unity of all things in nature, suddenly finds himself in a dangerous situation in which he must do all to survive.

517, on the other hand, begins with explicit references to the future. It is an urban dystopian novel in which grey uniform buildings dominate the centre of the city, while suburbia is marked with ruined concrete blocks—well-known dystopian characteristics. What is interesting and specific in this novel is the humorous treatment of the theme of the so-called "сегашизам" ("presentism"): the government forbids any reference to the past and demands that everyone lives only for and in the present, trying to adjust even language in such a way that it would be unable to express past tenses.

In terms of the structure, *517* explores ethical issues from three different perspectives represented in the three types of chapters that are intertwined in the novella. First, the brief chapters dedicated to the activities of an unnamed "сегашист" ("presentist"), narrated from a third person, yet subjective, point of view, which are aimed at ironically representing the "presentist." Second, the longer chapters narrated from a more detached third-person point of view, in which the reader follows the development of an underground resistance movement through its five members who rebel against society, also called "anti-presentists." And finally, the brief chapters told from a first-person viewpoint (given in italics) of one of the rebels.

The aspect that differentiates *517* from other grim dystopian visions is the humour with which it is told, despite the pessimistic perspective. Thus, mentioning that 517 may be a code name for a hero, the narrator adds that there have been stories of other "нумерички херои" ("numeric heroes") as well, offering as examples: 007, 501 and 99,102%.[10] The reference to 007 would be clear to many readers, but 501, for example, has a specifically Macedonian context: namely, in the 1980s, the most famous and, in fact, almost the only widely known foreign brand of jeans in Macedonia (part of Yugoslavia at the time) was Levi's 501, a brand that meant prestige among teenagers. The number 517 transforms in the novel into some kind of mysterious remnant of a past age, which might have some nostalgic value but, at the same time, mocks the obsession with this famous brand. Humour intermingled with irony also accompanies the chapters that describe instances of presentism: a presentist sonnet in which the master of presentism is praised, a presentist hairdresser who makes only presentist hairstyles, and so on, contribute to the humorous atmosphere; yet, interpreted seriously, they refer to the dominance of ideology in every sphere of life during dictatorial regimes.

Бродот. Конзархија takes place in the near future—2039. It includes many technological novelties, but the relations between people seem to largely resemble those of today's world. The setting is Skopje, and there is an ecological statement that is relevant for the city: plastic bottles and other instances of garbage appear on the surface of the river, which today is the case with the Vardar River, flowing through Skopje, so the novel discusses the extent to which today's ecological problem can extend in the future.

The dystopian elements are recognisable in the novel's advanced technological devices, the technological opportunities for people to upgrade their physical appearance, as well as perhaps the most convincing dystopian element in the novel: the changes made to the language, in which Macedonian terms have been replaced with English words. There is a modern tendency among Macedonian language speakers to replace Macedonian words with internationalisms. Osmanli brings this to a more extreme form, which is, at the same time, a very probable development of the language: in the novel, for example, the term "domicile speech" has replaced the term "mother tongue," which is used at present. The language spoken by the characters in the novel is regulated with a so-called "Директива за заштита на хигиената на домицилниот говор" ("Directive for the Protection of the Hygiene of the Domicile Speech"), which is an expression of the government's desire to control language and thus the communication between people.[11] Other instances of the "domicile speech" in *Бродот. Конзархија* are the English words outlook, involving, default, username, resetting, reply, frequent—all of them written in Cyrillic and pronounced just as they are in English, replacing existing Macedonian counterparts.

There are certain dystopian elements, however, that are not as convincing as the mentioned language tendencies. In the context of language, these include the conversations between the protagonists, who speak in highly formal and exceedingly polite language, using phrases such as "highly respected" and "revered sir/madam." Such formal conversations have never been part of Macedonian linguistic culture, so the probability of the domination of such speech in the near future Skopje is small. Another element that seems more to be inspired by Western science fiction and dystopian films than by the Macedonia context is the establishment of consarchies, which are territorial and financial districts whose business managers are also political leaders with vast political powers. Dystopian visions are often formulated out of fear that certain negative developments in the present moment may take an extreme form. Taking into consideration the levels of poverty and unemployment in Macedonia today, and the fact that the creation of large-scale corporations has never been part of its history, the existence of such consarchies in the near future appears improbable.

The Rulers' Concealment of the Past and the Protagonists' Nostalgia

The dystopian works discussed here do not focus on issues commonly addressed in utopian and dystopian fiction, such as the treatment of criminals, child care, eugenics, nature vs. urban environment, gender or euthanasia.[12] All three novels do, however, focus on mourning the lost ethical values of the past—a theme not uncommon in dystopian literature. This is strongly expressed both in the nostalgia of the main characters and in the implicit condemnation of the authorities who conceal the past from the citizens. The protagonists of all three works display reverence for compassion, respect, justice, integrity, truthfulness, faithfulness, trustworthiness, discernment, conscientiousness—qualities that are considered central in virtue ethics.[13] In the course of the novels, it becomes clear that the main characters are guided by a belief that such virtues could lead to the defeat of the authoritarian regimes and establishment of a more just society. In *Дворскиот поет во апарат за летање*, the young woman Estrela tries to explain to the somewhat naive protagonist Igor the importance of discernment and faithfulness. The five rebels in *517*, and especially the narrator referred to as The Withdrawn One, show trustworthiness, compassion and respect to one another from the moment they meet—traits that differentiate them from the rest of the society. Slobodan (whose name means freedom) and Tatjana in *Бродот. Конзархија* are the only characters in the novel who are faithful to each other and value integrity. They all see these virtues that they cherish as part of a system no longer in existence in their society.

It could be argued that this nostalgia has its roots in the socialist system that existed in former Yugoslavia and in Macedonia as one of its republics. This is recognisable not only in the mentioned novels but also in the public discourse, with one of the frequent arguments, expressed by sociologist, artists, writers and political thinkers alike, being that all ethical values have been lost. The dystopian novels seem to perpetuate this kind of discourse. In socialist Macedonia, there appeared to be a relative degree of economic equality—certainly not complete economic equality but greater than in the last 28 years since the disintegration of Yugoslavia when the gap between the few rich and the majority poor has greatly widened. In part due to unfair competition and corruption, money is a much greater issue today than previously. Also a subject of nostalgia is the clear hierarchical system that was in place in the past, according to which working experience and merit were an important criterion for obtaining a job—something undermined in recent decades as party membership became the greatest advantage in employment. Post-communist nostalgia has become a subject of numerous studies, and one such notable study is *Post-Communist Nostalgia* (2010) edited by Maria Todorova and Zsuzsa Gille, which discusses the complex reasons why it is still present in former communist and socialist territories despite the oppression of the communist regimes. As Todorova writes, practically "all capital cities of Eastern Europe have a monument to the victims of communism, and most have museums, not commemorating but condemning Communist rule."[14] Yet the novels discussed do not dwell on the negative sides of the socialist ideology but on the personal experience of the characters who are faced with the immoral behaviour of those around them: mainly, of the corrupted officials (the Duke, the Pakanskis, the presentists) who aim to destroy all signs of just and fair society and who have climbed the hierarchical ladder not on the basis of merit but on the basis of deceit.

In all three novels, nostalgia is also closely related to the idea that the government can retain its power only if it conceals the past from the citizens, so that they do not have a point of comparison that would presumably make them aware that their present is worse than the past. The authorities' insistence on hiding the past in order to retain their political power and eliminate all threats to the established order is a recurrent motive in dystopian fiction, as seen in Yevgeny Zamyatin's *Мы* (*We*, 1921), George Orwell's *Nineteen Eighty-Four* (1949) and Aldous Huxley's *Brave New World* (1932), among others. The Duke in Urošević's *Дворскиот поет во апарат за летање* does not allow the poet to view the telescope or the helicopters represented on the tapestry in the castle. The concept of presentism in Mihajlović Kostadinovska's *517* is based on the dominant ideology that tries to eliminate all references to time. In Osmanli's *Бродот. Конзархија*, foreign (mostly English) words replace pre-existing Macedonian synonyms. All assume that the authorities' main concern is to uphold the stability of an unethical system that places personal power over the wellbeing of others.

In *Дворскиот поет во апарат за летање*, the authorities' control over knowledge about the past is demonstrated when the Duchess and the astrologer do not reveal to Igor that what he accidentally saw illustrated on a tapestry were flying machines. It soon becomes clear that Igor has seen helicopters without knowing what they are since in this post-apocalyptic future, there is no more sign of them. It is significant that the tapestry also represents the image of a terrifying battle, which suggests that people's weapons of mass destruction had led them to obliterate the world as we know it, leaving almost no trace of the technological developments of the past. In order to prevent Igor from learning the truth, the Duke and his collaborator, Sesil B. de Smil, take Igor on an "expedition" to restore Papa-tema from the thieves who stole it.

In this post-apocalyptic world, where nature dominates, and in which there is no sign of civilisation, the Duke and his company meet various small and scattered groups of people: some look for gold with tools that Igor does not recognise, others have strange and simple clothes and treat the foreigners kindly. At one point they see an observatory, which is completely new to Igor: "Тоа беше нешто мошне различно од сето она што му припаѓаше на нему познатиот свет; нешто таинствено и моќно зрачеше од неа и тој ја гледаше со восхит" ("It was something very different from everything that belonged to the world known to him; something mysterious and powerful was shining from it and he was looking at it with admiration").[15] Such scenes, in which Igor is the only one confused and alienated from his surroundings, dominate the novel. Like the protagonists of many other dystopian novels, Igor seems unable to fit into the existing hierarchy.

The moment when Igor starts to transform by learning something about the past is in his encounter with a young woman named Estrela, who shows him remains of the technology from previous centuries. In a hotel, no longer functioning as a hotel but as some kind of gathering point, they see a group of young women and men in black formal clothes who make gestures with theatrical movements and call themselves the Lost Generation. On the river, there are numerous long, thin boats with people who row on each side, and with a platform on the first boat on which there is a throne and a man tended to by partly naked women—clearly a reference to ancient Egypt. At Igor's comment that he does not understand anything, Estrela explains that there are a lot of things that come from different times, numerous ways of life that coexist almost on the same location.

Не постои повеќе единствена цивилизација. Секој сега живее сам за себе. Сакам да речам – секоја заедница. Всушност – секој може да си одбере време, епоха, култура, обичаи, изглед на облеката, оружје – зависи од вкусот но и од тоа што ќе најде. Едни живеат во еден век, други во друг и како да не се забележуваат едни со други. И ништо заедничко повеќе не ги врзува.

(There is no longer a single civilisation. Now everyone lives for him/herself. I mean—every community. In fact—each one can choose time, epoch, culture, customs, dress code, weapons—depending on the taste and on what they find. Some live in one century, others in another and it is as though they don't notice each other. And they have nothing in common to connect them.)[16]

In this post-apocalyptic setting, all that remains after the wars are scattered, unconnected groups that barely survive and live in fear of being attacked, without having any opportunity to develop their potentials. Estrela laments that there is no longer anything connecting people. What she reveals to Igor about the different groups that emulate past communities demonstrates that she favours the empathy and closeness that have disappeared in the world of the novel, the lifestyle that no longer exists in which there was a sense of community and belonging. All these scenes express the longing for ethical values that, as Igor assumes, must have existed in the past.

In *517*, even mentioning the past is prohibited, so we learn about the principles of this future society not implicitly from the story itself but from an introduction that explicitly provides background information about the order of society and indicates the ethical position of the text. The introduction focuses on a legend that exists clandestinely in this future society, setting up a mystery that will be discovered by the protagonists:

Во градот каде што секое сеќавање беше забрането, каде што узеите беа срамнети до темел и сите книги беа изгорени, каде што немаше часовници, вони редови, работни часови, … таму, во подземјето на тој град, кружеше легендата за 517.

(In the city where every memory was forbidden, where museums were completely destroyed and all books were burnt, where there were no clocks, no bus or train schedules, … and all else that resembled the flowing of time was scorned, there, in the underground of that city, the legend of 517 was circling.)[17]

Various legends of the number 517 exist: that it may refer to one person who somehow gave sense to human life, that it is an enigmatic number which may be revealed with some secret formula, that it may be the name of a street where two lovers lived and which became a symbol of everlasting love—these are just a few of the numerous possibilities. The search for the meaning of the number remains persistent in the novel and efficiently moves the story forward.

The suppression of memory through burning books and destroying museums, as well as the existence of an underground resistance movement, are well-worn devices of dystopian novels (Ray Bradbury's *Fahrenheit 451* [1953] is an example), thus placing this novella in a familiar, if perhaps a bit outdated, context. However, *517* goes a step further by strengthening the power of the present over the past through the dominant concept of presentism that puts the present on a pedestal. Two groups of people dominate: those belonging to "the elite" who uphold presentism and for whom there were no disappointments, no birthdays, no memories or anything that may be associated with time, as the point of life was to live in the present moment; and the very small number of those who were powerless and invisible, who believed that all things can come back.

Although it is not certain when the novel is set, it is probably the not-too-distant future, as the first-person narrator still remembers things that are common today. At one point, speaking about her parents, she says: "*Се сеќавам дека порано си велевме добра ноќ, се бакнувавме, си посакувавме пријатен сон, собара за одмор ја викавме спална соба, но сето тоа е сега забрането*" ("*I remember that we used to wish each other good night, we used to kiss and wish each other pleasant dreams, we used to call the resting room a bedroom, but it is now all forbidden*").[18] What is it about the expected future that seems much grimmer? Alienation is perhaps the most noticeable concern. The five rebels have a much closer relationship among them than any other characters, even closer than family members. The distancing of people is also visible through the fact that any private contact that would bring about emotional involvement is discouraged: people can speak to each other loudly, conveying pieces of information, but they must not speak to each other quietly or show sympathy. Sex is freely practised but no deeper partnerships exist. Nostalgia thus dominates *517*, where the past is shown as encompassing much closer and more honest relationships among people. The protagonists are rather young and have mostly been exposed (except, as it is suggested, in their early childhood) to the presentist system. Therefore, the fact that they could qualify the facades of buildings as banal or the existing ideology as senseless, although they do not have points of comparison for either the architecture or the ideology, is an expression of their nostalgia—longing for an idealised past.

The architecture of Skopje is also one of the central concerns of the novella and is closely related to nostalgia and the desire to bring back the old look of the city. It has ethical implications as well as aesthetic, as could be expected given the "Skopje 2014" project. The government's purpose with the project was not only to give the city another aesthetic look but to promote a different identity, one not based on historical research but that would contribute to the rise of nationalism. This aspect of "Skopje 2014," along with its financial implications, contributed to the project becoming an ethical and identity issue as well as an aesthetic one. The protagonists of *517* are, therefore, trying to reconstruct the past look of Skopje by bringing together pieces of old photographs they manage to find.

Nostalgia in *The Ship. Consarchy* is part of the ethical system that differentiates clearly between positive or morally upright characters and negative or corrupted characters. Slobodan Savin and Tatjana Urova are undoubtedly presented as positive characters: they both have artistic tendencies (Slobodan is a holograph designer, Tatjana understands and appreciates art), are not interested in money, start a relationship after being single and are nostalgic for the past. Slaven Pakanski, Sofija Pakanska or Anton Poljakov, on the other hand, are presented as negative characters mostly through the fact that their priority in life is to be successful, wealthy and powerful. What goes along with these characteristics is sexual promiscuity—specifically, the Pakanskis and Poljakov are either unfaithful to their partners or do not mind having intercourse with someone who is married. According to the somewhat traditional moral perspectives that have dominated literature and society and are still present today, faithfulness and loyalty in marriage are signs of morally virtuous behaviour. The Pakanskis and Poljakov, contrary to Savin and Urova, do not adhere to such traditional norms, consequently the sexual relations each of these characters has outside of marriage are linked to their moral corruption in general. This aspect of the novel reflects, rather than questions, established ethical norms.

In this context, nostalgia represents an inclination towards the values of the past, when people were more empathetic and more dedicated to each other. "На ова место треба да се каже дека Слободан Савин носи една емпатиска, лирична и сентиментална личност" ("We should mention here that Slobdan Savin is an empathic, lyrical and sentimental person"), the narrator explains, "и дека него на чуден начин го допираа старите нешта:

сронатите или напуштените куќи, старите улици и маала … црно-белите фотографии кои се наоѓаа по антикварните дуќани, изветвените премети на кои неочекувано ќе наидеше" ("and that he is strangely touched by old things: ruined or abandoned houses, the old streets and mahalas … black and white photographs in the antiquarian shops, old object that he would unexpectedly encounter"), adding that such objects "за него претставуваа бивши вредности, и беа зафрлени или отфрлени сеќавања на некои одвеани времиња и веќе исчезнати луе, идеи, занеси, заложби" ("represented for him former values, and were thrown or rejected memories of some past times and of people, ideas, desires, strives that have already disappeared").[19] It is easy to recognise this kind of nostalgia as a reflection of the nostalgia frequently expressed in the discourse in present-day Macedonia, especially through the attachment to black and white photographs and the praising of old streets and old mahalas that have disappeared with the urbanisation of Skopje.

The novel also implicitly reacts to the "Skopje 2014" project when describing the look of the city in 2039:

> Таа мешаница од градби и стилови што кричеа од помпезност – и тоа Савин никако не можеше да го свари – всушност беа идеолошки знаци на распознавање на поранешниот неоантички доминион Аксиј чиешто официјално име сега, две децении подоцна, гласеше – *Конзархија Брод и Крајбрежје*
>
> (This mixture of buildings and architectural styles that was so visibly pompous—which Savin could not stand—were actually ideological signs of recognising the former neoantique dominion Aksij whose official name now, two decades later, was—*Consarchy Ship and Coastal Area.*)[20]

Along with the criticism of the mixture of incompatible architectural styles, this comment also assists in identifying Savin as a character whose attitudes are evaluated as moral since he does not approve of the way the old look of Skopje has been transformed.

Wars and Dehumanisation

In addition to the nostalgia that seems to be a common concern in Macedonian dystopian fiction, the three works in question treat various other ethical issues, each focusing on a different one. The main concern explored in *Дворскиот поет во апарат за летање* is war as a cause of the destruction of the world. In a conversation with one of the soldiers, Igor hears the rumours spread among the soldiers that there once used to be carriages that ran on fire and were used in wars, boxes in which there were living images, houses with a hundred floors and black fish-like boxes that used to go under water. Most soldiers do not believe such tales, but the soldier telling those present about such strange machines and buildings from the past insists: "И кај нас имало такви нешта, и машини што летаат, и кочии што одат со помош на оган, и сѐ, но потоа дошла војна..." ("There were such things, even in our region, including flying machines, carriages that run on fire, and everything, but then came war...").[21] This implies that he is talking about a large-scale war; a war so extensive that it wiped out most marks of the achievements of civilisation. Parallels can be made between the Duke and De Smil with the military leaders of the past, who have very likely contributed to the annihilation of civilisation. The Duke's threats to all the people they encounter on the expedition, alongside his evil and dark desires for war, are distinctly unethical: he puts his selfish and violent goals above the wellbeing of his people.

517 touches upon dehumanisation and the eradication of identity. This is why there are no names in the novel. Neither the presentists nor the anti-presentists have any identity, so that the narrator chooses the five rebels names that fit their character: The Persistent One, The Withdrawn One, Mixer, Picker and Blabber. There are also other aspects in the novella that refer to dehumanisation with bitter irony and touch upon the ethical concerns of present-day Macedonia, predicting they would take a more exaggerated form in the future. One of the most serious problems facing Macedonia today is its state-owned health-care system, from a lack of basic medications and hospital materials, to old hospital buildings and old medical equipment, to the lack of an organised system of dealing with patients. The most frequent question of a patient who comes to a doctor's appointment is: "Who is last?"—the purpose of the question being to know after whom to enter the doctor's office as there are usually a lot of people waiting. This question is forbidden in the society of *517* to avoid reminding patients of the passage of time—so although they wait for a long time, they do not identify any reason to be discontent. The problem is thus resolved (the patients do not complain of waiting) without being resolved (they still wait for a long time).

Dehumanisation is also addressed in *Бродот. Конзархија*, although in a very different way. The people inhabiting the world of the novel can use upgrades for their bodies (so they can look different, for example), and they can change their clothes electronically. These science fictional elements in Osmanli's novel indicate scepticism and occasionally resistance towards new technological developments. The novel thus reflects certain earlier dystopian narratives, such of *Brave New World*, which Evie Kendal describes as "philosophical thought experiments and ... socially conservative, technophobic cautionary tales."[22] In Osmanli's novel, the condemnation of technology is not very explicit, but the quick upgrading is suggestive of artificial beauty and preoccupation with external appearance as opposed to the intellect. *Бродот. Конзархија* shows corruption on the highest level, and it is only Slobodan and Tatijna, who show true solidarity, that are presented in a positive light. It is somewhat difficult to interpret them in a wholly positive way, however, since although they are not impressed by power or wealth, they are still part of the corrupted world and do not oppose it: Tatjana works at Pakanski's office, while Slobodan does not in any way disapprove of the system—he is just not interested in money himself.

Ambiguous Visions of the Future

The pessimistic ending is one of the standard elements of dystopian fiction, although this trend has altered somewhat since the 1970s. The last sentence of Offred in Margaret Atwood's *The Handmaid's Tale* (1985), for example, suggests that light is as possible as darkness, and in the concluding chapter, the reader is assured that Gilead is no longer in power even if the post-Gilead society is imperfect. Tom Moylan examines the various possibilities of interpretation that dystopian novels employ to convey their position towards reality, stating that although all dystopian texts present pessimistic social alternatives, "some affiliate with a utopian tendency as they maintain a horizon of hope (or at least invite readings that do), while others only appear to be dystopian allies of Utopia as they retain an anti-utopian disposition that forecloses all transformative possibility, and yet others negotiate a more strategically ambiguous position somewhere along the antinomic continuum."[23] Such ambiguous endings are present in various forms in the three Macedonian works discussed. The texts of Urošević, Mihajlović Kostadinovska and Osmanli exhibit a tone of pessimism, however they are simultaneously ambiguous about the persistence of the grim society in the future and "maintain a horizon of hope," as Moylan says. Among them, *517*, like Atwood's novel, includes a Post Scriptum, which functions as a concluding chapter and suggests a somewhat brighter future than the one described in the novella.

The ending of *Дворскиот поет во апарат за летање* is not only unexpected, but it also complicates the understanding of the novel, including how it is analysed and interpreted, and what its ethical assumptions may be. "Проклет идиот, ни ја расипа сцената. Од каде ли се појави!" ("What a fool, he destroyed our scene. Where did he come from!") shouts Sesil B. de Smil at the end of the novel, to which the assistant director adds: right, like the helicopter before.[24] In other words, the ending reveals that all previous developments in the novel were actually part of a film that was being shot, in which Igor is the only one with no idea of what was happening, believing that his life in the castle and the later expedition to search for Papa-tema were real. Looking back at the novel from this perspective, there are occasional allusions that may have given clues about the possibility that it is all a kind of reality show, but they are not clear until after the final reveal. In this context, the novel may be compared to the film *The Truman Show*, which was released in 1998—two years after *Дворскиот поет во апарат за летање* was published—and reflects similar concerns with the growing popularity of reality television. Taking this into consideration, the perspective of the novel changes from a post-apocalyptic future expressing fear about large-scale catastrophic wars, to a different kind of future that has also seen the elimination of the world as we know it: the old world of privacy has been destroyed, for better or for worse, to be replaced by one in which the media dominate so much that there is almost no place to hide from them. Comparison can be made with William Morris's *News from Nowhere* (1890), which ends with the revelation that the events described were a dream, although this does not deny the importance of the society imagined within the dream. Thus, despite the ending revealing it has all been a film, the developments within that film certainly invite us to explore the issues of an apocalyptic war and the responsibility of the citizens of the world to prevent such wars from happening.

Дворскиот поет во апарат за летање is certainly the most ambiguous novel among the three, both in regard to its ending and the ethical implications derived from it: Igor's end, considering that he was in danger minutes before the reveal of the film setting, may seem comforting for a moment, relieving the tension of what the protagonist may face. A few moments later, however, it already seems pessimistic: although everyone is just an actor or actress, and the dangers were not real, the standards of solidarity and justice that Estrela seemed to epitomise and fight for also suddenly vanish—they were just an act and it seems impossible that they would ever be part of the real world.

Although the ending of *517* is also ambiguous in regard to optimism or pessimism, it is less ambiguous in regard to the ethical values it advocates such as solidarity and rebellion against an authoritarian ideology. In the course of the story, optimism is built up as the five characters who barely knew each other in the beginning work together towards achieving their goal in a coordinated manner. They even manage to carry out coordinated guerrilla actions: they stop people on the streets to ask them where street 517 or building 517 is, they put the price of beer or of a concert at 517 denars, and so on, so that the number gets stuck in people's minds. The final discovery of 517 reveals the secret behind the presentist ideology, contributing to its undermining. When The Withdrawn One sees a photograph from the past with a long, straight street ending at a ruined building with a large clock, she remembers that she fell near that building when she was a child, wondering why the clock, which she imagined as a giant, did not come to rescue her. Her mother tells her that the ruined building was the old railway station in Skopje, destroyed in the great Skopje earthquake of 1963, and that the arrows of the clock had stopped at the moment of the earthquake—5:17. "*Значи… 517 е тој… тоа, тоа не е херој, тоа е само часовницкот на болката, часовникот кој не умрел, но останал да стои и да сведочи за една лузна во времето, во рпосторот, во нас…*" ("*So… that is 517… it, it is not a hero, it is just a clock of the pain, the clock that did not die, but remained standing in order to witness a scar in time, in space, in us…*").[25] Although this is the end of the rebellious company, which gets captured by the presentist police the following day, it is also the beginning of a wider and more organised anti-presentist movement, inspired by the five brave if unfortunate protagonists, as the narrator informs us in the post scriptum of the novella. The narrator ends by stating that "кога антисегашизмот дојде на власт времето се вгнезди во градот во својот вообичаен тек" ("when anti-presentism came to power, time re-entered the city with its usual flow)," implying afterwards that certain negative aspects continue in the new ideology as well.[26] People will again be obsessed with soap operas and political debates (an ironic reference to the two most popular TV shows in present-day Macedonia) and will soon forget the dangers of the totalitarian regime, thus making themselves susceptible to fall victims to a new regime that manipulates the truth because "никој веќе не обрнуваше внимание на такви детали" ("no one paid attention to such details any more").[27]

Бродот. Конзархија seems to provide an optimistic ending: the holographic image Slobodan Savin creates as his masterpiece engulfs him as a water wall, then withdraws and reveals an area coming out of "митскиот потоп" ("the mythical flood").[28] The river and the air seem clear and it appears to be easier to breathe, thus Savin plunges into pleasant memories. Although this may only be an illusion, it is one that brings joy to the protagonist. The optimism, however, is present only on the level of the individual, whereas, on the level of society, there do not seem to be any signs of change.

In the context of dystopian fiction in general, what these three Macedonian novels share is a reaction to the political and social circumstances faced by their authors, such as the threat of war or the lack of concern for the health-care system. They all express nostalgia for a better past and attempt to defy the governing structures that aim to conceal history. *Бродот. Конзархија* and *517* contain certain elements that resonate with the dystopian fiction of the first half of the twentieth century, including a totalitarian society ruled either by a strong political government or by a large corporation with ties to the governing elites. *Дворскиот поет во апарат за летање* and *517* introduce additional aspects to the dystopian genre: the post-apocalyptic warning of the dangers of a total war in *Дворскиот поет во апарат за летање* is intertwined with the ambiguous possibility of a post-reality future in which privacy is eliminated under the influence of media, while the humorous treatment of themes in *517* gives a fresh perspective to the pessimistic future.

Notes

1. Claeys, "Origins of Dystopia," 109.
2. Su, *Ethics and Nostalgia*, 2.
3. Su, *Ethics and Nostalgia*, 9.
4. Niemeyer, "Introduction," 1.
5. Niemeyer, "Introduction," 2.
6. Stableford, "Ecology and Dystopia," 273.
7. БИРН Македонија, "*Досие*." The BIRN (Balkan Investigative Reporting Network) Macedonia has, since 2014, published a series of articles under the title "Досие: Скопје 2014" (Dossier: Skopje 2014) that includes research, news and analyses of the project.
8. Капушевска-Дракулевска, *Во лавиринтите на фантастиката*, 91. All translations from Macedonian texts are my own.
9. Урошевиќ, *Доврскиот поет*, 15.

10. Михајловиќ Костадиновска, *517*, 6.
11. Османли, *Бродот. Конзархија*, 16.
12. Carey, *Faber Book of Utopias.*
13. Russell, *Cambridge Companion of Virtue Ethics.*
14. Todorova, "Introduction," 4.
15. Урошевиќ, *Доврскиот*, 53.
16. Урошевиќ, *Доврскиот*, 128.
17. Михајловиќ Костадиновска, *517*, 5.
18. Михајловиќ Костадиновска, *517*, 19.
19. Османли, *Бродот. Конзархија*, 7–8.
20. Османли, *Бродот. Конзархија*, 23.
21. Урошевиќ, *Доврскиот*, 58.
22. Kendal, "Utopian Visions," 90–91.
23. Moylan, *Scraps*, 147.
24. Урошевиќ, *Доврскиот*, 207.
25. Михајловиќ Костадиновска, *517*, 88.
26. Михајловиќ Костадиновска, *517*, 94.
27. Михајловиќ Костадиновска, *517*, 95.
28. Османли, *Бродот. Конзархија*, 197.

Works Cited

Atwood, Margaret. *The Handmaid's Tale.* Toronto: Emblem, 2017.

Bradbury, Ray. *Fahrenheit 451.* New York: Ballantine Books, 2008.

Carey, John. *The Faber Book of Utopias.* London: Faber and Faber, 2000.

Claeys, Gregory. "The Origins of Dystopia: Wells, Huxley and Orwell." In *The Cambridge Companion to Utopian Literature*, edited by Gregory Claeys, 107–131. Cambridge: Cambridge University Press, 2010.

Huxley, Aldous. *Brave New World.* New York: Harper Perennial, 2006.

Jameson, Fredric. *Archaeologies of the Future.* New York: Verso, 2005.

Kendal, Evie. "Utopian Visions of 'Making People': Science Fiction and Debates on Cloning, Ectogenesis, Genetic Engineering, and Genetic Discrimination." In *Biopolitics and Utopia: An Interdisciplinary Reader*, edited by Patricia Stapleton and Andrew Byers, 89–117. New York: Palgrave Macmillan, 2015.

Morris, William. *News from Nowhere and Other Writings.* London: Penguin Classics, 1993.

Moylan, Tom. *The Scraps of the Untainted Sky: Science Fiction, Utopia, Dystopia.* Boulder: Westview Press, 2000.

Niemeyer, Katharina. "Introduction: Media and Nostalgia." In *Media and Nostalgia: Yearning for the Past, Present and Future*, edited by Katharina Niemeyer, 1–23. Basingstoke: Palgrave Macmillan, 2014.

Orwell, George. *Nineteen Eighty-Four*. London: Penguin, 1998.
Russell, Daniel C., ed. *The Cambridge Companion of Virtue Ethics*. Cambridge: Cambridge University Press, 2013.
Stableford, Brian. "Ecology and Dystopia." In *The Cambridge Companion to Utopian Literature*, edited by Gregory Claeys, 259–281. Cambridge: Cambridge University Press, 2010.
Su, John J. *Ethics and Nostalgia in the Contemporary Novel*. Cambridge: Cambridge University Press, 2005.
Todorova, Maria. "Introduction: From Utopia to Propaganda and Back." In *Post-Communist Nostalgia*, edited by Maria Todorova and Zsuzsa Gille, 1–13. New York: Berghahn Books, 2010.
Zamyatin, Yevgeny. *We*. New York: Random House, 2006.
Капушевска-Дракулевска, Лидија. *Во лавиринтите на фантастиката*. Скопје: Магор, 1998.
Михајловиќ Костадиновска, Сања. *517*. Скопје: Темплум, 2015.
Османли, Томислав. *Бродот. Конзархија*. Скопје: Магор, 2016.
Урошевиќ, Влада. *Доврскиот поет во апарат за летање. Дива лига*. Скопје: Магор, 2004.

CHAPTER 13

Cairo in 2015 and in 2023: The Dreadful Fates of the Egyptian Capital in Jamil Nasir's *Tower of Dreams* and Ahmed Khaled Towfik's *Utopia*

Anna Madoeuf and Delphine Pagès-El Karoui

Immense, luxurious glass skyscrapers towering over endless miserable suburbs, inextricable traffic jams, pollution, terrorist attacks, followed by earthquakes... In a world where the inequality gap has grown wider by every measure, the West refuses Egypt credit, while its capital of 35 million residents is shaken by intense social and tectonic jolts... This is a glimpse of Cairo in 2015, according to Jamil Nasir's 1999 novel *Tower of Dreams.*[1]

Published nine years later, Ahmed Khaled Towfik's portrait of Cairo in 2023, in his novel *Utopia* (2008 as يوتوبيا), is no more uplifting. The Egyptian territory is split into two warring spaces: the rich westernised Egyptians live entrenched in a few gated residential communities on the Mediterranean coast (Utopia is the name of the community where one of the narrators lives)

A. Madoeuf (✉)
Université de Tours, Tours, France

D. Pagès-El Karoui (✉)
INALCO, Sorbonne Paris Cité University, Paris, France

Z. Kendal et al. (eds.), *Ethical Futures and Global Science Fiction*, Studies in Global Science Fiction,
https://doi.org/10.1007/978-3-030-27893-9_13

whereas the rest of the territory is reduced to an immense shanty town where the remaining, miserable population is left to rot. The image of the gated community, which reflects the country's current urban fragmentation, serves as a spatial metaphor for denouncing the worsening of social inequalities and the disappearance of the middle classes. This speculative, futuristic novel opens with the author warning Egyptians about the current state of their society, and the dark future that may result from it, and closes with a revolt against the rich. After the 25 January 2011 revolution, the novel was sometimes seen, with hindsight, as having been prophetic.[2]

These two narratives belong to a dystopian genre of literature—that is, exactly opposite of a utopia, when novelists imagine a world worse that their present society, emphasising all its shortcomings—and are set in Cairo, the largest city in the Middle East and on the African continent.[3] Urban settings are often the preferred backdrop for science fiction novels, in which the city serves as a stage on which the author can lay out a critical vision of urban and social change[4]:

> La ville de science-fiction, produit de sociétés injustes et inégalitaires, apparaît le plus souvent comme l'accomplissement inéluctable d'une série de problèmes et de dysfonctionnements qui hypothèquent l'avenir des métropoles contemporaines: congestion, pollution, déficience des services urbains, dissolution du lien social, accroissement des inégalités économiques et des disparités spatiales, violence quotidienne… De Babel à New York en passant par Babylone et Coruscant, la science-fiction fait le procès d'une ville insoutenable qui n'est qu'un reflet des villes d'aujourd'hui.
>
> (The unjust and inegalitarian cities of science fiction often appear as the inevitable consequence of a series of problems and dysfunctions that threaten the future of contemporary metropolises: congestion, pollution, deficient urban services, the dissolution of social bonds, the growth of economic inequalities and spatial disparities, daily violence, etc. From Babel to New York, Babylon to Curscant, science fiction frequently denounces the way tomorrow's cities reflect today's unsustainability.)[5]

We have chosen to compare these two novels for two reasons. First, because they are among the very few science fiction novels by Arabic-speaking authors and both are available in English. Second, because they represent some of the few attempts to imagine a city of the Global South and, what's more, an "Arab" or "Oriental" city, in the future, a space-time that fiction (literature and cinema, in particular) often reserves for cities in wealthy countries, whose representations, realities and landscapes seem more easily transposable to the future. Science fiction remains a fairly marginal and academically neglected genre in Arabic literature, having first appeared in Egypt in the 1950s. Since then, Egypt has remained the leading country for the production of Arabic science fiction, with *Utopia* the first of the genre to become a bestseller.[6] Over the past decade, there has been a surge in Arabic science fiction, following the disillusions of the 2011 revolutions.[7]

In light of the current context in Egypt and Cairo and the time when these novels were produced, we propose to analyse the two novels' parallel geographic representations of Egypt's capital city, as they imagine it in the future. We will examine the similarities, or lack thereof, between the two imaginaries of Cairo promoted by these authors, who have very different backgrounds and wrote ten years apart (the 1990s and the 2000s). The son of a Palestinian refugee, Nasir is an American author who writes in English and who spent a portion of his childhood in the Middle East. Born in 1962 and dying prematurely in 2018, Towfik is an Egyptian science fiction author who writes in Arabic. Prior to the publication of *Utopia,* he was considered a minor (even if prolific) author of fiction for adolescents. After the success of *Utopia,* he began to be celebrated as a significant author.[8] We will seek to identify some of the historical events that influence their imaginaries of Cairo and to evaluate the critical, even prophetic, dimension of their fiction, insofar as dystopias serve primarily to denounce the dysfunction of contemporary society. We will begin by considering the genre—science fiction—to which the novels belong and then examine the ethical issues of social justice, urban dysfunction and segregation. Finally, we will analyse economic inequality, environmental ethics and the interaction between ethics and global politics in each novel.

Parallel Futures, or Parallels in the Future?

Two Rare Science Fiction Imaginings of Cairo

In general, cities of the Global South are less commonly depicted than those of the North in science fiction and related genres or tend to be ignored in academic discourse when they do exist. There is an overabundance of work associating futurism, modernity, inventiveness, escape and imagination with cities of the North, including classic science fiction films such as *Brazil* (1985), *The Fifth Element* (1997) and *Blade Runner* (1982). Cities of the South are presented as, simply or presumed to be, too bogged down in the litany of their dysfunctions to fully take part in the process of imagining the future or to the extent that they do, it is to conjure up by way of continuities and extrapolations its problems. It is as though their movement to the future is blocked since the process of imagining the future amplifies their current problems, which are augmented exponentially by time. The cities of the "Orient," which are extreme cases of the South, are represented as being weighed down and held back by the inertia of their Orientalised past and struggle to acquire an imaginary future other than one of chronic ills staged to a backdrop of stereotyped changelessness.[9] If, as in *Tower of Dreams* and *Utopia*, Cairo is projected into some future time, it is along with its baggage of structural defects; imaginaries of its fate are not necessarily signs that it has a future.

Both Nasir's *Tower of Dreams* and Towfik's *Utopia* belong to a genre of social science fiction that projects the reader into a near and supposedly credible future. *Tower of Dreams* belongs more to the genre of science fiction in the strict sense, insofar as the scientific and technological innovations in *Utopia* are relatively minor (see below on Pyrol, Libidafro and Phlogistine), whereas in *Tower of Dreams*, references to virtual worlds and cyberspace are much more pervasive (supercomputers, holograms, etc.). Despite their fictitiousness, these speculative narratives are anchored in reality. Places, in particular, are specified and often retain some familiar features. This is the case in *Tower of Dreams*, which is set explicitly in Cairo in 2015—a time that was relatively close to the book's writing at the very end of the twentieth century. In *Utopia*, the future is no more distant since we learn in the book's first chapter that it is set 50 years after the Israel-Arab war of 1973, in other words, in 2023. This highly symbolic date serves as a clear reminder of the centrality in Egypt, like in most Arab countries, of the conflictual relationship with Israel. That is, in fact, one of the major novelties of the world Towfik invents, since in 2023, tensions with the Israelis have completely disappeared, along with all memory of the conflict: "لا أعرف لماذا في حقبة ما كان المصريون يكرهون الإسرائيليين؟" ("I don't know why—at some point in time—the Egyptians used to hate Israelis").[10] Unlike in *Tower of Dreams*, the urban landscapes of *Utopia* have nothing futuristic about them, even within the gated communities. Perhaps Towfik intends for this to signify the social regression of the two worlds he depicts. The only difference with today pertains to the bodies of the rich youth (the narrator has purple hair that stands up vertically in a Mohawk, is covered in piercings, dyes his teeth various colours and has a fake wound on his forehead).

Tower of Dreams is heavily spatially referenced. Several Cairo neighbourhoods appear in the novel, including Darb al-Ahmar (situated in the historic city), Shafa'i (the "city of the dead," Cairo's ancient Necropolis), Garden City (a pericentral neighbourhood that emerged from a villa subdivision at the beginning of the twentieth century) and Mohandiseen (an upmarket neighbourhood, developed since the 1960s). The famous Tahrir Square is also mentioned, and the novel's protagonist stays at the Nile Meridien hotel on Roda Island (today, the Hyatt). Towfik's geographical references are less abundant; *Utopia* is very short and devotes very little time to lengthy descriptions, remaining focussed on the action.

Orient of the Past, Orient of the Future: A Well of Images

Blaine, the American protagonist, collects dreamlike images from Cairo and other cities of the Global North, places where tourists no longer travel but about which people still dream. He comes to collect fragments of representations for a multinational corporation specialised in locating, collecting, sorting and selling these images that Westerners crave and which are necessary for their sensorial and psychological wellbeing, as well as for constructing their dreams.

Despite many ups and downs, the "Orient" remains a fertile and unique source of images for the rest of the world, and in that respect, at least, Cairo remains rich—indeed, this may be its only wealth. This is obviously a fiction, and a futuristic one, yet the theme strangely echoes the very essence of the Orientalist travel narrative, born almost two centuries earlier, and, in particular, one magisterial and incantatory phrase from the opening of François-René de Chateaubriand's *Itinéraire de Paris à Jérusalem* (1811), a work considered to be among the first travelogues of the Orientalist genre: "J'allais chercher des images, voilà tout" ("I went looking for images, nothing more").[11] Two centuries apart, and in literary registers as different as travel writing and science fiction, the same impetus that drove Chateaubriand towards the Orient drives the protagonist of Nasir's novel towards Cairo. Their shared motive is explicitly to capture images. What *Tower of Dreams* suggests is that the Orient remains, even in the imaginary of its future and its imagined future, a well of iconography that is both quantitatively inexhaustible and limited in variety.

There is no trace of Orientalism or of the Oriental city in *Utopia*, in which the urban environment is divided into two contradictory models, the gated community of the rich and the immense shantytown of the poor.

This binary, not to mention caricatured, vision of urban space is also reflected in the novel's structure, which is based on five parts alternatively titled *predator* and *prey*. The narrative is recounted, in turns, by two narrators, the first of whom lives in Utopia and whose name remains unknown (we will call him the Predator), and the second, named Gaber, who belongs to the other world, that of the poor.

Predator and Prey

Utopia's first narrator is a 16-year-old man, the egocentric and decadent son of a wealthy industrialist who, despite his numerous sexual relations and use of all kinds of drugs, is bored to death. As a break from his life's monotony, he decides to follow the example set by his friends and undergo their rite of passage, which consists of murdering a poor resident of the slums and bringing one of his limbs back to Utopia as a kind of trophy. This odious narrator does not experience redemption through his contact with the slums and commits several atrocities: he rapes Safeya, Gaber's sister, then kills Gaber once he has driven them home. He then cuts off one of Gaber's arms, which he embalms. He will now be able to brag about having completed his initiation ritual for his friendship circle. In contrast, Gaber appears like a kind of prophet of Egypt's downfall. Although he has no illusions about the personalities of the two young characters from Utopia, he chooses not to kill them, despite his hatred for them, since his refusal to employ violence is the only thing connecting him to his humanity. His death and Safeya's rape will be the spark that ignites the flames of revolt. Social inequalities are thus at the heart of the novel, focusing on the opposition of its two main characters: the Predator, a rich man from Utopia, and Gaber, one of the intellectual poor who lives in the world outside. To signify their absolute alterity, the rich refer to those who do not live in Utopia or other gated communities as "the Others."

Literary Fiction, Revealer of Urban Transformations

Urban Dysfunction

In *Tower of Dreams*, Cairo is "a collapsed Third World city," immense, sprawling, foetid and full of waste, inhabited by 35 million residents (of a total of 100 million in the country), the perimeter grows by a kilometre every year as poor suburbs, consisting of temporary, run-down and improvised housing built along transport corridors, surround and suffocate the city—to the extent one still even exists.[12] The population grows ceaselessly and every month 150,000 people move into the capital, which is described as a "sea of disasters."[13]

Traffic jams and various forms of pollution have grown, and in this suffocating city, with its leaden air, people doze during the day, then drive at breakneck speeds at night to move from one neighbourhood to another. The situation is particularly tense since the economy is in shambles and the West has just refused to grant additional aid, needed in the aftermath of an earthquake, on grounds that Egypt is overpopulated and poorly regulated. People invoke God constantly; terrorist attacks become more frequent; casinos, cinemas, alcohol and cabarets are targeted by violent attacks and constant criticism. People fear another earthquake that will doubtless be more intense—a final reckoning. These social and tectonic jolts subject the city to tumultuous and harmful times and make life so uncertain and vain that people seem to just drift from day to day.

Several key themes can be identified and developed from Nasir's inglorious portrait of Cairo in 2015. The first remark is that the general atmosphere and setting appear to be exaggerations of the catastrophic 1990s descriptions of large cities of the Global South. The following examples, although they seem like caricatures in hindsight, are representative of the dominant discourses of Western nations:

> Au Sud: des mégapoles disloquées. Des centaines de millions d'hommes, des dizaines de millions d'enfants, déracinés, abandonnés à leur sort; pollution et spéculation, infrastructures dégradées ou inexistantes, misère et violence: les villes du tiers-monde semblent annoncer la décomposition du modèle occidental de civilisation urbaine.
>
> (In the South: broken megalopolises. Hundreds of millions of men, tens of millions of children, uprooted, abandoned to their fate; pollution and speculation, run down or inexistent infrastructures, misery and violence: the cities of the Third World seem to herald the breakdown of the Western model of urban civilisation.)[14]

The passage devoted to Cairo in *Géographie universelle* was not particularly enthusiastic either:

> La situation du Caire est un avant-goût des difficultés à venir: 4 % de croissance annuelle, des densités exceptionnellement fortes dans le cœur même de l'agglomération qui rassemble 9 millions d'habitants en 1985, avec 46 %

> de logements sans eau, 53 % dépourvus d'assainissement, des conditions de transport désastreuses, une voirie inadaptée dans un tissu urbain particulièrement dense, des réseaux d'adduction d'eau et d'électricité à bout de souffle. Cependant, le pire n'est pas toujours certain, et la capacité d'ajustement reste considérable. Des associations peuvent prendre en charge la vie de leur quartier. La gestion urbaine est un immense bricolage, qui permet une adaptation aux vicissitudes d'un pays fortement endetté, en proie au terrorisme et aux mouvements insurrectionnels.
>
> (The situation in Cairo provides a taste of the difficulties to come: 4% annual growth, exceptionally high densities in the heart of an agglomeration of 9 million people in 1985, with 46% of households lacking running water, 53% without sewer hookups, disastrous transport infrastructures, an inadequate road network in a particularly dense urban fabric, water and electricity grids at their breaking point. However, disaster is not always inevitable, and people's capacity for adaptation remains considerable. Associations sometimes take on the role of managing their neighbourhood. Urban management is a vast process of improvisation that allows for adapting to the ups and downs of heavily indebted country beset by terrorism and insurrection.)[15]

Discussions of Cairo's demography never failed to mention very high population numbers and extreme growth, and the media discourse tended to be monolithic: "La capitale égyptienne, entraînée par une croissance de 4% par an, est aujourd'hui confrontée au défi que représentent chaque année 500 000 citadins supplémentaires" ("Egypt's capital, swept along by a growth rate of 4% per annum, today faces the challenge of 500,000 additional residents in the city each year").[16]

In *Utopia*, ten years later, Towfik's portrait of Cairo outside the gated communities remains dire. The description of the city of the poor, reduced to a giant slum, is apocalyptic. Public services have disappeared: electricity and sewer networks have stopped working, the metro has been abandoned and cinemas no longer exist—even the word "cinema" has disappeared from people's memories, a sign of the vanishing of culture. The inhabitants have been reduced to an animal-like state, struggling against one another to find food and avoid illness or seeking escape from their miserable condition through alcohol, drugs or religion. One of their only opportunities to find work is in *Utopia*, where they carry out the most drudgerous tasks and are tolerated during the day but must leave at night on a bus that carries them home. Despite their physical separation, the two worlds have one point in common: violence as a symbol of their dehumanisation and the total extinction of ethics—the disappearance of all moral principles. The only one character who is different from both the rich and the poor is Gaber, an intellectual, who delivers long speeches about how growing inequalities and the selfishness of the rich led to this severe segregation and state of violence. Gaber is unique in his non-violence, standing apart from the other members of his gang. Gaber's sense of moral superiority, however, leads to his death, although this triggers the revolution. Towfik thus condemns a general trend which has developed in Egypt since the 2000s, the rich escaping Cairo to live in gated communities that have flourished in the desert. If both rich and poor are depicted as evil in Towfik's *Utopia*, leaving no faith in humanity, the segregation of rich and poor is presented as the principal cause of this calamitous situation.

Vertical Segregation/Horizontal Segregation

While the two novels converge in the vision of urban dysfunction, their authors diverge as to the urban model they imagine to be predominant in the Cairo of the near future. Nasir develops the theme of vertical segregation, whereas Towfik envisages horizontal segregation.

The social geography of *Tower of Dreams* follows a centre-periphery model from the banks of the Nile towards the outlying developments but, above all, takes the form of vertical segregation that establishes a sharp social stratification. The rich live apart, perched atop giant secured towers between which they move in sophisticated flying limousines, an imported technology. The innumerable poor, meanwhile, are everywhere else, getting by as they can and occupying the interstitial spaces of this tumultuous city, living pressed against one another, in every sense of the term. The rich can always escape abroad, but the poor have no choice but to remain.

Socio-spatial segregation, which is a core theme in *Tower of Dreams*, clearly appears to have increased between 1990 and today but has followed a different pattern than that anticipated by Nasir, who imagined that the "anarchic" lower-class peripheries would make up the majority of the city in the long run and smother the centre. Indeed, the 1990s and 2000s witnessed the emergence of numerous vast residential quarters around Egypt's capital, some of which, in the form of gated communities, were closed to the outside. Today, the surface area of these compounds exceeds that of the lower-class suburbs, where the housing is much more compact, and they have contributed the most to the agglomeration's peripheral sprawl.[17]

Vertical expansion, for which the regional model would be Dubai, with its futuristic neighbourhoods prickling with towers and crowned with the highest skyscraper in the world (Burj Khalifa), never really took hold in Cairo. The *Cairo 2050* development project, launched in 2007, sought to transform Cairo into a green, global and connected city. Its aim was to endow Cairo with what are currently considered the elements of urban modernity (densification of the downtown through the construction of high towers for offices, luxury hotels, leisure activities, etc.) in order for it to preserve its place as the "leading capital of the Middle East."[18] This project was sharply criticised, particularly after the revolution, among other things because it reinforced socio-spatial inequalities instead of resolving them.

In *Utopia*, the rich choose to protect themselves from the poor by locking themselves in closed residential cities. The theme of enclosure is fundamental in the novel, the opening scene of which consists of the narrator and his girlfriend watching a poor wretch who tried to illegally enter the gated community, which is protected by barbed wire and gates, being shot dead by a US marine. Retired US marines provide surveillance in the city and prevent anyone without authorisation from entering. The internal organisation of the opulent city follows a rigorously zoned pattern: the Gardens quarter houses schools and places of worship (mosques, churches and synagogues), while giant shopping malls and large villas are located in the Malls quarter. In order to remain connected with the world, Utopia has an interior airport, so its residents can avoid having to cross the poor area and risk being lynched along the way. The community has its own newspapers, laws and courts and is governed by a council of wise men, presided over by the most eminent businessmen among them. Only one cleavage persists in *Utopia* between the generations: the spoiled youth of Utopia are idle and depraved, partaking openly and excessively in sex and drugs, while their parents still have the hypocrisy to hide their vices behind the veil of respectability and religion. Whereas the youth no longer believe in anything, the elderly cling to religion out of fear of losing their identity. More than just the city, these dystopian fictions also reinvent Egypt's environment, society and economy, as well as regional geopolitics.

Dystopias: The Reinvention of Worlds

Grand Projects and Black Clouds

In *Tower of Dreams*, since the flooding of the Qattara Depression in 1996 and the digging of a canal from the Mediterranean, Egypt's once mostly stable climate has become stormy and unpredictable. The theme of grand public works projects evoked by Nasir is, today more than ever, at the forefront of actual Egyptian current affairs. In August 2015, President-General Al-Sisi inaugurated, with great fanfare, the widened Suez Canal, presented as the "cadeau de la mère du monde au monde" ("the gift to the world from the Mother of the World").[19] The president had previously unveiled to the world a series of Pharaonic development projects, among which were the creation *ex-nihilo* of a new administrative capital between Cairo and Suez and the construction of a replica of the ancient lighthouse of Alexandria. These contemporary projects for a "New Egypt" are part of a national tradition of mega-projects which, like other grandiose actions still to come, are part of a process of staging and legitimating the established power.

Finally, although a possible causal connection between grand national projects and increased seismic activity remains in the realm of fiction, the theme of climactic disturbances and worsening pollution sadly revealed itself to be premonitory, as demonstrated by the dreadful "black cloud" that appeared in the Cairo sky in 1999 and has hung over the capital in the autumn (and sometimes spring) ever since, the origins of which are related to various forms of pollution and their interaction.[20] In Towfik's work, the question of pollution is much less present, but he imagines a radical transformation of the Egyptian economy and society.

The New Economic and Social Reality

In *Utopia*, Towfik prophesises the successive collapse of the four pillars of Egypt's rentier economy: tourism, hydrocarbons, the Suez Canal and emigrant remittances. At the time of Towfik's writing, these constituted the backbone of the Egyptian economy and its main sources of foreign currency. In the 2010 of the world of *Utopia*, an American chemist invented Pyrol, a petroleum substitute that immediately took the place of oil as an energy source. In order to guarantee a supply of Pyrol, the elites sold their past without a second thought, pawning the entirety of Egypt's Pharaonic monuments to the Americans in exchange for a guaranteed 50-year supply of Pyrol, reserved exclusively for the use of Utopia and other gated communities. As a result, tourism declined considerably. Another consequence of the discovery of Pyrol was that the Gulf countries, the wealth of which came from oil exports, became rapidly impoverished and expelled the Egyptian emigrants, ending their cash remittances. Finally, a further innovation consisting of a canal dug by Israel, short-circuiting the Suez Canal, deprives Egypt of its precious revenue. In parallel to these events, Towfik describes the disappearance of the middle classes under the effects of ultra-liberal capitalism, in which large firms dominate and the State is reduced to nothing (it no longer even fulfils its sovereign role related to defence since retired US marines protect the gated communities). It stopped all forms of social assistance, privatised everything. One day, it simply stopped paying salaries and running its services. The police disappeared. In the meantime, the rich continued getting richer Thus, the disappearance of the middle classes led to spatial segregation:

> لكن الحياة معنا صارت أمرًا مستحي لا لا... اضطرت هذه الطبقات إلى أن تعزل نفسها طلبًا للأمان في تلك المستعمرات على الساحل الشمالي، وقد استعملوا رجال المارينز لأنهم يضمنون ولاءهم بينما لا يضمنون ولاء «البودي جارد» المطحون بدوره.. إن فكرة أن يثور محيط الفقر هذا كانت تؤرقهم.. كل الثورات الشعبية في التاريخ بدأت بذبح الأثرياء.. هكذا تكون مجتمعان أحدهما يملك كل شيء والآخر لا يملك شيئًا.. أهمية المجتمع الثاني لا تزيد على كونه سوقًا استهلاكيَّة لا بأس بها..
>
> (Then living with us became an impossible matter, so your class was forced to isolate itself, seeking safety in those settlements along the North Coast. They employed Marines because they could guarantee their loyalty, unlike the loyalty of their wretched bodyguards. The idea that this ocean of poverty would rise up used to keep them awake at night. Since the dawn of time, all popular revolts have begun with slaughtering the rich.
>
> And now two societies have formed. One of them owns everything and the other owns nothing. The second society is only important as a consumer market, nothing more.)[21]

In the Egypt of the 2000s, sexual frustration is at a peak due to the prohibition on sexual relations prior to marriage and the delays in marriage resulting from the ever-increasing cost of the ceremony. However, in the Egypt of *Utopia*, sex has become commonplace, among both rich and poor. This has done little to improve the situation of women, who remain more than ever at the mercy of men and are forced to resort to theft or prostitution. Another change relative to the Egypt of the 2000s, and which concerns both worlds, is the easing of religious tensions since the three religions—Islam, Christianity and Judaism—coexist peacefully. This erasure of religious belonging goes hand in hand, for Towfik, with the dissolution of national identity:

> ستة عشر عامًا من عمرك وأنت لا تنتمي إلا إلى «يوتوبيا».. أنت مواطن «يوتوباوي» ذوبتك الحياة المترفة وذوبك الملل، فصرت لا تعرف الأمريكي من المصري من الإسرائيلي.. صرت لا تعرف نفسك من الآخرين ..

(Sixteen years old, and you don't belong anywhere except Utopia. You're a Utopian resident, softened by a life of luxury and boredom. You end up unable to tell an American from an Egyptian from an Israeli. You end up unable to tell yourself apart from other people).[22]

Here, Towfik describes a process of individualisation carried to the extreme: the dissolution of collective identity and the loss of the memory of national history produces monsters, especially among the young generations, totally deprived of ethics. The fact that the enemy par excellence for contemporary Egypt, Israel, is no longer *the* enemy (instead the poor are), is a sign, among others, that regional geopolitics have been profoundly altered.

Geopolitics of the Middle East: A World of Chaos and Dependence on the West

Towfik imagines a new Middle East, based on "مال خليجي..(قبل أن ينضب).. ذكاء إسرائيلي .. أيد عاملة مصرية رخيصة ...." ("Gulf money (before it dried up), Israeli know-how, and cheap Egyptian manual labour").[23] The residents of Utopia do not consider Israel an enemy. The narrator claims to have numerous Israeli friends, and he mentions the presence of synagogues in Utopia. The idea of increased dependence on the West is also present. This dependence is, above all, economic: in 2023, Egypt depends on the West for its imports of Pyrol, Phlogistine (a drug imported from Denmark that all Egyptians, rich or poor, are crazy over) and Libidafro (a kind of new Viagra from France). The dependence also extends to the realm of security, with the ex-US Marines mentioned previously.

Nasir's vision is even darker, since he depicts a world that is as dominated and controlled by the West as never before, in a regional context of chaos and instability in the Middle East:

The regimes in the area had never fully recovered from the prodemocracy *intifadahs* of '02 and '03, which had left a jumble of autonomous and semiautonomous regions in their wake, creating fertile ground for smuggling, gunrunning, and money-laundering, which the governments tried to combat by strict border control.[24]

This quotation has sad echoes with today's post-revolutionary Middle East.

Cairo, an Eschatological City: Earthquakes and Revolutions

In *Tower of Dreams*, Cairo is a restless society troubled by social upheaval but, in the end, mainly passive, whose constant agitation seems paradoxically to guarantee the impossibility of change, and which is in an endemic state of political torpor. The city's evolution, survival and destiny seem to primarily depend on unpredictable, deterministic factors. There are underlying convulsions and recurrent riots, associated with misery and shortage, but they are generated by pauperisation and lack and not a rebellious collective conscience or any real politicisation. In the end, the government is toppled but by a military coup, which occurs too late to change anything.

In fact, at the end of Nasir's novel, deliverance does come, in the form of a cataclysmic punishment by fate, which some consider to be a tectonic movement and others a divine retribution. This dual reading of the event calls to mind the polemical discourses that followed the earthquake that hit Egypt, and especially Cairo, on October 12, 1992:

> Ainsi politisé, le séisme s'est également "islamisé", puisque le Parti du travail (de tendance islamiste) y a vu le signe de la colère divine s'abattant sur un gouvernement corrompu et une société éloignée du droit chemin.
>
> (In addition to being politicized in this way, the earthquake was also "Islamized," since the Workers' Party (of Islamist tendencies) saw in it a sign of divine anger raining down on a corrupt government and a society that had strayed from the righteous path.)[25]

The fictitious catastrophe of 2015 takes eight million lives and destroys the city. Nothing is left of either Cairo's rich or poor neighbourhoods, with both equally destroyed. All that remains are camps of refugees aided by international organisations, a final image that is also recognisable from the contemporary reality of the Middle East. Perhaps everything will start again from there, perhaps not…

Utopia also closes with an apocalyptic scene but of another nature. The poor, outraged by Gaber's murder and Safiyya's rape, march towards Utopia to attack it. Towfik dots his narrative with revolutionary references (evocations of the Iranian revolution of 1979 and of the French revolution, with the recurrence of the image of the storming of the Bastille) but never appears very convinced. Revolution is either derided or relegated to the distant future. When the young man from Utopia asks Gaber why the poor do not revolt, he responds that the security apparatus is so powerful that it easily dissuades any would-be rioters.

The dystopia imagined here is more of a social critique than a strictly political one. Towfik principally denounces Westernised elites, whose selfishness has led them to sell their country and who have allowed the majority of the population to sink into poverty. Indeed, when revolution is envisaged, it is from a social angle: "لو حدثت الثورة يومًا ما فلسوف نبدأ بالتهام كلابهم المدللة السمينة كلها" ("If the revolution happens, we'll begin by devouring all their fat, pampered dogs").[26]

There are various divergences between the January 25, 2011 revolution and the scenarios imagined by Towfik and Nasir. What the experience of 2011 above all emphasises, for present purposes, is its definitive refutation of the representation of the Egyptian people as politically apathetic in structural way, which had become a kind of accepted wisdom and which is echoed in the obvious resignation of *Tower of Dreams*, despite the affordances of being imaginary and future-oriented.

Poverty, unemployment and social inequalities did rank among the main causes of people's exasperation, but the protest movement took an eminently political tone, focused on ousting President Mubarak and bringing down the regime. In addition, the middle class had not disappeared, even if it suffered greatly as a result of the economic crisis since it was youth belonging to this category who started the protests, following the Tunisian precedent. These youth, whom the media characterised as "Westernised" because they used the internet as a tool of contestation, considered themselves, above all, to be fiercely Egyptian, while simultaneously deeply anchored in modernity.

President Morsi's destitution by the military in July 2013, acting with the support of a massive people's movement, might lead us to think that Nasir was right in the end since Egypt is once again under the yoke of an authoritarian military regime. However, prior to 2015, in Egypt, there was 2011 and the overthrow of a regime, even if, afterwards, the previous regime returned to power. There is no doubt that something historic happened in 2011 on the banks of the Nile.

Conclusion

While their novels are anchored in specific realities of Egypt's megalopolis, Nasir and Towfik's narrative schema are drawn from the classic repertoire of futuristic literature (a devastating natural disaster destroying a great city, enclosure, surveillance and securitisation of urban spaces, obsession with terrorist threats, etc.). They make similar use of the genre to denounce injustices of the capitalist system of globalisation and the socio-spatial inequalities that result from it. However, they propose two different imaginaries of urban fragmentation since one of them envisions vertical segregation while the other's is horizontal.[27]

Towfik's *Utopia* focuses on the spatial metaphor of the gated community and devotes no attention to artificial urban landscapes, virtual universes or cyberspace, which are much more present in Nasir's *Tower of Dreams*. However, this projection of a kind of post-modernity onto Cairo is paradoxical since the city's Oriental character remains firmly imprinted upon it.

The Cairo of *Tower of Dreams* is a metaphor for the Orient to the extent that the city (its landscapes, its society, its characters, its street scenes) reveals a receptacle-space, which is like a repository of compatible times, since modernity is never achieved, nor even appropriated, its dynamics being exogenous.[28] The past-present-future triptych is emphatic, the city is polytemporal; although it is located in the future, it appears to be characterised by immobility, fatalism, archaism; its present is like a caricature of its past and both seem equally bogged down.

Perhaps Cairo cannot be conjugated in the future imaginary since its "fictionally" logical future, the one flowing from its cumulative and contrasting representations (a city of the Global South, Third-World megalopolis, Arab-Oriental city), the one that makes it a city that is practically ontological, could only have a prophetic and inescapable ending. These divergent visions of Cairo, as a city of the Global South in the not too distant future, lead in two different directions, one apocalyptic and the other revolutionary. In *Tower of Dreams*, Cairo is a reflecting pool of a restless but passive society, a metaphor for the impossibility of change or of any event other than an earthquake. In *Utopia*, the oppressed poor ultimately revolt against the rich, in a struggle that has no certain outcome.

Thus, the great originality of these two novels consists of producing fiction that escapes traditional literary representations of Cairo, as an Arab city with its labyrinthine alleys and impasses that was so dear to the great novelist Naguib Mahfouz, as a modern downtown that was long the heart of intellectual or artistic life, of its informal quarters, or finally of the upper class suburbs such as Heliopolis or Maadi.[29] With *Tower of Dreams* and *Utopia*, Cairo earns its stripes in the science fiction genre and joins other fictional metropolises of the future.

Notes

1. An earlier version of this chapter first appeared in French: Anna Madœuf and Delphine Pagès-El Karoui, "Le Caire en 2015 et en 2023: deux dystopies anticipatrices? Les avenirs funestes de la capitale égyptienne dans *Tower of Dreams* et *Utopia*," *Annales de Géographie*, no. 709–710 (2016): 360–377.
2. Pagès-El Karoui, "*Utopia* ou l'anti-Tahrir."
3. Towfik has described *Utopia* as a "post-apocalyptic dystopia." Towfik, "Ahmed Khaled Towfik Interview."
4. Desbois, "Présence du futur"; Musset, *De New-York à Coruscant*.
5. Musset, "Penser la domination." Translation by the authors.
6. Barbaro, *La fantascienza*; Khayrutdinov, "Ahmad Khaled Tawfik's Novel."
7. Greenberg, "Ahmed Khaled Towfik."
8. Jacquemond, "Yacoubian Building."
9. For more on the notion of the "Orient," see Said, *Orientalism*.
10. Towfik, يوتوبيا, 12; Towfik, *Utopia*, 6.
11. De Chateaubriand, *Itinéraire*, 7. Translation by the authors.
12. Nasir, *Tower of Dreams*, 6.
13. Nasir, *Tower of Dreams*, 47.
14. "La ville partout," 39. Translation by the authors.
15. Dollfus and Brunet, *Géographie universelle*, 496. Translation by the authors.
16. Pourtier, "Villes africaines," 26. Translation by the authors. Pourtier also uses the metaphor of a human and urban "tide."
17. Braud, "Derrière les murs"; Dénis and Séjourné, "Le Caire métropole privatise"; Florin, "Les quartiers."
18. Mahmoud and Abd Elrahman, "La planification."
19. Sallon, "Le président." Translation by the authors.
20. In *Tower of Dreams,* a causal relationship is posited between the flooding of the Qattara Depression (a project that was envisaged in the 1990s but never realised), climate change and the recurrence of earthquakes.
21. Towfik, يوتوبيا, 98; Towfik, *Utopia*, 82.
22. Towfik, يوتوبيا, 12; Towfik, *Utopia*, 6.
23. Towfik, يوتوبيا, 98; Towfik, *Utopia*, 82.
24. Nasir, *Tower of Dreams*, 30.
25. El Kadi, "Le tremblement," 7. Translation by the authors.
26. Towfik, يوتوبيا, 73; Towfik, *Utopia*, 60.
27. On the necessity of analysing Egyptian gated communities in terms other than those of urban fragmentation, see: Braud, "Derrière les murs," 89–102.
28. The Orient is, of course, elsewhere par excellence, but it is also a place where time and the relationship to time are considered different as well.
29. Heshmat, "L'évolution."

Works Cited

Barbaro, Ada. *La fantascienza nella letteratura araba*. Roma: Carroci, 2013.

Braud, Élise. "Derrière les murs, l'écho de Tahrir: Le Caire et ses compounds, une fragmentation à nuancer." In *Villes arabes, cités rebelles*, edited by Roman Stadnicki, 89–102. Paris: Éditions du Cygne, 2015.

De Chateaubriand, François-René. *Itinéraire de Paris à Jérusalem*. Paris: Flammarion, 1811.

Desbois, Henri. "Présence du futur. Le cyberespace et les imaginaires urbains de science-fiction." *Géographie et cultures*, no. 61 (2007): 123–140.

Denis, Éric, and Marion Séjourné. "Le Caire métropole privatisée." *Urbanisme*, no. 328 (2003): 31–37.

Dollfus, Olivier, and Roger Brunet. *Mondes Nouveaux*, vol. 1 of *Géographie universelle*, edited by Roger Brunet. Paris: Hachette/Reclus, 1990.

El Kadi, Galila. "Le tremblement de terre en Égypte." *Égypte/Monde arabe* 14 (1993): 163–196. doi: https://doi.org/10.4000/ema.584.

Florin, Bénédicte. "Les quartiers fermés du Grand Caire. Dimensions urbanistiques et idéologiques d'une forme de ville: nouvelle urbanité ou césure urbaine ?" *L'Espace Politique* 17 (2012). doi: https://doi.org/10.4000/espacepolitique.2393.

Greenberg, Nathaniel. "Ahmed Khaled Towfik: Days of Rage and Horror in Arabic Science Fiction." *Critique: Studies in Contemporary Fiction* (2018): 1–10. https://doi.org/10.1080/00111619.2018.1494130.

Heshmat, Dina. *L'évolution des représentations de la ville du Caire dans la littérature égyptienne moderne et contemporaine*. PhD diss., Université de Paris 3, 2004.

Khayrutdinov, Dinar Rafisovich. "Ahmad Khaled Tawfik's Novel *Utopia* as an Important Example of the New Wave of Science Fiction in Arabic Literature." *World Applied Sciences Journal* 31, no. 2 (2014): 190–192.

Jacquemond, Richard. "The Yacoubian Building and Its Sisters: Reflections on Readership and Written Culture in Modern Egypt." In *Popular Culture in the Middle East and North Africa: A Postcolonial Outlook*, edited by Mounira Soliman and Walid El Hamamsy, 223–236. New York: Routledge, 2013.

Mahmoud, Randa, and Ahmed Abd Elrahman. "La planification controversée du Grand Caire avant/après 2011." *Égypte/Monde arabe* 11 (2014): 177–201. Doi: https://doi.org/10.4000/ema.3305.

Musset, Alain. *De New-York à Coruscant: Essai de géofiction*. Paris: Presses universitaires de France, 2005.

———. "Penser la domination et l'émancipation dans la ville à travers la science-fiction." *La forge numérique*. June 8, 2015. http://www.unicaen.fr/recherche/mrsh/forge/3372.

Nasir, Jamil. *Tower of Dreams*. New York: iUniverse, 2009.

Pagès-El Karoui, Delphine. "*Utopia* ou l'anti-Tahrir: le pire des mondes dans le roman de A. K. Towfik." *EchoGéo* 25 (2013). doi: https://doi.org/10.4000/echogeo.13512.

Pourtier, Roland. "Villes africaines." *Documentation photographique*, no. 8009 (1999): 1–63.

Said, Edward. *Orientalism*. New York: Vintage Books, 2003 [1978].

———. *L'orientalisme. L'Orient créé par l'Occident*. Paris: Seuil, 1980.

Sallon, Hélène. "Le président Sissi met en scène sa « nouvelle Egypte »." *Le Monde*, August 4, 2015. https://www.lemonde.fr/proche-orient/article/2015/08/04/le-president-sissi-met-en-scene-sa-nouvelle-egypte_4711198_3218.html.

Towfik, Ahmed. "Ahmed Khaled Towfik Interview," with Cheryl Morgan. *The World SF Blog*. June 11, 2012. https://worldsf.wordpress.com/2012/06/11/monday-original-content-ahmed-khaled-towfik-interview/.

———. *Utopia*. Translated by Chip Rossetti. Doha: Bloomsbury Qatar Foundation, 2010.

———. يوتوبيا: رواية. Cairo: Dar Merit, 2008; reprinted Doha: Bloomsbury Qatar Foundation, 2010. Page numbers refer to the Bloomsbury edition.

"La ville partout et partout en crise." Le Monde diplomatique: *Manière de voir*, no 13 (1991).

CHAPTER 14

Post-Capitalist Futures: A Report on Imagination

Nick Lawrence

In the midst of the ongoing and systemic crisis—world-political, world-economic and world-ecological—the century's second decade has witnessed not one but two golden ages: a golden age of crisis theory, together with a flowering of dystopian realisms. While this may seem to confirm the point that (critical) theory and (creative) practice tend to develop in tandem with the moving contradictions of capital itself, it's worth examining further the link between the drive to theorise epochal crisis and the drive to write it. For some thinkers on the left, the years since 2008 have seen a reinvigoration of debates over key categories such as value, labour, class and social reproduction, bringing in their wake an upsurge of formerly dormant utopian imaginings involving workless futures and full automation.[1] For others, the serial irresolution of world-systemic weakness has prompted a tonal shift from what Wolfgang Streeck calls "wishful demonstrations of the possible" to "a realistic accounting of the real."[2] Given this response, the galvanic charge of the May 1968 slogan—"Be realistic: demand the impossible"—may seem to invite recuperation as a matter of ethical necessity. Yet both theorists and imagineers of the bad new days confront a seismic shift in the landscape of the real that is their departure point.

N. Lawrence (✉)
University of Warwick, Coventry, UK

Z. Kendal et al. (eds.), *Ethical Futures and Global Science Fiction*, Studies in Global Science Fiction,
https://doi.org/10.1007/978-3-030-27893-9_14

Before presenting its diagnoses, for example, the new sobriety in theory has had to grapple with the full scope of current challenges to what counts, or might count, as "realism." These include the cardiac frailty of a global economy that routinely posts new highs on the major financial indices; the scale of endemic underemployment indexed against increasingly coercive labour conditions in core and peripheral regions alike; map-altering levels of mass migration and the resurgence of xenophobic nationalism; ramped-up applications of racialised state violence targeting populations already subject to historic levels of disciplinary policing; the profound alterations to social environments occasioned by online and digital media; the acceleration of climate breakdown coupled with a kamikaze embrace of fossil-fuel extractivism; and alongside all these developments, a circumstance that Mark Fisher identified years ago as the generalised atrophy of any sense of futurity beyond continuation of the status quo.[3]

The extremity of such conditions serves to up-end prior assumptions about the commonsensical "real," to suggest that current "reality" is increasingly improbable, uncanny and removed from what were once taken to be routine expectations of regularity, order and stability. Amitav Ghosh notes that these expectations form the tissue of a self-conscious modernity consolidated during the nineteenth century and grounded in the experience of the everyday, in which conditions of predictability set the parameters of the realist outlook. "Quite possibly," he adds, in acknowledgement of widespread disattention to the implications of present-day climate chaos, "this era, which so congratulates itself on its self-awareness, will come to be known as the time of the Great Derangement."[4] That is, of course, if there is a future era from which to look back and make such judgements.

For many writers, meanwhile, these conditions have occasioned less a return to older models of realist representation and more an embrace of the generic protocols of utopia's twin shadow. Fredric Jameson's much quoted aphorism—it is easier to imagine the end of the world than the end of capitalism—may no longer hold true, but only because both of these prospects are so easily conflated into a single looming terminus. (In this respect, we could recall the less-remarked outcome of epochal class struggle predicted by the *Communist Manifesto*: either "a revolutionary reconstitution of society at large" or "the common ruin of the contending classes."[5]) Getting real in crisis conditions such as these—a world, for example, where Royal Dutch Shell assumes in its corporate planning a global temperature rise of 4–6°C—means, for writers and other artists, pressing against the limits of even the dirtiest realisms.

The Return of Realism?

When what is termed the realist outlook enters one of its recurrent crisis periods, it tends to be accompanied by a host of morbid symptoms. These are apparent not only in baroque strategies of avoidance, denial, myopia and wishful thinking with respect to the climate crisis itself, but in the shifts and slippages framing the question of cultural form. The difficulties in tracking realism's shifting ground are apparent particularly in recent critical debates over the adequation of earlier literary forms to current realities. In theory circles, the historic opposition of realism and modernism is now often augmented with a new antithesis between realism and speculative fiction.[6] Yet this formulation is deceptive, insofar as it masks a more fundamental struggle over the terms and stakes of realism itself.

Jed Esty's recent identification of a millennial turn to literary realisms after the exhaustion of a postwar modernism associated with both "minimalist" (Kafka, Beckett, Coetzee) and "maximalist" (Joyce, Borges, Pynchon, Rushdie) modes suggests a revival of the realist-modernist dichotomy that shaped mid-twentieth-century debates in critical theory. Taking in a range of examples from contemporary global fiction, Esty argues that "worldly realisms are emerging as central to newly forming literary canons insofar as they appear to move us beyond the stale paradigms of the late twentieth century such as postmodernism or magical realism and to offer more direct access to problems of social and economic justice at the global scale."[7] This development is not restricted to novelistic practice alone; as he suggests, "new kinds of reality-based forms have challenged the social and entertainment value of fiction—its ability to sift and condense experience into aesthetic form, to reorganize the kaleidoscopic real into a legible pattern."[8]

Esty concedes that "the problem of contemporary realism begins with the pressure of the recirculated, mediated, and curated 'real' bearing down even on traditional realisms."[9] But a further problem arises in the attempt to assimilate to a previous era's critical opposition between "realism" and "modernism"—singular terms in Esty's usage—the recent turn towards what David Shields terms "reality hunger," a cultural tendency bypassing the socially mimetic ambitions of realist fiction (as consolidated from the nineteenth century onward) in favour of a more direct route towards a presentation of the "real."[10] In aesthetic terms, this phenomenon is in fact closer to collage (or sampling) than the artisanal rendering of social detail associated with classic realism. Rather than depend on a laboriously artful reconstruction of "social reality" within a single medium, in other words, the collage principle's transposition of social materials from one medium to another throws a further complication into any attempt to cleanly dichotomise realism and its opposites—not least since collage remains the modernist principle par excellence, yet is mobilised most frequently for the purpose of introducing indexical jolts of the "real." As Walter Benjamin was to observe in his essay on art's auratic decay under the conditions of capitalist reproducibility, the modern desire to "'get closer' to things" entails their multiplication as copies detached from their original context and deployed in a new "alignment of reality."[11] The social materials of collage are in this case less tools of representation, and closer to aspects of a fundamentally environmental process of assemblage and dispersal that envelopes producers and consumers alike. The permanent availability of these materials, within the paradigmatic model of database archives rather than sequential narrative,[12] is in part what distances the present-day "return of the real" from Lukácsian models of historical realism.

Cli-Fi and Crisis

But if older realisms and neo-modernist collage are alike stymied in fundamental ways by the challenge of responding critically to epochal crisis, where then is Marx's "musician of the future" to be found?[13] One answer is hinted at in Esty's observation that "if new realist novels find ways to represent 'combined and uneven development' in the global frame where it cannot be mediated into the destiny of a single people, this may well explain the rising force of apocalyptic and Anthropocene models as ways to identify collective problems operating at planetary scale."[14]

As the spectre haunting all attempts to provide Streeck's "realistic accounting of the real," climate change poses threats not only in terms of global overheating, rising sea levels, desertification, super-storms, flooding and ecocide, but also to literature's capacity to grasp and make sense of these developments in narrative form. An extreme example of what Rob Nixon describes as slow violence—the processes and consequences of anthropogenic planetary heating—raises questions concerning the potential of fiction as well as other forms of cultural production to adequately register the scale, complexity and dynamic of what is happening and about to happen.[15] Is Ghosh thus right to argue that world literature's failure to meet the challenge of the Anthropocene reflects a "broader imaginative and cultural failure that lies at the heart of the climate crisis"?[16] How far does this purported failure of the imagination resonate with critical debates on the failures of realism, in an age of capitalist realism?[17]

For forecasts of catastrophes to come, we turn to our cultural meteorologists. A sample of contemporary realism from an interview with Canadian science fiction (SF) author Peter Watts:

> The system is a jumbo jet, overbooked, overweighted, out of fuel over the Atlantic and already ten thousand feet below cruising altitude. Short-term economic concerns led to the overbooking; profit margins dictated skimping on the fuel. But nobody gives a shit about those things now. Now, the guys in the cockpit are just trying to keep the nose up in the forlorn and desperate hope that by some miracle, everyone won't be killed on impact. Now, the best-case scenario involves being alive when the sharks find you.[18]

Watts's genre credentials draw attention to a basic feature of the contemporary culture of crisis, that is, its increasing recourse to modes, registers and representational strategies associated with speculative fiction (fantasy, SF, supernaturalism in horror and the paranormal). As noted, these registers are overwhelmingly cast in the dystopian key. A quick survey of recent bestselling fiction—from YA to adult, genre to literary—demonstrates the ubiquity of the new dystopian realism: this is the age of *The Hunger Games*, the return of *The Handmaid's Tale*, *The Road*, *The Water Knife*, *Divergent*, *American War*, not to mention countless iterations of the zombie and post-apocalyptic franchises that dominate the multiplex and cable TV. Yet if such titles offer at least a partial response to Ghosh's charge of literary silence, it's worth asking, in light of this apparent hegemony, what forms of inoculation, screening, shielding and evasion might characterise the readiness with which the cultural rheostat is set to doom, especially in the higher income nations. When the Global North looks anxiously to the future for signs of dystopia, for example, to what degree does this mask the dystopian actuality of wide zones across the Global South? As Australian author Claire G. Coleman writes, "We, the Indigenous people of this continent, now live in a dystopia"—the end-times have come and gone.[19] And as cyberpunk author William Gibson noted some time ago, "The future is already here, it's just not very evenly distributed."[20]

At the same time, the dangers of a rote and routinised catastrophism in public discourse—and with it, the apocalypse fatigue that once again distances representation from the real—are also becoming more widely recognised among theorists and cultural producers alike. In her essay "Great Chaos Under Heaven," Sasha Lilley notes problems with catastrophist rhetoric not only on the reactionary right, but on the left as well.[21] Two tendencies in particular, one determinist and the other voluntarist, converge in responses to the conjunction of the Great Financial Crisis and the threat of global overheating. The former, in adopting the formula "the limits to capital are capital itself," posit a self-triggered systemic collapse without the need, or indeed capacity, of human agency to bring it about; the latter heralds the acceleration of economic disaster, climate change and intensifying state oppression as the necessary, indeed only, conditions enabling the prospect of revolutionary transformation. Both responses mask, as Lilley points out, an underlying structure of feeling: that of despair. "Such political despair is understandable," she notes; "it needs to be resisted nonetheless."[22] Her collaborator Eddie Yuen is more direct: "The politics of failure have failed." And Doug Henwood adds, even more directly: "Dystopia is for losers."[23]

Post-Capitalism: Theory and Practice

A rejection of routinised dystopia lays the way open for a return of more explicit forms of utopian thinking: in cultural terms, the revival of radical fantasy, SF and other forms of speculative fiction; in sociological terms, a renewed interest in models of post-capitalist transition, organisation and planning. Yet the dominance of dystopia in literary circles shows little sign of giving way to the efflorescence of utopian imaginaries that marked the last spring tide of speculative experimentation during the crisis period of the late 1960s to early 1970s, when socialist-feminist authors such as Ursula K. Le Guin, Samuel R. Delany, Marge Piercy and Joanna Russ took up the brief of imagining radical alternatives to the status quo in fictional form. Then, the emancipatory energies of the civil rights and new social movements, coupled with the crisis in American imperialism marked by the war in Vietnam and stirrings of a worldwide response to environmental degradation, led such writers to map alternative projections of a social and natural order beyond the settled determinations of mid-century modernisation. But although the veteran SF writer Kim Stanley Robinson retains a continuing interest in heterotopian modelling (as in his *New York 2140*), the overwhelming majority of contemporary SF, particularly in work falling under the banner of cli-fi, is resolutely set to the key of apocalypse. In contemporary sociological theory, however, the outlook has seemed healthier. A brief survey of twenty-first-century titles indicates a current upsurge in post-capitalist speculation, from David Schweickart's *After Capitalism* (2002) to Michael Albert's *Parecon: Life After Capitalism* (2003), Harry Shutt's *Beyond the Profits System: Possibilities for the Post-Capitalist Era* (2010), Erik O. Wright's *Envisioning Real Utopias* (2010), Nick Srnicek and Alex Williams's *Inventing the Future: Postcapitalism and a World Without Work* (2015) and Paul Mason's *PostCapitalism: A Guide to our Future* (2015).

As with other "post"-prefixed terms, "post-capitalism" has a mixed and often prevaricating history, most usually indicating a desire to avoid the naming of socialism or communism as possible futures.[24] In usage dating from the 1950s, when the consolidation of welfare state societies in the advanced industrial world prompted weak speculation that an evolution beyond monopoly capital was already under way, this history led C. B. Macpherson to wonder: "How much has capitalism changed? Are we in an era of post-capitalism?"[25] In the 1990s, by contrast, management theorists such as Peter Drucker read the tea-leaves of a post-Cold War order as indicating a techno-determinist supersession of class struggle, in favour of the inevitable historical transition to an artificial intelligence (AI) future: "The real and controlling resource and the absolutely decisive 'factor of production' is now neither capital, nor land, nor labour," he writes in *Post-Capitalist Society* (1994), "it is knowledge."[26]

While this cheerleading of so-called cognitive or creative capitalism[27] under the guise of a paradigm shift is increasingly discredited today—if still hugely influential in policy circles—left theorists have shown themselves to be not immune to the tendency of detaching the technological from the political. The embrace of accelerationist and technologically driven programmes for a transition out of capitalism's terminal contradictions has engendered further intractable problems of its own. According to Paul Mason's argument in his 2015 survey *PostCapitalism*, "The ultimate market signal from the future to the present [is] that an information economy may not be compatible with a market economy."[28] As David Runciman put it in a largely positive review of Mason's book, "This tension between knowledge (which is limitless) and ownership (which is limited) represents the basic contradiction of capitalism. Earlier thinkers caught sight of it from various different angles. Now the digital revolution has laid it bare."[29] But the digital revolution has in fact worked much along the lines of previous disruptions to business as usual, generating new sources of profit through a combination of appropriation (of common goods) and exploitation (of labour, both paid and unpaid). The book's argument for epochal change triggered by a technological paradigm shift overlooks the history of previous such transitions as exemplary of capitalist development.

Mason assumes that the existence of "sharing economy" exchanges represents an encroachment against capitalist commodification, when these may simply support the development of commercial applications as a form of unpaid work. The degree to which the sharing of information actively feeds the profit margins of the tech sector, for example, is sufficiently recognised to pass for received understanding; the category of "prosumption" in studies of the new landscape of labour, erasing the dividing line between production and consumption, further attests to the assimilability of the information economy to capitalist imperatives.[30] What Mason effectively ignores in his analysis is the attention paid by social reproduction theory to the symbiosis of paid and unpaid work as drivers of capitalist accumulation. This oversight points to the necessity of caution when it comes to interpreting as insurmountable contradictions what turn out instead simply to be capitalist opportunities. The climate crisis fuelled by a carbon economy is of course one of these.

Nick Srnicek and Alex Williams's *Inventing the Future*, meanwhile, at least recognises the potential importance of cultural imaginaries in augmenting the task of social-scientific speculation: "From predictions of new worlds of leisure, to Soviet-era cosmic communism, to Afro-futurist celebrations of the synthetic and diasporic nature of black culture, to post-gender dreams of radical feminism, ... the popular imagination of the [postwar] left envisaged societies vastly superior to anything we dream of today," they write.[31] Inheriting this legacy, today's left should thus "mobilise dreams of decarbonising the economy, space travel, robot economies—all the traditional touchstones of science fiction—in order to prepare for a day beyond capitalism."[32] Srnicek and Williams's reading of the present, as with the autonomist tradition from which they draw, finds utopian potential in the very symptoms of apparently terminal malaise: from the spectre of mass unemployment following the relentless expansion of automation into service and professional sectors, to the crisis of profitability signalled by open-source software and copy-left erosion of intellectual property, to the easy availability of 3D printing systems for the production of formerly restricted consumables. Yet the most striking feature of *Inventing the Future* is the basic modesty of its proposals and, as with Mason's *PostCapitalism*, its silence on the specifics of transitional struggle. Sam Kriss notes that Srnicek and Williams's critique of current left politics as incoherent conceals its own form of incoherence: the authors have "seen a deficiency in the means the left uses, and propose to correct it with a new set of aims. This is a category error—it's like saying that we're not walking quickly enough, so we should decide on a different destination."[33] And, as Owen Hatherley remarks:

> In the end postcapitalism, like postmodernism, is the name of an absence, not a positive programme. Like the anticapitalism of the early 2000s, it tells you what it's not: in this case, the old left, folk politics, social democracy or Stalinism, with their hierarchies and lack of cool free stuff … Postcapitalism tells you that the forces of production make something possible, then suggests either that you demand it, or that you're already doing it.[34]

An old and threadbare model of historical change underpins this emphasis on technology as productive force over and against the fetters of social relations. Above all, the question of how a "productive force" can be imagined outside the matrix of capitalist determinations goes unasked.

Reading (Post-)Capitalist Possibilities

This conceptual short-circuiting continually stymies the efforts of post-capitalist theorists. Moishe Postone has framed the problem, deeply embedded in positivist strands of the Marxist tradition, as follows: "The difficult task is to conceptually separate out the emancipatory dimension of the possibilities generated by capitalism from the non- or anti-emancipatory forms in which they have been generated."[35] In his insistence that any critical account of capitalism must grasp the basic forms and categories of its analysis—value, labour, commodity, capital itself—as historically specific, not transhistorical and neutral, Postone argues for a different understanding of capital's contradictions than autonomist or rigidly determinist interpretations, one that emphasises the logic of the growing gap between what is and what could be. This interpretation suggests that such a gap can be conceptualised in implicitly utopian terms, not as the inevitable conflict between forces and relations of production, but as "a gap between social labor as it is presently structured and social labor as it could be structured."[36] The contingency of capitalist arrangements can be discerned in the surplus of subjective forms, including cultural figurations, that are associated with different phases and conjunctures of capitalist history. "These forms are neither completely contingent nor are they preprogrammed ... That is to say, capital can generate the conditions of possibility of a society beyond capital, but the dialectic of capital is not a transhistorical dialectic of history. Capital will not change itself into something else."[37] In his differently inflected account of what Fredric Jameson made theoretically notorious, for example, Postone argues that postmodernism—or present-day capitalist culture in the high-income nations—can be understood "as a sort of premature post-capitalism, one that points to possibilities generated, but unrealized, in capitalism. At the same time, because postmodernism misrecognizes its context, it can serve as an ideology of legitimation for the new configuration of capitalism, of which it is a part."[38] "I agree with the image of capitalism as a runaway train," Postone concludes, nodding to Benjamin's revision of Marx's revolutionary locomotive of history, "although I think revolution entails more than just pulling the cord."[39] It is striking, however, how often struggles against a fossil-fuelled climate crisis reach for the emergency brake in precisely these terms. The suspension of questions of agency—who pulls the cord, under what conditions—too often assumes a basic continuity in the exercise of political power.

The problem of revolutionary transition is deliberately bracketed in what is one of the more interesting recent examples of post-capitalist theory. In his little book *Four Futures*, Peter Frase lays out what he terms four "ideal types" of post-capitalist scenario, each responding to two determining factors, economic and ecological: the emergence of a jobless future by means of the fully automated workplace and the threat of scarcity triggered by global overheating.[40] "Two specters are haunting Earth in the twenty-first century," he writes, "the specters of ecological catastrophe and automation."[41] Frase's thought experiment posits distinct outcomes for each factor. In the case of an automated future, the result at one end of the spectrum of possibilities would be an egalitarian society, in which all benefit from the new leisure afforded by machines doing the work for us, and, at the other end, a hierarchical society in which those who control the technology call the shots. As for ecology, the threat of climate change could bring about a decisive turn to renewable energy and consequently a new level of material abundance, or else it could usher in a new age of scarcity, in which finite resources dictate parsimony rather than wasteful consumerism as the only way forward. The resulting society types present four possible combinations: of abundance and equality, which Frase nominates as communism (and illustrates with reference to the *Star Trek* series); of scarcity and equality, or socialism, in which the collective guardianship of precious resources leads to what Ivan Illich calls a "convivial austerity"[42]; of abundance and hierarchy, or rentism, in which the few who own and control the intellectual property on which social organisation depends become near-absolute in their power; and scarcity and hierarchy, or exterminism, in which the increasing disposability of human populations reaches its logical endpoint in their wholesale liquidation at the hands of a militarised elite. In this scenario, the current forever wars of the Middle East, western Asia and central Africa share a commonality with the wars against immigrants and the poor—boats sunk in the Mediterranean, black teenagers shot down on the streets of the US—in what Christian Parenti calls "the politics of the armed lifeboat."[43] Exterminism is then one version of the barbarism that Rosa Luxemburg predicted would be the inevitable outcome of capitalist decay.

It should be stressed that for Frase these futures represent ideal types rather than the messy and hybrid actuality of any conceivable transition to a post-capitalist state; he is interested in exploring the intersecting logics of societies in which the capital-labour relation is abolished, or in which monopoly over the means of existence becomes absolute. In so doing, he teases out the fault lines defining present-day social and political struggles and exposes how provisional and historically contingent many of our hard-wired scepticisms and prejudices concerning possible futures turn out to be. Throughout, as well, Frase makes liberal use of illustrative examples drawn from speculative fiction, from Kurt Vonnegut's *Player Piano* (1952) to Kim Stanley Robinson's Three Californias trilogy (1984–1990) and Mars trilogy (1992–1996), to Andrew Stanton's *WALL-E* (2008) and Neil Blompkamp's *Elysium* (2013). Interleaved with references to Marx, André Gorz, Luxemburg and an array of contemporary social scientists, the effect is to highlight the overlap between speculative science fiction and speculative social science. As models for grasping aspects of a present reality, Frase's futures draw inspiration from the work of the SF canon as much as the social-theoretical one.

At the same time, Frase is careful to avoid any neat classification of his futures into static utopias or dystopias. As he points out, any of the scenarios he envisages could conceivably mutate into another; the road to communism may, as in classic Marxist analyses, emerge from capitalist contradictions, but it may also lead through the horrors of an exterminist phase of history:

> What's left when the "excess" bodies have been disposed of and the rich are finally left alone with their robots and their walled compounds? The combat drones and robot executioners could be decommissioned, the apparatus of surveillance gradually dismantled, and the remaining population could evolve past its brutal and dehumanising war morality and settle into a life of equality and abundance—in other words, into communism.
>
> As a descendant of Europeans in the United States, I have an idea of what that might be like. After all, I'm the beneficiary of a genocide.[44]

"We don't necessarily pick one of the four futures: we could get them all," he argues.[45] It is, in other words, not the abstract process of working towards a future goal, however complex and contradictory, that determines the likely course of human action, but rather the nature of the collective power built in the present. That power is the result of social struggles that are anything but speculative.

The Politics of Time (Travel)

In light of recent critical theorising, as well as SF writing, it may well seem that the stumbling blocks to living otherwise than the dystopian status quo—developing agency, building practical power, extending and connecting struggles—operate in speculative work as anamorphic blots or smears in the landscape of the present. Their blurred focus, that is, introduces a necessary lack of clarity to visualisations of the continuum linking present and future, which otherwise so easily lends itself to a conception of the "homogeneous, empty time" against which Benjamin's figure of the historical materialist must contend.[46] A revolutionary task: making time malleable again, restoring contestation over the time not just of wage labour or social reproduction in the abstract, but the innumerable live connections joining historical and contemporary struggles in such a way as to discompose the teleologies and narrative assumptions bringing "history" into alignment. For this, the trope of time travel, long a leading motif in speculative fiction but in a sense intrinsic to literary production generally, can offer useful guides.

We might think, for example, of the role of counterfactuals as a matter of reading as well as writing fiction: as in Roberto Bolaño's hallucinatory renditions of the Pinochet era, written partially under the influence of SF works such as Philip K. Dick's *Man in the High Castle* (1962), which serve as a reminder that, for much of Latin America, fascism—not liberal democracy—was actually victorious after the Second World War, and thus that counterfactual histories are embedded in the combined and uneven dynamics of the world-system. Raymond Williams's "resources of hope" might emerge, from this perspective, not simply in moments of past history, any more than in blueprints of the future, but instead in forms of connection or articulation between nodes of past, present and future conflicts: fuel for the future in imaginative reconstruction of our baseline infrastructures.[47] We can learn, as Jameson has suggested with respect to all utopian experiments, from past failures as well.

Looking Backward at *Looking Backward*

Time travel is of course the vehicle for many of the utopian SF works of the last two centuries. Of these, Edward Bellamy's speculative romance *Looking Backward* is by some measure the most successful in terms of circulation and historical impact. Published in 1888, it sold slowly in its first months, but by the end of the following year was selling 10,000 copies a week. Reviewed favourably by literary adjudicators, including William Dean Howells and Mark Twain, it eventually became the third bestselling American novel of the nineteenth century, after Harriet Beecher Stowe's *Uncle Tom's Cabin* (1852) and Lew Wallace's *Ben Hur* (1880).

Its impact on late Victorian culture is hard to measure. On one level, it provoked an entire genre of "answer" novels:

Arthur Dudley Vinton, *Looking Further Backward* (1890)
Richard Michaelis, *Looking Further Forward* (1890)
Arthur William Sanford, *Looking Upwards* (1892)
J. W. Roberts, *Looking Within* (1893)

Nor were these just topical responses by unknowns and amateurs. Some of the best-known authors of the day, including those celebrated for their realism, were prompted by Bellamy's book to try their hands at utopian romance:

William Morris, *News from Nowhere* (1890)
William Dean Howells, *A Traveller from Altruria* (1894)

H. G. Wells, *A Modern Utopia* (1905)
Charlotte Perkins Gilman, *Herland* (1915)

In assessing this influence, it's worth noting the different stances on the scope of individual agency in Stowe's anti-slavery novel and Bellamy's ostensibly anti-capitalist one. Whereas Stowe urges her readership to accept both collective and individual responsibility for implication in the system of slavery, Bellamy appeals to a middle-class sense of personal guilt over the inequities of the nineteenth-century class system, while essentially bypassing the possibility of individual or group redress. Based as it is on an evolutionary model devoid of conflict, the social change in *Looking Backward* appears to take place without the agency of any individual or social collectivity whatsoever.

A closer look at how this works reveals paradigmatic difficulties in the utopian enterprise. In Chapter XV of the novel, the nineteenth-century narrator, Julian West, is given a book to read as an example of what twentieth-century literature is like. He comments: "At the first reading what most impressed me was not so much what was in the book as what was left out of it," and goes on to note the absence of "all effects drawn from the contrasts of wealth and poverty, education and ignorance, coarseness and refinement, high and low, all motives drawn from social pride and ambition, the desire of being richer or the fear of being poorer, together with sordid anxieties of any sort for one's self or others."[48] This chimes with what Bellamy's reader may be thinking as well.[49] It's hard to imagine a book in which there is "love galore," as West puts it, but which lacks completely all the conflict and social contextualisation that help spark love in the first place.[50] It's hard not to conclude that such a book could only be colossally boring.

In terms of fictional satisfaction, we're constrained to admit that *Looking Backward* can get quite boring itself. With very little dramatic action and long stretches of patient, tedious explanation, it's less a novel than a series of static monologues. Moreover, this is not the tedium of perfection, the productive boredom of a vision of utopian leisure, but something more uncanny and unsettling—a transformation that transforms nothing, a non-redemptive redemption, in a kind of reverse optic from the messianic power that was so important for thinkers such as Benjamin and Adorno.

But if, like Julian West reading the twentieth-century masterpiece "Penthesilia," we ask ourselves what is left out of *its* vision, it becomes more interesting—not as a utopian blueprint for the twenty-first century, but as a cultural document of nineteenth-century failures of imagination. Consider, for example, what's missing in the following description of the brave new world of Boston in the year 2000:

> Miles of broad streets, shaded by trees and lined with fine buildings, for the most part not in continuous blocks but set in larger or smaller inclosures, stretched in every direction. Every quarter contained large open squares filled with trees, among which statues glistened and fountains flashed in the late afternoon sun. Public buildings of a colossal size and an architectural grandeur unparalleled in my day raised their stately piles on every side … Raising my eyes at last towards the horizon, I looked westward. That blue ribbon winding away to the sunset, was it not the sinuous Charles? I looked east; Boston harbor stretched before me within its headlands, not one of its green islets missing.[51]

The topography may be all there, but what *is* missing is any sign of people. There is, in *Looking Backward*, an almost spooky lack of social life. The novel contains only three speaking characters other than the narrator, all of them members of the Leete family. For the rest, we encounter a silent if efficient waiter and an equally efficient clerk; but even when the Leetes (a name that rhymes suspiciously with "elites") take Julian out to eat at the local public dining house, they're ushered into a private room where they eat by themselves. It's as if Bellamy can't imagine the benefits of a nationalised economy without sneaking in the old values of privatised space, possessive individualism and personal seclusion through the back door.

The lack of sociality in *Looking Backward* points to more than just a weakness of representation. It shows a crisis in the public-private divide that goes to the heart of Bellamy's utopian vision. For though the changes brought about by nationalisation and the creation of the industrial army are enormous, they leave curiously intact many of the structures of late Victorian life, including separate spheres for men and women, marriage, church, the nuclear family with a patriarch at its head, a clear hierarchy in the organisation of occupations, the defining split between mental and manual labour, privatised enjoyment of high culture within the sanctum of the home, and so on.

By the novel's end, when we learn that twenty-first century Edith Leete is the descendant of Julian's old fiancée from the nineteenth century, also named Edith, we begin to suspect that what *Looking Backward* really wants to do is reassure us that, no matter how great the changes effected by nationalisation, things will basically stay the same. We won't have to give up too much and we won't have to face serious change at either an individual or a class level.

This may explain why capitalists and financiers were among the most enthusiastic supporters of Bellamy's scheme. The horror driving Bellamy's utopian vision is ultimately not the suffering of the poor, but the class conflict that results from it. The strikes and labour agitation of the late nineteenth century are what alarm Julian West and keep his house from being built; only when the source of this conflict is removed, suggests the novel, will the nation be able to rebuild its postwar house on a peaceful foundation. But the means of removing conflict remain as mysterious by the novel's end as they are at its beginning.

The word "utopia" famously pivots on a pun on the Greek for both "good place" and "no place." In light of Bellamy's experiment in utopian fiction, what can we conclude about the fortunes of the idea of utopia, as well as of the genre of speculative romance? *Looking Backward* spawned an entire industry in the project of imagining society otherwise. Hundreds of thousands of copies of utopian novels were sold during the last third of the nineteenth century. Yet today, two decades on from Bellamy's benchmark year 2000, there is no outstanding example of utopian thought in the twenty-first century that has achieved success on a mass scale. Representations of dystopia abound, including scenarios built on environmental catastrophe, terrorist destruction, totalitarian takeover, foreign or alien usurpation. Utopia is reserved for the margins and islands within darker speculative science fiction scenarios, if it has any place at all in contemporary culture.

Utopia as Redemptive (Class) Struggle

In the second of his theses from "The Concept of History," Benjamin cites the now largely forgotten metaphysician Hermann Lotze: "'It is one of the most noteworthy particularities of the human heart,' writes Lotze, 'that so much selfishness in individuals coexists with the general lack of envy which every present day feels toward its future.'"[52] It is possible to read this citation as darkly satirical: only our complete lack of imagination, the philosopher might be suggesting, saves us from coveting a better world than the one we live in; or, conversely, our present course of destructiveness ensures that there won't be any future worth envying. But

Benjamin reads Lotze differently: "This observation indicates that the image of happiness we cherish is thoroughly colored by the time to which the course of our own existence has assigned us. There is happiness—such as could arouse envy in us—only in the air we have breathed, among people we could have talked to ... In other words, the idea of happiness is indissolubly bound up with the idea of redemption."[53] And this capacity for happiness, grounded in the present, thereby lays a (weakly messianic) claim on us, as surely as do the defeats and suffering of past generations: we are charged, like Benjamin's historical materialist, with a redemptive task in confronting the ruin of both past and present.

Similarly, Adorno and Horkheimer note that "whatever abundant anguish men suffered in their primal history, they are still incapable of imagining a happiness which does not live off the image of that history."[54] The distinction to be made concerns what Ernst Bloch differentiates as "abstract" and "concrete" utopias—those that are based in compensatory fantasy in isolation from present conditions, and those that develop out of real potentialities and tendencies in the present.[55] Even so, Adorno doesn't subscribe to Bloch's anticipatory understanding of art's utopian content: art doesn't, in his view, offer snapshots of the utopian future already residing within the present. Instead, as might be expected, he argues that it can, in its engagement with the materials of its time, both preserve and transform their negatively charged potential according to the procedures it undertakes. His own vision of redemption involves an altered way of seeing that depends on a necessary impossibility: "The only philosophy which can be responsibly practised in face of despair is the attempt to contemplate all things as they would present themselves from the standpoint of redemption ... Perspectives must be fashioned that displace and estrange the world, reveal it to be, with its rifts and crevices, as indigent and distorted as it will appear one day in the messianic light."[56]

No less a social documentarian of the new dystopias than Mike Davis chimes with Adorno in this regard:

> Only a return to explicitly utopian thinking can clarify the minimal conditions for the preservation of human solidarity in face of convergent planetary crises ... To raise our imaginations to the challenge of the Anthropocene, we must be able to envision alternative configurations of agents, practices and social relations, and this requires, in turn, that we suspend the politico-economic assumptions that chain us to the present.[57]

Alternative configurations, estranging perspectives, suspended assumptions: the imaginative tasks required for a world subject to anthropogenic climate heating are, it turns out, precisely those required for practical organisation in the fight against capitalist dystopia. At the same time, a politics of the imagination in the absence of active struggle—even more, motivated by avoidance or minimisation of struggle—reveals itself as paralysed from its inception.

Why Utopia Is So Hard

Fredric Jameson has recently suggested that "utopian thinking demands a revision of Gramsci's famous slogan, which might now run: cynicism of the intellect, utopianism of the will."[58] In a previous essay, he elucidates one reason why utopian thinking is so little in vogue these days:

> Is it not possible that the achievement of utopia will efface all previously existing utopian impulses? For as we have seen they are all formed and determined by the traits and ideologies imposed on us by our present condition, which will by then have disappeared without a trace. But what we call our personality is made up of these very things, of the miseries and the deformations, fully as much as the pleasures and fulfillments. I fear that we are not capable of imagining the disappearance of the former without the utter extinction of the latter as well, since the two are inextricably and causally bound together.[59]

That is, if conflict is simply what gets in the way of perfecting ourselves and our situation, then removing social conflict becomes the royal road to utopia, conceived of as a condition of harmonious stasis. But if conflict is what makes us who we *are*, constitutes us at both the collective and the individual level, then only a utopia that retains change, heterogeneity, variousness, unpredictability and conflict—a heterotopia, in other words—can have any purchase on our imaginations at all. At the same time, such a heterotopia must be able to excise the cancer of a value regime premised on endless growth from the body of human (and extra-human) history. Which remains considerably harder to imagine than looking backward from a steady-state nirvana.

The thinker most identified with the project of utopia and its fictional forms in recent years is of course Jameson himself, whose career-long investment in the modalities of social science fiction reached a culmination of sorts in the 2016 publication of *An American Utopia*. It came as no surprise to those familiar with *Archaeologies of the Future*, *Valences of the Dialectic* and *Representing Capital* that for Jameson, the fundamental bifurcation in utopian possibility concerns the question of labour, in which either a future relieved of the need to work for a living or a future in which all must do so on equal terms beckon as alternatives. Nor is it surprising that for Jameson, the latter option forms the most compelling utopian horizon. Full employment, for a child of the 1930s, must always retain a degree of lustre as an ideal that later generations, for whom paid work is simultaneously degraded and unavoidable, find harder to detect. And the notion of a nationally conscripted army is less foreign to one who came of age in the aftermath of the Second World War than to those who inherit the legacies of Vietnam, Afghanistan and Iraq.

What might appear counterintuitive, however, is that Jameson's vision of a national (but globally ubiquitous) army ensuring full employment is so much closer to Bellamy's industrial state than, say, to Morris's pastoral craft utopia. For Jameson, the utopian value of full employment is that it ensures a break with a capitalist system that relies on a reserve army of the unemployed and the permanent availability of new horizons for appropriating cheap labour. At the same time, his vision's checklist of difficulties to be overcome en route to this transformation, ranging from the political to the organisational to the psychological, echoes that of Bellamy's thought experiment, in which class struggle becomes a bad (nineteenth century) dream from which the utopian sleeper awakes into a state of well-marshalled harmony. An imaginative inability to confront the logic and logistics of historical transition bedevil *An American Utopia* as much as it does *Looking Backward*.

According to Jameson's well-known interpretation, however, utopia is precisely what marks, even celebrates, an essential negation of imaginative capacity. "Its function," he writes, "lies not in helping us to imagine a better future but rather in demonstrating our utter incapacity to imagine such a future—our imprisonment in a non-utopian present without historicity or futurity—so as to reveal the ideological closure of the system in which we are somehow trapped and confined."[60] Here we might note that the failure to imagine is a constant motif in Jameson's work, not least in the notorious pronouncement concerning capitalism's end. What bears investigating, however, is the conception of utopian capacities—"seeds of time," in Jameson's own phrase—encrypted within the very artefacts of (neo-) realism and (neo-)modernism in our own time, dystopian or otherwise.[61]

Why Utopia Is So Necessary

For the structure and potential of utopian thought, experiments retain their power—particularly in their catalysation of the time-travelling potential intrinsic to future imaginaries. One utopian novel written a century after Bellamy's offers an eloquent reminder of this power when *its* visitors from the future explain their reasons for making contact with the present:

> You may fail us … You individually may fail to understand us or to struggle in your own life and time. You of your time may fail to struggle altogether … [But] we must fight to come to exist, to remain in existence, to be the future that happens. That's why we reached you.[62]

Or, as poet Kevin Davies puts it in sci-po, if not sci-fi, terms:

> What gets *me* is
>
> the robots are doing
> *my* job, but I don't get
> the *money*,
> some extrapolated node
> of expansion-contraction gets
> my money, which *I* need
> for *time travel*.[63]

In a landscape simultaneously petrified and molten with accelerating catastrophe, it is precisely the heat-induced ripples of another time that require collective attention. An ethics of such attention may be one aspect of the political struggle—in solidarity with other times, other places—in our present moment of climate emergency.

Notes

1. Aside from those works examined below, see, for example, Weeks, *Problem with Work*; Bhattacharya, *Social Reproduction Theory*; Larsen et al., *Marxism and the Critique of Value*; "Misery and the Value Form."
2. Streeck, "Social Democracy's Last Rounds."
3. Fisher, *Capitalist Realism*, 2ff.
4. Ghosh, *Great Derangement*, 11.
5. Marx and Engels, *Manifesto of the Communist Party*, chap. 1.
6. See the perceptive discussion in McNeill, "Reading the Maps."
7. Esty, "Realism Wars," 323.

8. Esty, "Realism Wars," 318.
9. Esty, "Realism Wars," 318.
10. Shields, *Reality Hunger.*
11. Benjamin, "Work of Art," 105.
12. See Manovich, *Language of New Media.*
13. Marx, *Capital*, chap. 6.
14. Esty, "Realism Wars," 336.
15. Nixon, *Slow Violence.*
16. Ghosh, *Great Derangement*, 8.
17. As David Graeber notes, "We seem to be facing two insoluble problems. On the one hand, we have witnessed an endless series of global debt crises, which have grown only more and more severe since the seventies, to the point where the overall burden of debt—sovereign, municipal, corporate, personal—is obviously unsustainable. On the other, we have an ecological crisis, a galloping process of climate change that is threatening to throw the entire planet into drought, floods, chaos, starvation, and war. The two might seem unrelated. But ultimately they are the same." Graeber, "Practical Utopian's Guide."
18. Watts, "Wildlife, Natural and Artificial," 609.
19. Coleman, "Apocalypses."
20. Gibson, "Science in Science Fiction," 11: 22.
21. Lilley et al., *Catastrophism.*
22. Lilley, "Introduction," 8.
23. Lilley et al., *Catastrophism.*
24. See Nathan Brown's review of Srnicek and Williams: Brown, "Avoiding Communism."
25. Macpherson, "Post-Liberal-Democracy?" 13.
26. Drucker, *Post-Capitalist Society*, 5.
27. See, for example, Brouillette, *Literature and the Creative Economy.*
28. Mason, *PostCapitalism*, 81.
29. Runciman, "*PostCapitalism* by Paul Mason Review."
30. See, for example, the O'Neil and Frayssé collection *Digital Labour and Prosumer Capitalism.*
31. Srnicek and Williams, *Inventing the Future*, 11–12.
32. Srnicek and Williams, *Inventing the Future*, 183.
33. Kriss, "Future Has Already Happened."
34. Hatherley, "One Click at a Time."
35. Postone, *History and Heteronomy*, 106.
36. Postone, "Labor," 325.
37. Postone, "Labor," 329.
38. Postone, *History and Heteronomy*, 106.
39. Postone, "Labor," 329.

40. Frase, *Four Futures*, 27.
41. Frase, *Four Futures*, 1.
42. Illich, *Right to Useful Unemployment*, 36.
43. Parenti, *Tropic of Chaos*, 11.
44. Frase, *Four Futures*, 252–253.
45. Frase, *Four Futures*, 149.
46. Benjamin, "On the Concept of History," 395.
47. See Williams, *Towards 2000*, 241–269.
48. Bellamy, *Looking Backward*, 145.
49. See Jameson's seconding of Robert C. Elliott's thesis: "a utopia can be judged by the quality and position of the works of art it foretells; most, like More's, are either mute on the subject or distinctly unsatisfying, as in Morris and Bellamy alike." Jameson, *American Utopia*, 37.
50. Bellamy, *Looking Backward*, 114.
51. Bellamy, *Looking Backward*, 66.
52. Hermann Lotze, *Mikrokosmos*, vol. 3 (Leipzig: Hirzel, 1864), 49, quoted in Benjamin, "On the Concept of History," 389.
53. Benjamin, "On the Concept of History," 389.
54. Horkheimer and Adorno, *Dialectic of Enlightenment*, 64.
55. See Bloch, *Principle of Hope*.
56. Adorno, *Minima Moralia*, 247.
57. Davis, "Who Will Build the Ark?," 45.
58. Jameson, *American Utopia*, 11.
59. Jameson, "Politics of Utopia," 52.
60. Jameson, "Politics of Utopia," 46.
61. Jameson, *Seeds of Time*.
62. Piercy, *Woman on the Edge of Time*, 213.
63. Davies, *Comp*, frontispiece.

Works Cited

Adorno, Theodor W. *Minima Moralia: Reflections on a Damaged Life*. London: Verso, 2005.

Bellamy, Edward. *Looking Backward: 2000–1887*. New York: Dover, 1996 [1888].

Benjamin, Walter. "On the Concept of History," translated by Harry Zohn. In *Walter Benjamin: Selected Writings*, vol. 4, *1938–1940*, edited by Howard Eiland and Michael W. Jennings, 389–400. Cambridge: Harvard University Press, 2003 [1940].

———. "The Work of Art in the Age of its Technical Reproducibility (Second Version)," translated by Harry Zohn. In *Walter Benjamin: Selected Writings*, vol. 3, *1935–1938*, edited by Howard Eiland and Michael W. Jennings, 101–133. Cambridge: Harvard University Press, 2002 [1936].

Bhattacharya, Tithi, ed. *Social Reproduction Theory: Remapping Class, Recentering Oppression*. London: Pluto Books, 2017.
Bloch, Ernst. *The Principle of Hope*. 3 vols. Cambridge: MIT Press, 1995.
Brouillette, Sarah. *Literature and the Creative Economy*. Stanford: Stanford University Press, 2014.
Brown, Nathan. "Avoiding Communism: A Critique of Nick Srnicek and Alex Williams' *Inventing the Future*." *Parrhesia* 25 (2016): 155–171.
Coleman, Claire G. "Apocalypses Are More Than the Stuff of Fiction: First Nations Australians Survived One." *ABC News*. December 7, 2017. http://www.abc.net.au/news/2017-12-08/first-nations-australians-survived-an-apocalypse-says-author/9224026/.
Davies, Kevin. *Comp*. Washington, D.C.: Edge Books, 2000.
Davis, Mike. "Who Will Build the Ark?" *New Left Review* II, no. 61 (2010): 29–46.
Drucker, Peter F. *Post-Capitalist Society*. London: Routledge, 1994.
Esty, Jed. "Realism Wars." *Novel: A Forum on Fiction* 49, no. 2 (2016): 316–342.
Fisher, Mark. *Capitalist Realism: Is There No Alternative?* Winchester, UK: Zero Books, 2009.
Frase, Peter. *Four Futures*. London: Verso, 2015.
Ghosh, Amitav. *The Great Derangement: Climate Change and the Unthinkable*. Chicago: University of Chicago Press, 2016.
Gibson, William. "The Science in Science Fiction." Interview by Brooke Gladstone. *Talk of the Nation*. National Public Radio. November 30, 1999. http://www.npr.org/templates/story/story.php?storyId=1067220.
Graeber, David. "A Practical Utopian's Guide to the Coming Collapse." *The Baffler*, no. 22 (2013). http://thebaffler.com/salvos/a-practical-utopians-guide-to-the-coming-collapse.
Hatherley, Owen. "One Click at a Time." *London Review of Books* 38, no. 13 (2016): 3–6. https://www.lrb.co.uk/v38/n13/owen-hatherley/one-click-at-a-time.
Horkheimer, Max, and Theodor W. Adorno. *Dialectic of Enlightenment: Philosophical Fragments*. Edited by Gunzelin Schmid Noerr. Translated by Edmund Jephcott. Stanford: Stanford University Press, 2002 [1947].
Illich, Ivan. *The Right to Useful Unemployment and Its Professional Enemies*. London: Marion Boyars, 1978.
Jameson, Fredric. *An American Utopia: Dual Power and the Universal Army*. Edited by Slavoj Žižek. London: Verso, 2016.
———. "The Politics of Utopia." *New Left Review* II, no. 25 (2004): 35–54.
———. *Postmodernism: Or, the Cultural Logic of Late Capitalism*. London: Verso, 1992.
———. *The Seeds of Time*. New York: Columbia University Press, 1996.
Kriss, Sam. "The Future Has Already Happened." *Viewpoint*. June 1, 2016. https://www.viewpointmag.com/2016/06/01/the-future-has-already-happened/.

Larsen, Neil, Mathias Nilges, Josh Robinson and Nicholas Brown, eds. *Marxism and the Critique of Value*. Chicago: MCM' Publishing, 2014.

Lilley, Sasha. "Introduction: The Apocalyptic Politics of Collapse and Rebirth." In *Catastrophism: The Apocalyptic Politics of Collapse and Rebirth*, edited by Sascha Lilley, David McNally, Eddie Yuen and James Davis, 1–14. Oakland: PM Press, 2012.

Lilley, Sasha, David McNally, Eddie Yuen and James Davis, eds. *Catastrophism: The Apocalyptic Politics of Collapse and Rebirth*. Oakland: PM Press, 2012.

Macpherson, C. B. "Post-Liberal-Democracy?" *New Left Review* I, no. 33 (1965): 3–16.

Manovich, Lev. *The Language of New Media*. Cambridge: MIT Press, 2001.

Marx, Karl. *Capital: A Critique of Political Economy*, vol. 1. Translated by Samuel Moore and Edward Aveling, Edited by Frederick Engels. 2015 [1867]. Accessed January 8, 2019. https://www.marxists.org/archive/marx/works/1867-c1/.

Marx, Karl, and Friedrich Engels. *Manifesto of the Communist Party*. Translated by Samuel Moore in cooperation with Frederick Engels. 1987 [1848]. Accessed August 5, 2018. https://www.marxists.org/archive/marx/works/1848/communist-manifesto/index.htm.

Mason, Paul. *PostCapitalism: A Guide to Our Future*. London: Allen Lane, 2015.

McNeill, Dougal. "Reading the Maps: Realism, Science Fiction and Utopian Strategies." *Imagining the Future: Utopia and Dystopia*, ed. Andrew Milner, Matthew Ryan and Robert Savage. *Arena Journal New Series*, no. 25/26 (2006): 63–79.

"Misery and the Value Form". *Endnotes*, no. 2 (2010). Accessed 5 August 2018. https://endnotes.org.uk/issues/2.

Nixon, Rob. *Slow Violence and the Environmentalism of the Poor*. Cambridge: Harvard University Press, 2011.

O'Neil, Mathieu, and Olivier Frayssé, eds. *Digital Labour and Prosumer Capitalism: The US Matrix*. Basingstoke: Palgrave Macmillan, 2015.

Parenti, Christian. *Tropic of Chaos: Climate Change and the New Geography of Violence*. New York: Nation Books, 2011.

Piercy, Marge. *Woman on the Edge of Time*. London: Del Rey, 2016 [1976].

Postone, Moishe. *History and Heteronomy: Critical Essays*. Tokyo: University of Tokyo Center for Philosophy, 2009.

———. "Labor and the Logic of Abstraction: An Interview," by Timothy Brennan. *South Atlantic Quarterly* 108, no. 2 (2009): 305–330.

Robinson, Kim Stanley. *New York 2140*. New York: Orbit, 2017.

Runciman, David. "*PostCapitalism* by Paul Mason Review—A Worthy Successor to Marx?" *The Guardian*. August 15, 2015. https://www.theguardian.com/books/2015/aug/15/post-capitalism-by-paul-mason-review-worthy-successor-to-marx.

Shields, David. *Reality Hunger: A Manifesto*. New York: Alfred A. Knopf, 2010.

Srnicek, Nick, and Alex Williams. *Inventing the Future: Postcapitalism and a World Without Work*. London: Verso, 2015.

Streeck, Wolfgang. "Social Democracy's Last Rounds: An Interview with Wolfgang Streeck." *Jacobin*. February 22, 2016. https://www.jacobinmag.com/2016/02/wolfgang-streeck-europe-eurozone-austerity-neoliberalism-social-democracy/.

Watts, Peter. "Wildlife, Natural and Artificial: An Interview with Peter Watts." By Imre Szeman and Maria Whiteman. *Extrapolation* 48, no. 3 (2007): 603–619.

Weeks, Kathy. *The Problem with Work: Feminism, Marxism, Antiwork Politics, and Postwork Imaginaries*. Durham: Duke University Press, 2011.

Williams, Raymond. *Towards 2000*. London: Chatto and Windus, 1983.

Index[1]

A
Africa, 199, 212, 213, 215, 313
  African Futurism, vii, 211, 213–216, 229
Afrofuturism, 211, 213–216, 230n8
Aldiss, Brian W., vi, 54, 57, 81, 85
Algeria, vi, 212, 215
Algiers, 218
Alterity, vi, 5, 11, 12, 17, 22, 288
Animals, v, vi, 13, 16, 22, 49–67, 90, 91, 101–103, 105, 106, 111, 112, 114, 131, 143, 147, 149, 150, 156–158, 159n10, 174, 241, 264, 266, 290
  human-animal binary, 101, 102
  See also Ethics, animal ethics
Anthropocene, 103, 114, 159n8, 175, 176, 306, 319
Apocalypse, 109, 111, 131, 141, 142, 144–146, 150, 277, 290, 296, 298, 306, 307, 309
  ecopocalypse, vii, 142, 144, 149, 152, 158
  post-apocalyptic, 91, 114, 242, 262, 263, 265, 270, 271, 277, 279, 307
Appadurai, Arjun, 168
Aridjis, Homero, vii, 142, 144, 146, 147, 149–151, 154, 155, 157, 160n23, 160n36
  Apocalipsis con figura, 150
  ¿En quién piensas cuando haces el amor?, vii, 142
  La Leyenda de los soles, vii, 142, 146, 160n23
Asimov, Isaac, vi, 3–22
  Foundation trilogy, vi, 3–22

[1] Note: Page numbers followed by ‘n’ refer to notes.

Z. Kendal et al. (eds.), *Ethical Futures and Global Science Fiction*, Studies in Global Science Fiction,
https://doi.org/10.1007/978-3-030-27893-9

Astounding Science-Fiction (magazine), v, 5, 9, 10, 21, 22n4, 22n9, 23n35
Atwood, Margaret, vii, 61, 65, 66, 79, 81, 84, 86–92, 99–114, 276
The Handmaid's Tale, 276, 307
MaddAddam, vii, 61, 65, 66, 81, 84, 90, 91, 99–114
Oryx and Crake, 65, 66, 81, 90, 91, 109, 110, 112, 113
The Year of the Flood, 66, 90, 91, 109, 112
Australia, vi, vii, 77, 85, 92n2
Automation, 303, 310, 312
Azikiwe, Ben N., 168

B
Baccolini, Raffaella, 83
Bacigalupi, Paolo, 81
Bacon, Francis, 52, 67
Bagchi, Barnita, 32, 33, 39
Ballard, J. G., 84
Barthes, Roland, 19
Baxter, Stephen, 57
Belasco, Warren, 53, 59
Bellamy, Edward, 49, 55, 56, 67, 92, 315–318, 321, 322, 324n49
Benford, Gregory, 179
Benjamin, Walter, 133, 134, 256n5, 306, 312, 314, 316, 318, 319
Besson, Bernard, 81, 248
Biggar, Nigel, 167, 168
Bignall, Simone, 195
Bisson, Terry, 88
Bloch, Ernst, 123, 169, 319
Heimat, 169
Bloom, Daniel, vi, 80
Bolaño, Roberto, 314
Böttcher, Sven, 81
Boyle, T. C., 84
Bradbury, Ray, 193, 196, 272
Bradley, James, 61, 81
Brantenberg, Gerd, 36
Britain, vi, vii, 85, 190
Bryant, Dorothy, 78
Buckingham, James Silk, 55
Bulwer-Lytton, Edward, 54, 67, 68n34
Butler, Octavia E., vii, 166, 168–173, 176, 183
Xenogenesis trilogy (*Lilith's Brood*), 166
Butler, Samuel, 53

C
Callenbach, Ernest, 63, 64, 78, 90
Campanella, Tommaso, 52, 67
Campbell, John W., Jr., v, 9, 10, 19, 21, 22n4, 23n35, 212
Canada, vi, vii, 85
Cannibalism, 56, 62, 69n63, 155
Capitalism, 109, 122, 144, 188, 189, 196, 216, 217, 294, 304, 309–312, 321
See also Post-capitalism
Care, vi, vii, 30, 34, 38, 39, 41–43, 131, 148, 156, 174, 201, 213, 215–220, 223–229, 238, 251, 264, 268
Caribbean, 142, 144, 152, 212, 220, 226
Carson, Rachel, 131, 150
Catastrophe, 86, 110, 122, 127, 128, 131, 133, 134, 141, 144, 151, 160n36, 264, 296, 307, 312, 318, 322
catastrophism, 144
Charnas, Suzy McKee, 78
China, vi, vii, 83, 85, 120, 123–126, 129, 131, 137n51
Cultural Revolution, 125, 126, 131
Christianity, 252, 294
Clarke, Arthur C., 84

Climate change, vii, 77, 80–83, 85–87, 110, 131, 142–144, 150, 157, 158, 299n20, 306, 308, 312, 323n17
denial, 80, 81, 85, 305
fatalism, 80, 81, 85
Gaia, 80, 85
imperialism, 158
mitigation, 80, 81, 84, 85
negative adaptation, 81, 84, 85
positive adaptation, 81, 85, 86
Clute, John, 5
Coleman, Claire G., 307
Colonialism, vii, 142, 145, 153, 157, 158, 159n8, 165, 167, 170, 173, 176, 178–180, 187–189, 191–193, 195, 205, 212
neocolonialism, 205
Communism, 8, 10, 18, 121, 122, 125, 126, 131, 255, 269, 304, 309, 310, 313, 314
post-communism, 269
Crichton, Michael, 81
Cyborg, vii, 102, 104, 187–205

D

Davies, Kevin, 322
Delany, Samuel R., 4, 63, 64, 78, 308
Derrida, Jacques, 102
Determinism, 10
Dick, Philip K., 84, 206n30, 315
Diderot, Denis, 237, 238
Disch, Thomas M., 184n13
Djebar, Assia, vii, 211–230
Ombre sultane, vii, 211–230
Dürbeck, Gabriele, 90
Dworkin, Ronald, 176
Dystopia, vi, viii, 4, 12, 14, 18, 21, 22, 59, 62, 66, 78–85, 87, 90, 100, 108, 110, 119, 126, 129, 143–145, 149, 150, 154, 155, 158, 236, 239, 240, 242, 245, 248–250, 254, 255, 261–263, 266–270, 272, 274, 276, 279, 284, 285, 292, 297, 303, 307, 308, 313, 314, 318–321
classical, 79–81, 83, 84
critical, 78–81, 83, 84

E

Egypt, vi, 190, 270, 283, 285, 286, 288–297
Cairo, 283, 285, 286, 290, 291, 296
Esty, Jed, 305, 306
Ethics
animal ethics, vi, 50, 67
care ethics, 30, 34, 41–43
environmental ethics, 143, 157, 158, 159n8, 159n10, 285
postcolonial ethics, vii, 195–196
relational ethics, 177, 182
Eutopia, 78–80, 84, 86–88, 90–92, 239
classical, 85
critical, 79, 81, 85, 90, 91
Extraterrestrials (or aliens), 8, 10, 12, 20, 121–123, 127, 130, 165–167, 179, 181, 193, 196, 200

F

Fascism, vi, 3–5, 8–10, 21, 82, 315
Feminism, v–vii, 29, 30, 32–34, 36, 39–44, 51, 60, 61, 106, 116n37, 213, 216, 217, 224, 225, 229, 308, 310
ecofeminism, 103, 104
heteropatriarchy, 38
second wave, 34, 40, 42, 229
third wave, 42, 225

Flammarion, Camille, 55, 240
Fleck, Dirk C., 81, 86–92
  Das Südsee-Virus, 88
  Das Tahiti-Projekt, 88
  Feuer am Fuss, 88
  GO! Die Ökodiktatur, 88
  MAEVA!, 81, 88, 92
France, vii, 85, 90, 92n2, 105, 219, 235–255, 295
  Paris, 235–242, 244, 246, 249, 251, 253, 254
Franklin, Alfred, 87, 241
Frase, Peter, 312, 313
Futurians, 9

G

Gearhart, Sally, 78
Germany, vi, vii, 85, 88, 188
Gernsback, Hugo, v, 82
Ghosh, Amitav, vii, 84, 133, 150, 187–205, 304, 306, 307
  The Calcutta Chromosome, vii, 187–205
  The Great Derangement, 150, 304
Gibson, William, 307
Gilman, Charlotte Perkins, 29, 60, 61, 316
Global North, 142, 187–189, 193, 195, 205, 218, 287, 307
Global South, vii, 188, 189, 205, 219, 284, 285, 289, 298, 307
Gore, Al, 87
Graham, Elaine L., 101, 102, 108, 159n8, 189
Gunn, James, 6, 13, 19

H

Habermas, Jürgen, 176
Haiti, vi, 142, 144, 151, 152
Haraway, Donna J., 102, 104, 203
Hayles, N. Katherine, 101, 102
Heinlein, Robert A., 184n13
Herzog, Arthur, 81, 84
Hopkinson, Nalo, vii, 165, 192, 200, 211–230
  Midnight Robber, vii, 211–230
Hossain, Begum Rokeya Sakhawat, vi, 29–44, 61
  Padmarag (পদ্মরাগ), vi, 30
  The Secluded Ones (Abarodhbasini), 35
  Sultana's Dream, vi, 30, 32–44, 61
Houellebecq, Michel, vii, viii, 81, 99–114, 116n37, 235–255
  Les Particules élémentaires, vii, 99–114
  Soumission, viii, 236, 246, 247, 249, 250, 254
Hoy, David Couzens, 169
Hubbard, L. Ron, 10
Huxley, Aldous, 12, 236, 242, 262, 269

I

Imperialism, vii, 5, 9, 11, 23n22, 52, 142, 145, 152, 158, 159n8, 165, 167–169, 183, 187, 188, 191–193, 196, 215, 308
India, 190
  Bengali literature, 35
  colonial India, 34
Islam, 32, 218, 246, 249, 250, 252, 253, 294

J

Jamaica, vi, 223
Jameson, Fredric, 3, 12, 49, 50, 52, 63, 82, 122, 128, 304, 312, 315, 320, 321
Jeschke, Wolfgang, 81
Judaism, 294

K

Kingsolver, Barbara, 81, 84
Kurtz, Malisa, 84, 189, 194–196, 205, 313

L

Lake, David J., 57
Lane, Mary E. Bradley, 61
Langer, Jessica, 165, 192, 193, 199
Langford, David, 8
Le Guin, Ursula K., 12, 30, 62–64, 78, 179, 308
  Always Coming Home, 63
  The Dispossessed, 62, 63, 78
  The Left Hand of Darkness, 30
Lem, Stanisław, 12
  Solaris, 12
Lessing, Doris, 84
Levinas, Emmanuel, vi, 4, 5, 7, 11, 16, 19, 21
Lewis, C. S., 78, 82, 83
Ligny, Jean-Marc, 81
Lister, Umoya, 81
Liu, Cixin, vii, 81, 119–135
  The Three-Body Problem, vii, 83, 119–135
Liu, Ken, 120
López-Lozano, Miguel, 153, 159n17
Lovelock, James, 80

M

Macedonia, vi, viii, 261–279
  Skopje, 264, 267, 273, 274, 278
Marshall, Alan, 65
Martin, George R. R., 82, 83, 106, 152, 153
Marx, Karl, 7, 123, 128, 217, 306, 312, 313
Marxism, vi, 3, 7, 8, 21, 125, 311, 313
Mason, Paul, 309–311
McCarthy, Cormac, 84, 257n28
McCoy, John F., 55
McEwan, Ian, 81, 84
Mehan, Uppinder, 192
Mercier, Louis-Sébastien, viii, 235–255
  L'An 2440: Rêve s'il en fut jamais, viii, 235
Merleau-Ponty, Maurice, 103
Mexico, 142, 144, 151, 154, 160n36
  Hernán Cortés, 151, 153
  La Malinche, 153, 154
  Mexico City, 146, 147, 151, 160n33
Middle East, 284, 285, 292, 295, 296, 313
Miéville, China, 83
Mihajlović Kostadinovska, Sanja
  517, viii, 262, 265, 269, 276
Milner, Andrew, vii, 7, 62, 68n27
Moorcock, Michael, 10
More, Thomas, v, 49, 51, 52, 66, 67, 67–68n12, 78, 102, 135, 255, 324n49
Moretti, Franco, 85
Morrey, Douglas, 104, 106, 116n37
Morris, William, 55, 69n40, 92, 277, 315, 321, 324n49
Moylan, Tom, 62, 78–80, 83, 88, 122, 276

N

Nasir, Jamil, viii, 283–298
  Tower of Dreams, viii, 283–298
Nationalism, viii, 8, 9, 38, 56, 125, 153, 213, 214, 262, 264, 273, 304
Neuhaus, Nele, 81
Neville, Henry, 52

O

Olerich, Henry, 55
Orientalism, 197, 198, 224, 225, 286, 287
Orwell, George, 12, 236, 242, 262, 269
Osmanli, Tomislav, viii, 262, 265, 267, 269, 276
  Бродот. Конзархија, viii, 262, 267, 269, 275, 276
Otherness/the Other, 5, 6, 10–12, 14, 16–19, 21, 22, 179, 193, 196, 205, 288

P

Pan-Africanism, vii, 212–216
Phallocentrism, 42, 219
Piercy, Marge, 36, 61, 62, 64, 78, 308
Plato, 51
Plumwood, Val, 103, 104, 106, 108, 110, 143, 157, 159n9
Poland, 85
Politics, viii, 3, 8, 10, 12, 18, 21, 23n35, 31, 83, 85, 86, 88, 123, 124, 131, 135, 142, 144, 195, 212, 213, 215, 223, 224, 226, 229, 236, 238, 246, 255, 262, 311, 313
  Green politics, 85, 86
Post-capitalism, viii, 308–313
Postcolonialism, v, vii, 142, 143, 158, 202–205
  Green postcolonialism, 159n8
  postcolonial science fiction, 165–183, 188, 189, 191–196, 199, 200, 205
  postcolonial theory, 169, 189
Posthumanism, 92, 99–114, 166, 170
  ecological posthumanism, vii, 99–114
  techno-posthumanism, 101, 102
Postone, Moishe, 311, 312
Pratt, Mary Louise, 180
Pullman, Philip, 82, 83

R

Realism, 19, 20, 86, 87, 120, 121, 124–126, 131, 132, 211, 213, 223, 303–307, 315
Religion
  laïcité, 242, 243, 246, 255
  secularism, 236, 238, 239, 242–244, 246, 250, 252, 255
  See also individual religions
Rieder, John, 165, 187, 188, 191, 192
Robinson, Kim Stanley, 38, 65, 81, 84, 86–92, 94n43, 120, 308, 313
  Aurora, 81, 84
  Green Earth, 81, 86–88
  Mars trilogy, 88, 313
  New York 2140, 65, 81, 84, 87, 88, 90, 92, 308
  Pacific Edge, 88
  Red Mars, 88
  Science in the Capital trilogy, 81, 86
  Three Californias trilogy, 313
  2312, 65, 81, 84
Ross, Ronald, 190, 191, 194, 198, 199, 201, 203
Roumain, Jacques, vii, 142, 144, 145, 147, 149–153, 155, 157
  Gouverneurs de la rosée, vii, 142, 145
Rousseau, Jean-Jacques, 237, 238, 240, 243
Russ, Joanna, 62, 63, 78, 308
Russia, vi, 12, 17, 18, 85, 92n2, 188
  Soviet Union, 12

S

Sanders, Bernie, 87
Santesso, Aaron, 3, 8
Sargent, Lyman Tower, 29, 50–52, 55, 78, 80
Sawyer, Andy, 192
Schätzing, Frank, 81
Science Fiction (SF)
  genre SF, 4, 5, 8–11, 19, 22n4, 23n22, 84

golden age, 4, 22n9
Science Fiction Magazines, 3
pulps, v, 3, 4, 6, 8–12, 22n4, 22n9
See also individual titles
Shelley, Mary, vi, 54, 82
Frankenstein, vi, 54
The Last Man, 54
Shelley, Percy, 53, 54
Simon, Joel, 160n36
Skrimshire, Stefan, 144
Socialism, 7, 18, 20, 119–126, 129, 131, 132, 135, 218, 240, 242, 247, 251, 255, 268, 269, 308, 309, 313
South Africa, vi, vii, 85
Spivak, Gayatri Chakravorty, 201
Srnicek, Nick, 309–311
Stapledon, Olaf, 59
Su, John J., 262
Suvin, Darko, 3, 50, 82

T
Tolkien, J. R. R., 82, 83
Towfik, Ahmed Khaled, viii, 283–298
Utopia, viii, 283–298
Turner, George, 81

U
Uchronia, 235, 237, 239, 240, 242, 249, 254, 255
United States of America (USA), vii, 79, 94n43, 125, 142, 184n12, 188
New York, 9, 65, 87, 189
Urošević, Vlada, viii, 261, 262, 265, 269, 276
Дворскиот поет во апарат за летање, viii, 261, 269
Utopia, v, viii, 29, 41, 50–56, 58–63, 65, 78–80, 86, 106, 108, 110, 119, 120, 123, 124, 126, 134–135, 240, 242, 262, 276, 283–298, 318–322

V
Vegetarianism, vi, 49–67, 68n34, 69n40, 90, 172
Verne, Jules, 82, 240, 241, 250, 255
Voltaire, 238
Vonnegut, Kurt, Jr., 84, 313

W
Watts, David, 142
Watts, Peter, 307
Weber, Max, 80, 81
Wells, H. G., 50, 56–60, 62, 66, 196, 236, 242, 262, 316
First Men in the Moon, 56
The Island of Doctor Moreau, 57, 58
Men Like Gods, 58, 62
Modern Utopia, 59, 60, 316
The Shape of Things to Come, 58
The Time Machine, 56–58, 66, 236
The War in the Air, 56
The War of the Worlds, 57, 58
Williams, Alex, 309–311
Winterson, Jeanette, 81, 83
Wolfe, Cary, 102
Wolfe, Gene, 184n13
Wollheim, Donald, 7, 9, 10, 23n35
Wong, Brian, 177
World systems theory, 85
Wright, Alexis, 81

Y
Yuen, Eddie, 141, 142, 308

Z
Zamyatin, Yevgeny, vi, 3–22, 236, 242, 269
Мы (*We*), vi, 3–22, 236, 269